Earth's climate response to a changing Sun

Coordinated by: Jean Lilensten, Thierry Dudok de Wit, Katja Matthes

Editors: Thierry Dudok de Wit, Ilaria Ermolli, Margit Haberreiter, Harry Kambezidis, Mai Mai Lam, Jean Lilensten, Katja Matthes, Irina Mironova, Hauke Schmidt, Annika Seppälä, Eija Tanskanen, Kleareti Tourpali, Yoav Yair

17, avenue du Hoggar
Parc d'activité de Courtabœuf, BP 112
91944 Les Ulis Cedex A, France

COST (European Cooperation in Science and Technology) is a pan-European intergovernmental framework. Its mission is to enable break-through scientific and technological developments leading to new concepts and products and thereby contribute to strengthening Europe's research and innovation capacities. It allows researchers, engineers and scholars to jointly develop their own ideas and take new initiatives across all fields of science and technology, while promoting multi- and interdisciplinary approaches. COST aims at fostering a better integration of less research intensive countries to the knowledge hubs of the European Research Area. The COST Association, an International not-for-profit Association under Belgian Law, integrates all management, governing and administrative 10 functions necessary for the operation of the framework. The COST Association has currently 36 Member Countries.
http://www.cost.eu/

COST is supported by the EU Framework Programme Horizon 2020

Cover illustration: Sunset observed from the International Space Station - expedition 15 (NASA). Aurora at Svalbard, by Cyril Simon Wedlund (Aalto-University (Finland) and IPAG-CNRS).

Printed in France
ISBN: 978-2-7598-1733-7
DOI: 10.1051/978-2-7598-1733-7

This work is subject to copyright. All rights are reserved, whether the whole or part of the material is concerned, specifically the rights of translation, reprinting, re-use of illustrations, recitation, broad-casting, reproduction on microfilms or in other ways, and storage in data bank. Duplication of this publication or parts thereof is only permitted under the provisions of the French Copyright law of March 11, 1957. Violations fall under the prosecution act of the French Copyright law.
© EDP Science, 2015

Contents

Foreword   VII

Preface   XI

PART I. INTRODUCTION TO THE SUN-CLIMATE CONNECTIONS....... 1

    1.1 The Earth's atmosphere: an introduction 3

    1.2 The impact of solar variability on climate 13

    1.3 The Sun-Earth connection, on scales from minutes to millennia 19

    1.4 The role of the Sun in climate change: a brief history 27

    1.5 The role of the Sun in climate change: a societal viewpcint............. 35

    1.6 The debate about solar activity and climate change 41

    References of Part I 49

PART II. SOLAR AND SPACE FORCING 53

    2.1 Basics of solar and heliospheric modulation 55

    2.2 Solar radiative forcing 67

    2.3 Variability of solar and galactic cosmic rays 77

    2.4 Variability and effects by solar wind 85

    2.5 Variations of solar activity 91

    2.6 Understanding solar activity 97

INFOBOX 2.1 Orbital forcing of glacial - interglacial cycles 103

INFOBOX 2.2 Grand minima and maxima of solar activity................... 111

INFOBOX 2.3 A practical guide to solar forcing data 113

    References of Part II...121

PART III. DETECTING SOLAR INFLUENCE ON CLIMATE 127

   3.1 Observations on paleoclimatic time scales 129

   3.2 Ground-based observations 139

   3.3 Satellite observations 155

   3.4 Reanalysis data 165

   3.5 Uncertainties and unknowns in atmospheric observations:
How do they affect the solar signal identification? 171

   3.6 Numerical models of atmosphere and ocean 179

   3.7 From climate to Earth system models 187

   3.8 Uncertainties in the modeling of the solar influence on climate 195

   3.9 Detection and attribution: How is the solar signal identified and
distinguished from the response to other forcings? 203

INFOBOX 3.1 Why are models needed in the first place, and can they be
trusted? 211

INFOBOX 3.2 Model Equations and how they are solved 213

   References of Part III 221

PART IV. IMPACTS ON THE EARTH SYSTEM 227

   4.1 Direct impact of solar irradiance variability 229

   4.2 'Top-down' versus 'bottom-up' mechanisms for solar-climate
coupling 237

   4.3 Interactions of different sources of variability 247

   4.4 Impact of solar variability on the magnetosphere 255

   4.5 Atmospheric ionisation by solar energetic particle precipitation 261

   4.6 Impact of energetic particle precipitation on atmospheric chemistry
and climate 267

   4.7 The impact of cosmic rays on clouds 273

   4.8 Impact of solar variability on the global electric circuit 281

INFOBOX 4.1 Modeled impact of total solar irradiance (TSI) forcing 289

INFOBOX 4.2 Lightning, cosmic rays and energetic particles 293

INFOBOX 4.3 The influence of solar variability on extreme weather 297
   References of Part IV 301
PART V. CONCLUSION 309

Conclusions   311

   References of Part V 320

Glossary   321

The authors   341

FOREWORD

Roger-Maurice Bonnet[1]

1 Why did I accept to write this foreword?

As former solar astronomer, and member of the Service d'Aéronomie of CNRS in France (now renamed LATMOS), I am since long deeply involved in – and motivated to – understanding the role of the Sun on the Earth's climate. However, when Thierry contacted me for writing this foreword, I had some hesitations before answering positively. The response of our climate to the changing Sun has been passionately debated at length among the concerned scientists, solar physicists, meteorologists, climatologists, and many non-specialists, and these debates often made me uncomfortable. The topic deserves indeed a lot of care and its scientific analysis an utmost rigor, both too easily overwhelmed by irrationalism and passion. Why that passion? The issue is not the Sun, a modest G-type star fairly well understood, but rather the climate of our Earth, the most complex of all planets in the Solar System because it is a water planet, because it has an atmosphere and hosts life, particularly human life, the whole set resulting in an incredibly complex system permeated by complex interactions, which make other planets, like Mars for example, look very simple. That is where the problem lies!

The 20th century has witnessed a quasi-exponential growth of the world population followed by a similar growth of the demand on energy necessary to sustain its needs, and of course on fossil fuels, the cheapest source and admittedly the largest producer of CO_2, whose efficient greenhouse effect might be the most powerful cause of the changes observed in the recent evolution of the Earth's climate. Hence the debate! Those who trust in a solar effect are right: nobody would argue that without the Sun, life would not exist on Earth in its present form at least. Nobody would argue against Milanković's theory that describes the collective effects of changes in the Earth's orbital movements upon the climate, and that winters are colder than summers. On the other side, few would argue against paleoclimate studies concluding that without the greenhouse effect of methane and carbon dioxide in the atmosphere of the young Earth, when the Sun output was 70% fainter than at present, our planet would have remained a cosmic snowball

---

[1] International Space Science Institute, Bern, Switzerland

for ever. Those who doubt that the present growth in the concentration of atmospheric greenhouse gas is largely responsible for the observed changes, naturally receive the support of those who do not like the idea that Man's activities might be the cause of the present climate problem, because the remedy to anthropogenic forcing would responsibly imply adopting profound changes in our dealing with industrial activities, fossil fuel production, economy, and in our social systems. The "business-as-usual" attitude is indeed less disturbing in the short range than accepting the necessity of such changes!

In the course of last century and also recently, nearly all the wars on the planet have oil in their background. The fight for oil is indeed a serious matter: oil is relatively cheap and a source of immense revenues. The scientists therefore bear a crucial responsibility in placing the debate between natural and anthropogenic causes of climate change on solid irrefutable rationale grounds. As a scientist accepting that responsibility, I consider that careful attention should be given to the reasoning of the skeptics on both sides, in their expressing doubts about the other camps arguments. Not only because Thierry is a dear friend and colleague but also a man of high scientific rigor whom I fully respect, and because of what precedes, I eventually accepted to write these lines.

2 A surprising document

This monograph, built from some 30 contributions involving more than 60 authors, is surprisingly diverse and non-uniform. That is probably the usual form of reporting that characterises European COST actions, which intend to offer means of intellectual exchanges, be they of a scientific, technological, sociological or political nature using all possibilities of networking between selected members of the respective communities. The topic of this work responds quite exactly to that definition! Its final success would be reached when those involved would eventually concur that solar variability in all its forms does have effects on the Earth's climate, agree on what these effects are, and what are their intensity. To reach that goal, interdisciplinarity is essential and that might not have been so easy to implement, remembering the idiosyncrasy that for a long time has characterised parts of the communities involved. Undoubtedly, this report testifies that this goal was achieved and all the participants to this impressive effort must be congratulated. In that respect, it is important to appreciate the participation of a relatively large number of young scientists whose opinions are remarkably open and far from being biased by corporate protective behavior. They undoubtedly contributed in no small way to approach the Earth not anymore as a set of independent layers, an onion, but rather, as the great French poet Paul Eluard would say, like a "blue orange" inside which all layers interact, forcing the fall of the tight barriers that for a long time made the science of solar variability influences on Earth not fully effective. That courageous approach persuaded me to look closer at the diverse pieces of that colorful stain glass window and to appreciate the result of that confrontation between individual view points, opening new avenues for further discussions, and potentially leading to practical progress.

3 What do I get from this book?

The four main parts of this book follow a logical – nearly Cartesian – order, from a general introduction, followed by a documented analysis of the factors related to solar forcing, and a discussion of detection methods, data analysis techniques and related uncertainties, to terminate with a concluding review of the impacts of solar variability on the Earth System. It is probably the first time that we do have in hand a most detailed analysis so far of the mechanisms potentially contributing to the natural variability of our climate. Even though none of these seems yet to trigger a profound revolution in our understanding of the causes of climate change, many contribute to progress and advances, which justify the genuine value of this interdisciplinary effort. Improving the accuracy in reconstructing the variability of solar phenomena, and more particularly the still-too-low accuracy of UV spectral irradiance measurements, can be recognised as an important message to the scientists and their funding agencies for establishing a network of space-based instruments certainly reinforced by the perspective of a necessary long-term continuity. Recognising the importance of the Global Electric Circuit in modeling the atmospheric phenomena linking the ionosphere to the Earth's surface is another important output which opens a set of research perspectives to answer many questions related to the influence of solar activity in the generation of solar high energy particles and in the modulation of galactic cosmic rays that influence the circuit. Meanwhile, reaching the conclusion that Galactic cosmic rays are unable to significantly alter global cloud cover, nor explain recent global warming, is pleasant to read, remembering the rather long-lasting controversy raised by that possibility. Generating an open scientific discussion to address controversial issues is fine (that is exactly the rationale for creating the International Space Science Institute, in Bern), provided however that at some stage, conclusions should eventually be drawn and if not, agreements be reached on what should be done to later achieve that goal.

4 Reaching the readers, the political world and the social targets

Therefore, I cannot end this preface without raising a question and a concern. The question is: to whom is this book addressed? Is it just a tool, a step in the understanding of the Earth's climates natural variability, a motivation for scientists to initiate more specific studies, furthering their analysis and calibrating their progress? The concern has to do with how will policy makers interpret the document. As a positive response to the question above, or as a justification for them to conclude that because science has not yet delivered its last word about the main causes of climate change, it is too early to take the necessary decisions on capping CO_2 emissions? Because the complexity of the factors intervening in climate forcing is so large, and has not allowed yet to properly quantifying their effects relatively to other sources of forcing, it might indeed be interpreted by some that there is no urgency to act, thereby further delaying the necessary remedies to the deterioration of our atmosphere and of the climate, sacrificing for centuries to come the future of our descendants. That would be regrettable to say it mildly, and using stronger words, highly irresponsible making the present work a very counterproductive and dangerous exercise.

Without any doubt, a positive answer to the question is exactly what should be the conclusion of this work, opening a set of new research avenues leading to a more precise quantification of the relative importance of solar variability with respect to the most powerful anthropogenic forcing mechanisms, and to new discoveries in the understanding of our changing climate. To further engage in both theoretical and observational researches, aiming at improving the precision of the data and of their use in modeling the various forcing mechanisms, while ensuring their continuity over the remaining decades of this century, would be highly gratifying for all those involved in this important work, and would undoubtedly represent the most positive breakthrough of this COST action. It would fortunately bring to a halt the skepticism and "wait-and-see" attitude that in the recent past has often resulted from an improper communication from the scientists involved in this research to explaining their science while identifying the limits of their convictions. This book does offer a rich menu of possible actions that should be pursued with utmost vigor and determination. It possesses all the tools and justifications to add an essential and strong support to the understanding of climate change. The readers cannot interpret its conclusions in a different way.

PREFACE

It's the Sun, stupid! The Sun has indeed come under close scrutiny as a possible cause for global warming observed since the 1950s. Over 99.96% of the energy entering the Earth's atmosphere comes from it, which would make the Sun a natural culprit for the changing terrestrial climate.

Although the Sun is a remarkably stable star over the timescale of a human life, many scientists have wondered whether its variability could affect climate. These studies really gained interest in the 1980s only. This growing awareness coincided with the first direct satellite observations of the amount of energy radiated by the Sun, which indeed turned out to vary.

The Sun–climate connections are one of the most challenging problems faced today in climate science. There are two reasons for this. The first one has to do with the high complexity of the underlying physical and chemical processes at the Sun, and in the Earth's atmosphere. One may imagine these processes as multiple links of a long chain that extends all the way from the solar interior to the Earth's surface. For many years, it was believed that this chain could be approximated by one single quantity, ironically called the solar constant, whose main impact is a direct heating of land and oceans. Today we know that this chain has numerous more subtle aspects: the impact of changing ultraviolet radiation on ozone production, changes in upper atmospheric composition induced by high-energy particles, fluctuations in the solar wind that eventually affect aerosol production by means of the Earth's global electric field, and more. All these mechanisms are strongly interrelated, and affect all layers of the Earth's atmosphere, thereby making it extremely difficult to investigate them one by one, and quantify their impact. A paradigm is shifting: instead of understanding a system by dissecting it into its individual parts, we need a more holistic approach that focuses on the interactions between the different elements of the chain.

The second challenge encountered in the study of Sun–climate connections is the fragmentation of the scientific communities. For many decades, scientists were organised as communities that were specialised on specific aspects, with limited consideration for the full picture. The authors of this handbook have, individually, strong expertise on specific parts of the Sun-climate connections, such as solar physics, high-energy particle physics, geomagnetism, climate modelling, atmospheric chemistry, and more. They have tried to overcome this fragmentation of their communities by interacting and working more closely together. As they were doing so, a global picture of the chain gradually took shape.

The framework that has enabled such multidisciplinary interactions is a pan-European COST action named TOSCA (Towards a more complete assessment of the impact of solar variability on the Earth's climate[1]), which lasted from 2011 until 2015. COST stands for Cooperation in Science and Technology[2], and is a framework supporting cooperation among scientists and researchers across Europe. TOSCA involved over one hundred scientists, both young and senior, and this handbook is its main outcome. COST actions are fully open to participation, and the texts indeed reflect the diversity of the scientists involved. We deliberately allowed the contents of this handbook to reflect this diversity.

Our prime objective with this handbook is to show *science in action* on a highly challenging topic with major societal consequences. Today, there is strong evidence for solar variability to represent a minor contribution to global warming, as observed since the 1950s. However, small as the solar forcing may be, many open questions remain regarding the mechanisms by which it may affect climate, and the magnitude of their forcing. For that reason, this handbook concentrates on the impact of solar variability on climate, rather than on the role of the Sun in global warming.

The results collected represent the state of the art of what we know today about the Sun–climate connections, also raising many open questions. Some of the results shown here may be refuted tomorrow, or new mechanisms may be discovered. However, progress on such a highly multidisciplinary topic will be achieved only if the new evidence is supported by facts that can be tested, and explained in terms of physical mechanisms, and not just beliefs or speculations.

How to read this handbook

This handbook consists of five parts, each of which is subdivided into chapters: part one is a general introduction to Sun–climate connections, part two addresses solar and space forcing, part three is about detecting solar influence on climate, part four addresses impacts on the Earth's system, and part five ends with a conclusion, and a glossary. We intentionally structured this handbook as a mosaic of short chapters that reflect the multiple facets of Sun–climate connections. Each chapter may be considered as a piece of stained glass that fits in a multi-coloured window. We encourage the reader to exercise his/her judgment, and imagine what the global picture may look like.

The chapters are self-contained and can be read independently. Some recurrent technical words are explained in the glossary. Most chapters provide some

---

[1] http://www.tosca-cost.eu
[2] http://www.cost.eu

references for further reading, while more specific and technical references are all grouped at the end of each part.

The number of authors involved in this handbook is large. At the end of it, you will find for each of them a short biography, enabling you to appreciate the diversity of the communities involved.

Acknowledgements

This handbook would never have seen the light of day without an active editorial team that played a major role in harmonising all the different chapters. We gratefully thank the members of this team: Ilaria Ermolli, Margit Haberreiter, Harry Kambezidis, Mai Mai Lam, Irina Mironova, Annika Seppälä, Hauke Schmidt, Eija Tanskanen, Kleareti Tourpali, and Yoav Yair.

We also thank Agnès Henri (EDP Sciences) for her assistance in helping us make a handbook out of such a large ensemble of chapters, each of which has a history.

Finally, COST, by funding our Action TOSCA made this whole endeavour possible.

Thierry Dudok de Wit, Jean Lilensten. and Katja Matthes,
August 2015

Part I

INTRODUCTION TO THE SUN-CLIMATE CONNECTIONS

© ISS, NASA

CHAPTER 1.1

THE EARTH'S ATMOSPHERE: AN INTRODUCTION

Kleareti Tourpali[1], Jean Lilensten[2] and Roxana Bojariu[3]

1 Introduction

The Earth's atmosphere, the thin envelope of air that surrounds our planet, consists of a mixture of gases, extending from the surface of the Earth to the edge of space and tied to the Earth mainly by its gravitational force, and partly by the magnetic field at high altitude. It comprises mainly molecular nitrogen (78% in mass) and molecular oxygen (21%), with the other 1% composed of trace gases, named so because their concentrations are very small. Despite their small amount, carbon dioxide, ozone and water vapor are very important to the Earth's climate and can have a large impact on atmospheric processes. Above about 200 km, atomic oxygen, although tenuous, becomes the major species while atomic hydrogen is the most abundant species above typically 1000 km. Depending on the approach taken, there are various ways one can consider a distinction between atmospheric layers: composition, temperature, electromagnetic properties.

Although pressure and density decrease with altitude in the atmosphere, temperature remains relatively constant or even increases with altitude at certain regions, so it may well be used to distinguish between atmospheric layers. Based on temperature, the atmosphere is divided into four layers: the troposphere, stratosphere, mesosphere and thermosphere, as illustrated in Figure 1. Based on composition and mixing, the atmosphere can be divided into two major layers, the homosphere and heterosphere, defined by whether the atmospheric gases are well mixed. The homosphere, starting from surface, includes the troposphere, stratosphere, mesosphere and the lowest part of the thermosphere. From there on lies

---

[1] Lab of Atmospheric Physics, Aristotle University, Campus Box 149, Thessaloniki 54124, Greece
[2] Institut de Planétologie et d'Astrophysique de Grenoble (IPAG), Bâtiment D de physique, BP 53, 38041 Grenoble cedex 9, France
[3] National Meteorological Administration, Sos. Bucuresti-Ploiesti 97, 013686 București, Romania

the heterosphere where the chemical composition varies with altitude, and gases are stratified due to their molecular weight. In this section, we discuss the properties of the atmospheric layers distinguished by both composition and temperature profiles.

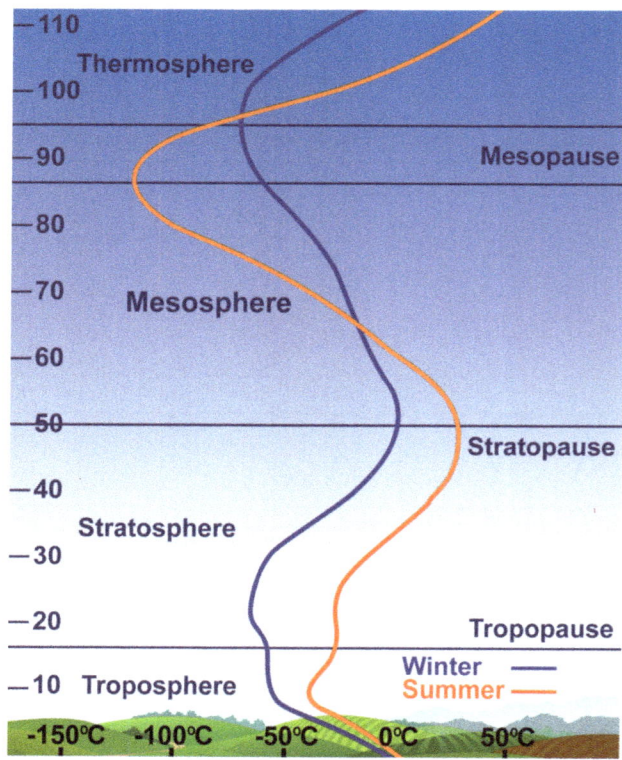

Fig. 1. Structure of the atmosphere and its layers based on temperature distribution during summer (orange line) and winter (blue line). (Adapted from http://www.athena-spu.gr/).

2 The homosphere

The amount of solar radiative energy that arrives perpendicularly to the Earth is on average 1361 W m^{-2}; this quantity is called Total Solar Irradiance (TSI). However, taking into account the spherical shape of the Earth, this value must be divided by 4 (i.e., 340 W m^{-2}) to obtain the overall mean amount of energy that reaches the Earth.

About 50% of this energy is used to heat the surface, 20% is absorbed in the atmosphere and 30% is reflected back to space (the albedo is the ratio between the amount of energy reflected or diffused and the amount received. It approaches 1 for snow, for instance, and therefore amounts to 0.3 on average on the surface of the Earth).

The energy absorbed on the surface of the Earth heats the ground and then, by conduction and convection, the air in the lowest layer of the atmosphere. It also leads to considerable evaporation from the oceans and lakes. This phenomenon, which is amplified by the transpiration of plants, results in a rate of relative humidity of more than 50% on average on the globe in the troposphere, which contains 99% of atmospheric water vapor.

2.1 The troposphere

The term troposphere is a Greek-derived term (*tropein* – to change, circulate or mix), indicating the turbulent nature of this layer where weather systems circulate air masses and move around almost all content of atmospheric water vapor. Water vapor concentration varies from trace amounts in polar regions to nearly 4% in the tropics. Most clouds are present in this layer, too. The troposphere contains around 75–80% of the total mass of the Earth's atmosphere. As mentioned above, the main constituents of the atmosphere are nitrogen, oxygen, argon and traces of ozone, and other gases and particles. Carbon dioxide is also present in small amounts, but its concentration has increased by more than 40% in comparison with the pre-industrial level (e.g., 1750) due to human activities. Like water vapor, carbon dioxide is a greenhouse gas, which traps a part of the Earth's heat in the troposphere.

The troposphere extends upward to about 16–20 km above the Equator, and to about 7–8 km above the poles. The tropospheric thickness depends on latitude, season, and diurnal cycle. It is highest in summer and lowest in winter. This seasonal effect is strongest at mid-latitudes.

The troposphere is mainly heated from the Earth's surface warmed by solar radiation. The uneven heating of the Earth surface by the Sun (which warms land and ocean at the Equator more than at the poles) leads to large-scale patterns of winds that move heat and moisture around the globe. Atmospheric circulation is deflected by the Earth's rotation as it goes between the poles and Equator, creating weather systems and belts of surface winds. The warmed air tends to rise leading to convection which is the mechanism responsible for the vertical transport of heat in the troposphere while horizontal heat transfer is due to advection. Air convection generates clouds and ultimately precipitation when liquid or solid water particles grow large enough in size to fall toward the surface. Temperature, pressure, air density and water vapor content in the troposphere decrease with height. Air temperature decreases with height at a mean lapse rate of about -6.5 C km^{-1}, up to the tropopause (the boundary between the troposphere and stratosphere). However, sometimes temperature can increase with height within limited tropospheric layers leading to temperature inversions, which prevent the vertical mixing of air. Such atmospheric stability can lead to air pollution episodes at ground level. The temperature is constant in the tropopause. In the troposphere, winds increase with height leading to jet streams which usually meander in the upper parts, moving air at speeds of about 100 km h^{-1}, just below the tropopause. Atmospheric circulation in the upper troposphere strongly influences weather systems in the lower troposphere, too.

The troposphere influences the stratosphere – the next layer beyond tropopause – through atmospheric waves that propagate upward. A number of studies show that this wave forcing leads to slowly evolving changes in stratospheric circulation, affecting polar areas (polar vortex; see next paragraph), which in turn can feed back to tropospheric phenomena (such as the Arctic/North Atlantic Oscillation) that shape anomalous weather regimes in winter.

2.2 The stratosphere

The stratosphere is a region where the air temperature increases gradually to the stratopause (found at an average height of 50 km), the area marking the border to the mesosphere (Figure 2). Due to the increase of temperature with height through the whole stratosphere, warmer air lies above colder air. Hence, convection is not favored. The atmospheric conditions are stable and vertical mixing is suppressed. Only small amounts of water vapor enter the stratosphere (due to their freeze-drying at the tropopause), so the stratosphere is a very dry region.

Ozone plays the major role in regulating the thermal regime of the stratosphere. The stratospheric region with the highest ozone concentration is known as the "ozone layer" or "ozonosphere", located at heights roughly between 15 and 35 km extending over the entire globe, with some variation in altitude and thickness. Its peak concentration occurs at heights from 26 to 28 km (or slightly higher) in the tropics and from about 12 to 20 km toward the poles. Typical ozone concentrations at its peak are about 10 parts per million by volume (ppmv), reaching to higher values over the tropics. Temperature increases with ozone concentration, as solar energy is converted to kinetic energy when ozone molecules absorb ultraviolet radiation, resulting in heating of the stratosphere. Temperature increases also due to thermal energy released during the ozone formation cycle. The reaction of oxygen (O) and oxygen molecules (O_2) that produces ozone is exothermal; heat is released and the stratosphere warms. As ozone is the main gas involved in radiative heating of the stratosphere, its solar-induced variations can directly affect the radiative balance of the stratosphere with indirect effects on circulation. A detailed account of solar-induced variations in the stratosphere and the ozone layer is given in Chapter 4.1.

Intense interactions take place among radiative, dynamical and chemical processes, in which the horizontal mixing of gaseous components proceeds much more rapidly than in vertical mixing. The horizontal structure of the stratosphere and the stratospheric circulation present a number of interesting features. Air rises in the tropical region from the troposphere into the stratosphere, is transported upward and poleward in the stratosphere, and descends in stratospheric middle and polar latitudes. This circulation, called the Brewer-Dobson Circulation, is strong in the respective winter hemisphere and transfers ozone and other trace constituents from the tropics to the polar regions, as illustrated in Figure 2. The polar vortex is a large-scale circumpolar cyclonic circulation (strong west-to-east winds that circle the polar regions). The polar vortex extends from the upper levels of the troposphere through the stratosphere and into the mesosphere (above 50 km) and is associated with low temperature and ozone values. Polar vortices occur over both hemispheres, strongest during the respective winter season, when the pole-to-equator temperature gradient is strongest. They play an important role in the coupling between the troposphere and stratosphere (e.g., Arctic Oscillation) and in ozone depletion (e.g., Antarctic Ozone Hole).

Other important features of the stratosphere are the quasi-biennial oscillation (QBO) and the Sudden Stratospheric Warmings (SSWs), both of which affect circulation and transport of tracers, such as ozone and water vapor (see the Glossary for their brief descriptions).

2.3 The mesosphere

The mesosphere's (middle sphere) average location is from about 50 to 90 km above the Earth's surface, and is characterised by temperature decreasing with height. Thus, the mesosphere is warmer at its lowest level (just above the stratopause), and becomes coldest at its highest level. The upper limit of the mesosphere, the mesopause, is at the altitude at which the temperature reaches a minimum. The mesopause, a region that experiences large height variations, is located at altitudes between 85 and 100 km, depending on latitude and season. The average mesopause temperature is about −90 °C (Figure 1), but the lowest temperatures

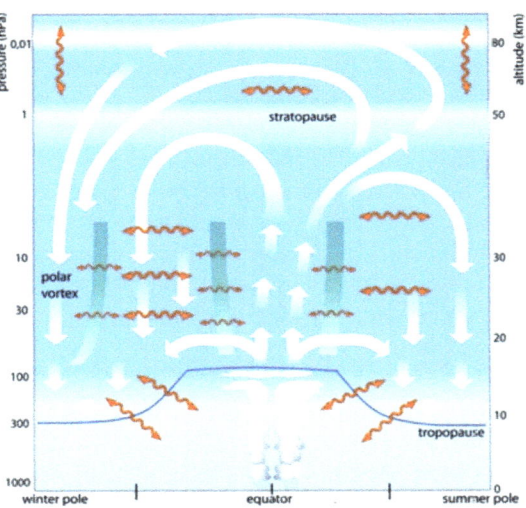

Fig. 2. Schematic representation of the Brewer-Dobson Circulation and Middle Atmosphere Transport. Arrows show the different pathways of air mass transport in the stratosphere and the stratosphere–troposphere exchange. Retrieved from Brewer-Dobson Circulation, (2011) http://www.eoearth.org/view/article/150779/.

are found over the poles during the summer season, where the temperature may often drop below $-100\,°C$ (at the same time, the other (winter) pole may have higher temperatures by several °C), so the summer polar mesopause is the coldest level in the whole atmosphere.

The percentage of oxygen, nitrogen, and carbon dioxide in the air in the mesosphere is almost the same as that in the lower atmospheric levels, but the density of air is much less. The water vapor content is very small, and the mesosphere contains higher percentages of ozone than the lower levels. The mesosphere is the coldest among the atmospheric layers, so that the small amounts of water vapor that are present there are frozen into ice crystal clouds. These are the highest clouds in the Earth's atmosphere, called Noctilucent Clouds and can be observed after sunset at high latitudes during summer.

The mesosphere is a region where solar radiation and energetic particles contribute to downward energy transfer, and gravity and planetary waves along with the atmospheric tides transfer energy upwards from the stratosphere. Strong variations occur in the mesosphere as a function of height, both in the physical processes as well as in the chemical composition, especially in the vicinity of the mesopause.

The mesosphere together with the lower thermosphere (sometimes referred to as Mesosphere-Lower Thermosphere or MLT) is the region where vertical turbulent mixing is gradually replaced by molecular diffusion as we move to higher altitudes. The amount of solar radiation available to photolyse atmospheric molecules increases rapidly; molecules dominate the atmospheric composition at lower levels and atomic species, which are lighter, are abundant in the upper layers.

The major contributions to the energy budget are the absorption of solar radiation by ozone and molecular oxygen, the emission of infrared radiation by CO_2, dissipative processes from the interaction with gravity waves, and transport of heat through advection. The mesosphere is also the layer in which many meteors burn up when they enter the Earth's atmosphere, as a result of the collision with some of the gas molecules present in this layer.

The stratosphere and mesosphere together are in many cases referred to as the "middle atmosphere". In all the "low" layers we have just described, molecules and atoms mingle, producing a homogeneous gas. The concentration is roughly 10^{25} particles per cubic meter at ground level and 10^{19} particles per cubic meter at the homopause (or turbopause), which is its upper limit.

3 The heterosphere, the thermosphere, the ionosphere

The heterosphere begins above the homosphere. It only became possible to explore the properties of this part of the atmosphere with the advent of radio communications in the twentieth century. Subsequently, sophisticated radar techniques and measurements by satellite revealed a complex, dynamic medium, a gas consisting of a mixture of electrically charged particles and neutral particles. This sheath still raises many questions about the part it plays in the ecosystem of the Earth and in the emergence of life on Earth.

In the heterosphere, the concentration of molecules and atoms becomes very low and each component behaves as if it were alone. Here, the perfect gas behaviour of the whole of the homosphere now applies separately to nitrogen, oxygen, and hydrogen, with a fundamental difference: each has its own scale height. The immediate result is a variation in their exponential concentration but with different decrease rates. At about 80 km, molecular nitrogen is predominant, followed by molecular oxygen. Above about 250 km , atomic oxygen prevails. However, at around 1,000 km, hydrogen becomes the most abundant element: since there is no longer a convection mixture, the heaviest elements stay in the lower layers and the lighter elements "float" above them.

One fundamental feature of the heterosphere is that it constitutes a filter for radiation in the extreme ultraviolet (EUV). These emissions, strongly correlated with solar activity, are absorbed in ionisation, excitation, dissociation and heating. We are therefore in a medium that is very different from the matter with which we are familiar, a mixture of neutral gas, ions that are more or less energised and electrons. The neutral gas has been given the name of thermosphere. The combination of ionised gas, ions and electrons is the ionosphere. The mixture of the ionosphere and thermosphere is a plasma called the "upper atmosphere". Its properties are quite different from those of a classic gas consisting of neutral particles, since the movement of the charged particles is sensitive to the electrical and magnetic fields. However, the proportion of charged particles remains low in comparison with that of the neutral particles: about one billionth at an altitude of 100 km and one tenth at around 1,000 km.

The heating of the upper atmosphere is indirect: the temperature of the atmosphere is increased by the friction of energised particles against those that have not been energised and by chemical reactions, not by direct interaction between the atmosphere and solar radiation. The temperature rises considerably above the mesosphere, roughly 8 to 10 °C per kilometer between 100 and 150 km. This is a positive temperature gradient: hot air no longer rises since it is already on top. Convection is inhibited and only conduction can transfer energy from one layer to the next.

This heating is effective as far as 200–300 km up; above this altitude, the atmosphere is too thin to conduct heat. Higher up, the temperature becomes constant and is then called exospheric temperature T_∞. Its typical value is between 1,000 and 1,200 K during a quiet period but it can exceed 2,000 K during periods of high solar activity[4]. This explains why this part of the atmosphere is called the thermosphere.

When the Sun is quiet, the temperature is about 1,000 K for the ions at 400 km and 1,500 K for electrons. However, these are low values compared with those that occur when the Sun is particularly active: the temperature of the ions can then be as high as 2,000 K and that of the electrons 3,000 K at the same altitude. The four effects of solar radiation on the high atmosphere (ionisation,

---

[4] Temperatures in the lower atmosphere are usually expressed in units of °C, whereas higher up, Kelvin (K) are used. Both temperatures are related by $T[°C] = T[K] - 273.15$.

dissociation, excitation, heating) depend on the intensity of the radiation and therefore on solar activity.

4 The magnetosphere

Above an altitude of about 600 km, and up to several terrestrial radii, particle concentration becomes so low that their behaviour is no longer a function of collisions but is a consequence of the configuration of the magnetic field. The nature of the environment changes and becomes the magnetosphere. The magnetosphere is not, strictly speaking, a part of the atmosphere, but rather part of the Earth space environment. However, as this book is dealing with the impact of the solar variability on the climate, the magnetosphere cannot be ignored.

The solar wind and the geomagnetic field interact in various ways. One is described in terms of both magnetic pressure (of the Earth's field) and kinetic pressure (of the solar wind). Furthermore, the magnetic field influences the charged particles of the solar wind, creating a large series of currents inside of the magnetosphere. We cannot describe all of them, and will only mention some that are relevant for this book.

The front of the magnetopause is located at approximately 10 terrestrial radii from the surface of our planet (the Earth's average radius is 6371 km). This distance increases to approximately 13 terrestrial radii during periods of low solar activity, and may occasionally drop to 6 radii during periods of high activity, i.e., when massive perturbations called "coronal mass ejections" are ejected from the Sun. On its flanks, the magnetopause is at about 15 terrestrial radii from the Earth. On the night side, it is shaped like the tail of a comet. On the outside, the space is subjected to the solar wind and the interplanetary magnetic field. Inside the magnetosphere, the terrestrial magnetic field is the controlling force. The magnetopause stands like a porous barrier between the two. On the night side, the magnetopause closes at several dozen of terrestrial radii.

A small fraction only of the solar wind can penetrate the magnetosphere. When its charged particles come closer to the Earth, they encounter an increasingly intense magnetic field. They end up being stopped, typically at a distance ranging from two to ten terrestrial radii, thereby creating a ring of current, also called Van Allen Belt, which surrounds the Earth at low latitudes. This ring is located between four and seven terrestrial radii; its current density is typically 10^8 A m^{-2}, and 90% of its particles have energies ranging from 10 keV to 250 keV, which is about 4 orders of magnitude more than in the solar wind. The influence of the ring current on Earth is noticeable at low latitudes, in equatorial regions. These particles are finally trapped and follow the magnetic field lines toward the south or the north magnetic poles. The rings plotted by the feet of the belt are located at high latitudes, between 65° and 75° North or South. They are called the auroral ovals, and coincide with the regions where auroras are most frequently observed.

Below five to six terrestrial radii from the surface of the Earth, the atmosphere is carried along by the rotation of the planet; this is known as "corotation". This area is called the plasmasphere. Here the density varies between 10 billion particles per cubic meter (10^{10} m^{-3}) at the top of the ionosphere, at approximately 1,000 km, and 100 billion (10^8 m^{-3}) at its outer boundary, the plasmapause.

Inside the plasmasphere there is another radiation belt, the first Van Allen belt, or inner belt. It is in a compact region centered above the equator, at an altitude of a few thousand kilometers. The origin of the belt is not to be found in the solar wind but primarily in the ionisation triggered by cosmic rays.

5 Conclusion

In the description of the Earth's atmosphere presented in this chapter, we have tried to give the reader a brief insight of the basic properties, composition, chemistry and processes that govern the atmospheric layers. This may seem as describing an onion: a series of layers that do not communicate with each other. However, this is not the case, as, and in order to understand the impact of solar activity on climate on its full extent, one should keep in mind that the Earth and its atmosphere constitute an integrated complex system of interlinked and interacting physical, chemical and biological processes. Our introduction here may serve as the background information for the understanding of the following chapters. Many processes occur at the interfacing boundaries between layers, and the physics of the system lies in the interactions. Moreover, this description is static, while the atmosphere is dynamic: it is constantly changing and interacting in a multitude of ways that connect layers to one another.

Further reading

Andrews, D. G., Holton, J. R., and Leovy, C. B.: 1987, *Middle Atmosphere Dynamics*, New York, Academic Press.

Brasseur, G. P. and Solomon, S.: 2005, *Aeronomy of the middle atmosphere: Chemistry and Physics of the Stratosphere and Mesosphere*, 3rd edition, Berlin, Springer.

Danielson, E. W., Levin, J. and Abrams E., 2003: *Meteorology*, New York, McGraw-Hill.

Holton, J. R., Curry, J. A., and Pyle, J. A. (editors): 2002, *Encyclopedia Of Atmospheric Sciences*, Ney York, Academic Press.

Lilensten, J., and Bornarel, J., 2006: *Space weather, environment and societies*, Berlin, Springer.

Peixoto, J. P., and Oort, A. H., 1992: *Physics of Climate*, New York, American Institute of Physics.

Wallace J. M. and P V. Hobbs, 2006: *Atmospheric Science: An Introductory Survey*, New York, Academic Press.

CHAPTER 1.2

THE IMPACT OF SOLAR VARIABILITY ON CLIMATE

Katja Matthes[1], Joanna Haigh[2] and Arnold Hanslmeier[3]

1 Introduction

Many observational studies have found solar influences on climate on decadal and longer timescales (see part 3 for details). There is little evidence that solar variations are a major factor in driving recent global climate change but considerable evidence for a solar influence on the climate of particular regions as well as throughout the atmosphere. The Sun impacts the stratosphere (15–50 km), the troposphere and at the Earth's surface as well as in the ocean (see part 3). During high solar activity, higher temperatures and larger ozone concentrations are observed in the tropical stratosphere, and stronger pressure gradients are observed over the North Atlantic and Europe. During solar maxima, the stratospheric winter polar vortex is stronger with positive zonal wind anomalies extending down, through the troposphere, to the ground. At the surface this can be expressed as a positive phase of the North Atlantic Oscillation (NAO), which is a feature of the natural variability of the climate characterised by the difference in surface pressure between the Icelandic Low and Azores High. A positive phase of the NAO is associated with stronger westerlies, more northerly storm tracks and milder winters in Europe and North America (Figure 1). During solar minima, the reverse picture is true: the stratospheric winter polar vortex is weaker and negative zonal wind anomalies extend to the surface. This represents a negative phase of the NAO and hence weaker westerlies and cold, snowy winters in Europe and North America. Recent work suggests that the NAO response peaks a few years after the solar cycle maximum, possibly due to atmosphere–ocean interactions over the North Atlantic (see Chapter 4.2 for details). Solar cycle signals have been also observed in the tropics where stronger trade winds, a stronger overturning (Walker) circulation

---

[1] GEOMAR Helmholtz Centre for Ocean Research Kiel, Düsternbrooker Weg 20, 24105 Kiel, Germany; and Christian-Albrechts Universität zu Kiel, Kiel, Germany
[2] Grantham Institute, Imperial College London, Exhibition Road, London SW7 2AZ, United Kingdom
[3] Universität Graz, Universitätsplatz 5, 8010 Graz, Austria

and shifts in precipitation, as well as changes in sea surface temperature patterns, occur during solar maxima (Figure 2). There is some observational evidence for a connection between geomagnetic activity and surface air temperatures. During active geomagnetic activity, which are characterised by energetic particle precipitation towards the Earth and which peak a few years after a solar cycle maximum, typical regional surface temperatures pattern have been reported (see Chapter 4.6 for more information).

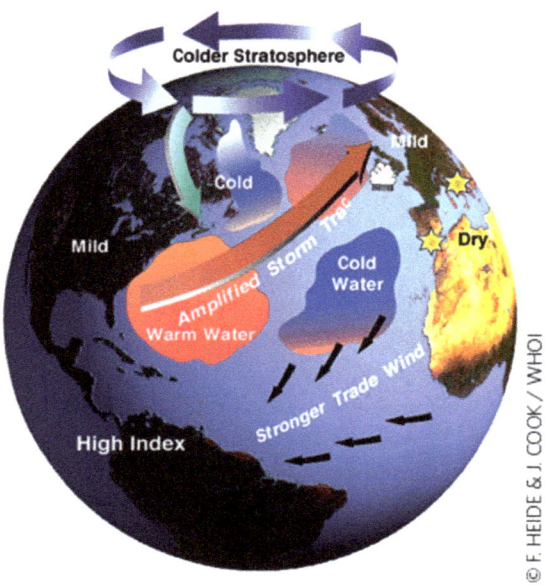

Fig. 1. The positive phase of the North Atlantic Oscillation and its relation to the stratospheric as well as oceanic circulations. When the NAO index is high, the North Atlantic storm track is stronger, and northern Europe experiences mild winters. During solar maximum years, the winter stratosphere is colder, the polar vortex is stronger and this associates with a more positive NAO at the surface. Copyright ©2007 Woods Hole Oceanographic Institution, All Rights Reserved.

The coincidence of an unusual long, deep solar minimum around 2008 together with the occurrence of a few cold and snowy winters in Europe since 2009 revived the discussion of solar influence on climate. On longer timescales, the Sun is currently declining from a grand maximum and moving towards a new grand minimum (see Chapter 2.5 and Infobox 2.2) although predictions of solar activity are very difficult so that the date of the next grand minimum is very uncertain. If the Sun were to enter a new Maunder-like minimum, it would make only a small compensation for human-induced global warming but it might well affect regional climate patterns. Over the solar cycle of about 11 years, solar variability offers a degree of predictability in particular for regional climate, especially in winter in the Northern hemisphere. Understanding solar variability might therefore help reduce the uncertainty of future regional climate predictions on decadal time scales.

The identification of solar signals in climate observations is often difficult because the available observational records are shorter than 100 years which would be at least needed to reliably detect robust solar signals (see part 3 for more details). Sufficient observations covering the atmosphere from the surface to the stratosphere and mesosphere only exist for the last three solar cycles and often solar signals are "hidden"/overshadowed by other signals, such as volcanic eruptions or ENSO or man-made climate change. For these reasons, the extraction of solar signals from data using statistical techniques needs to be very thorough and interpretation of the results should be carried out with caution.

There are a number of mechanisms proposed for how solar variability could impact climate. Climate models in combination with findings from observations are useful to test the proposed mechanisms and understand the pathways. A brief review of existing mechanisms is now presented (see part 4 for an in-depth overview).

2 Mechanisms for solar influence on climate

Figure 3 gives a schematic overview of how solar variability can influence climate. The influences can be categorised into two groups: the first is due to changes in the Sun's radiant output (TSI: Total Solar Irradiance; UV: Ultra Violet) and the second due to the Sun's influence on the energetic particles reaching Earth (SEPs: Solar Energetic Particles, GCRs: Galactic Cosmic Rays).

Total Solar Irradiance (TSI), which is the sum of all irradiance coming from the Sun, varies by only about 0.1% with the solar cycle (see Chapter 2.2 for more details). The irradiance is directly absorbed by the Earth and ocean surface and would lead to a small warming of about 0.1 K particularly in the tropics (see arrow with "TSI" in Figure 3) but is thought to be modulated through air–sea coupling; it changes precipitation and vertical motions which in turn influence trade winds and ocean upwelling. During solar maximum, this so-called "bottom-up" mechanism is proposed to lead to stronger Hadley and Walker circulations and associated colder Sea Surface Temperatures (SSTs) in the tropical Pacific (Figure 2). However, the details of the bottom-up mechanism and even the sign of SST response in the tropical Pacific are still under debate (see Chapter 4.2 for more details).

Fractional variations in the UV part of the Sun's radiative output (Solar Spectral Irradiance: SSI) are much larger than those in the Visible (VIS) and Near Infrared (NIR) reaching 5 to 8% in the region important for ozone chemistry (see Chapter 2.2 for more details). Enhanced UV radiation during solar maximum leads to a warming in the tropics around the stratopause and to greater ozone formation down below. The ozone absorbs UV radiation at longer wavelengths and gives an additional warming (see arrow with "UV" in Figure 3). The warming

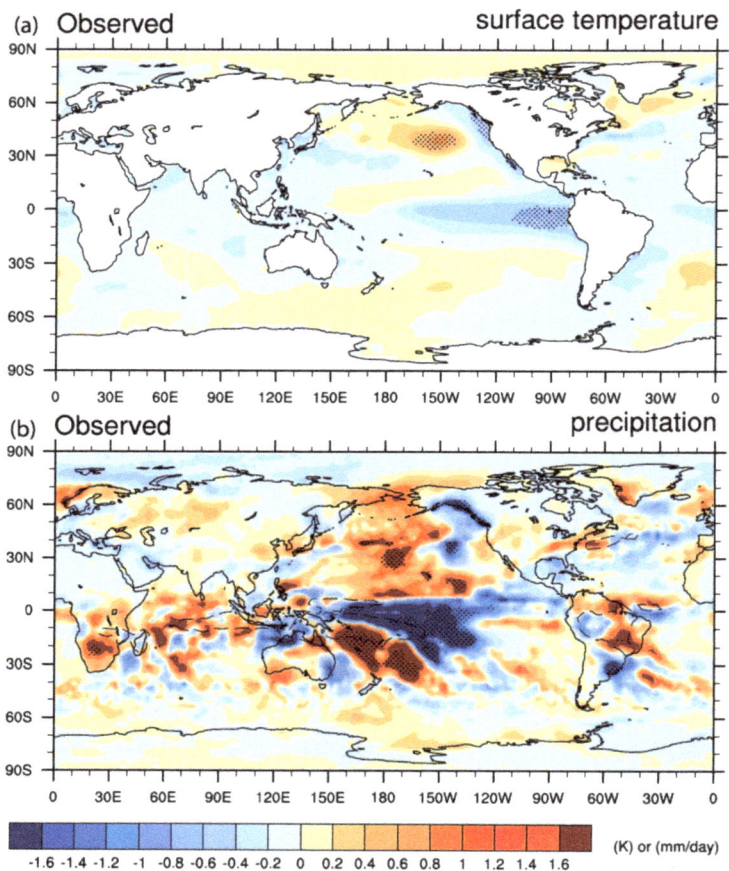

Fig. 2. (a) Composite average sea surface temperature anomaly in the Pacific sector for December, January, and February (DJF) for 11 peak solar years (C). (b) Same as (a) but for composite average surface precipitation anomaly from three available peak solar years (mm s^{-1}). Adapted from Meehl et al. (2009) by Gray et al. (2010) (with permission).

in the tropics leads to stronger winds in the subtropical upper stratosphere (ΔU in Figure 3), which influence the background state for planetary waves propagating upward from the troposphere (thick black arrows in Figure 3). Through complex interactions between the atmospheric flow and planetary waves, the signal in the middle atmosphere is transmitted down to the troposphere, where it modifies e.g., the NAO and leads to measurable regional effects (Figure 1). In addition to the effects in the polar region, there is also a warming in the tropical lower stratosphere which affects the vertical propagation of synoptic scale waves (thin black arrows in Figure 3). These small scale waves deposit momentum in the tropopause region and thus influence the strength and position of the sub-tropical

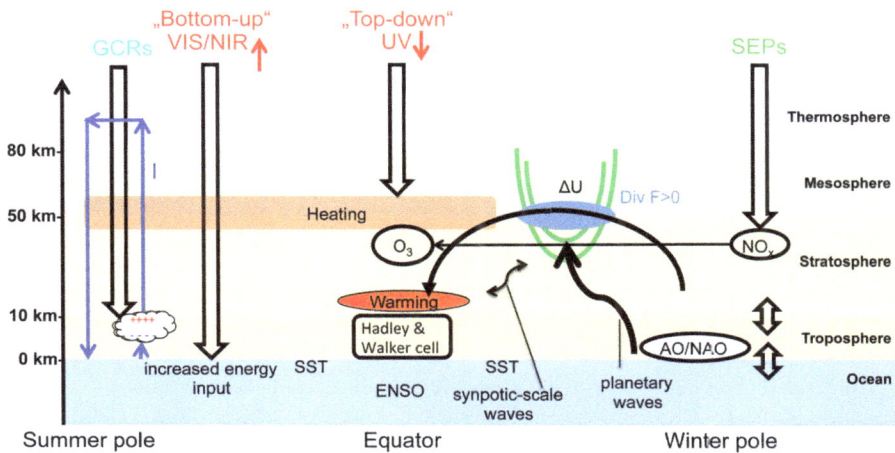

Fig. 3. Schematic representation of possible mechanisms for solar influence on climate based on Kodera and Kuroda (2002) and redrawn from Gray et al. (2010). Shown are direct and indirect effects through solar irradiance changes (VIS/NIR and UV) during solar maxima as well as energetic particle effects (SEPs: Solar Energetic Particles and GCRs: Galactic Cosmic Rays). The two double arrows denote the coupling between different atmospheric layers (troposphere and stratosphere) or the atmosphere and the ocean. See text for details.

jets. The changes in the vertical structure may also influence convection in the tropical troposphere. Since a number of other factors, such as ENSO, the QBO, or volcanic eruptions, influence the temperature of the lower stratosphere, the detection of solar signals is complex here (see Chapter 4.3 for more details on this). The processes, involving stratospheric heating by solar UV and subsequent dynamical adjustment, are sometimes characterised as a "top-down" mechanism and have been confirmed in a number of climate modeling studies (for a review, see Gray et al., 2010; see Chapter 4.2 for more details).

Besides the two solar irradiance effects, which are likely to work in combination, the effects of energetic particles need to be considered. Solar Energetic Particles (SEPs), such as solar protons or energetic electrons, which enter the atmosphere at polar latitudes during high geomagnetic activity, ionise atmospheric particles and form radicals, such as NO_x, which are transported down into the stratosphere during winter (arrow with "SEPs" in Figure 3; for more details see Chapters 4.5 and 4.6). These radicals play an important role in ozone chemistry and there is evidence that air depleted in ozone following an SEP event is transported over a period of months from the polar upper stratosphere to lower altitudes and latitudes. Some climate models already include particle effects (in particular solar proton events and auroral particles), but the magnitude of the different effects, as well as their possible long-term climate impact, are still under investigation (see Chapter 4.6).

Even less well established are the possible climate effects of Galactic Cosmic Rays (GCRs). In response to the strength of the heliospheric magnetic field, the incidence of GCRs on Earth is inversely correlated with solar activity (see Chapter 2.3). GCRs ionise the lower atmosphere with a peak in ionisation rates near the tropopause (arrow with "GCRs" in Figure 3). It is proposed that ionised aerosols can act preferentially as cloud nuclei and hence a higher incidence of GCRs might increase cloud cover. The processes necessary for this to take place are under critical revision, specifically with cloud chamber experiments at the CERN particle accelerator in Geneva (see Chapter 4.7 for more details). GCRs also contribute to variations in the Global Electric Circuit (GEC) and another area of research concerns whereby cloud formation could be affected (see Chapter 4.8 for more details).

The impact of solar variability on climate has been studied for a long time and some of the mechanisms involved are now being clarified. Climate models are continuously improving but new and continuous measurements of solar activity, as well as of atmospheric parameters, are needed in order to improve our understanding. The inclusion of all important aspects in climate models may contribute to the improvement of regional climate predictions.

Further reading

Gray, L.J. et al.: 2008, *Solar Influences on Climate*, Rev. Geophys., 48, RG4001.

Kodera, K. and Kuroda, Y.: 2002, *Dynamical response to the solar cycle*, J. Geophys. Res., 107, 4749.

Meehl, G.A., J.M. Arblaster, K. Matthes, F. Sassi, and H. van Loon: 2009, *Amplifying the Pacific Climate System Response to a Small 11-Year Solar Cycle Forcing*, Science, 325, 1114.

CHAPTER 1.3

THE SUN-EARTH CONNECTION, ON SCALES FROM MINUTES TO MILLENNIA

Thierry Dudok de Wit[1]

1 Sun-climate interactions: a tale of many mechanisms

The Sun affects the terrestrial environment in many different ways, and is its main energy source, continuously providing $2 \cdot 10^{17}$ W. The main mechanisms by which this energy enters the atmosphere are illustrated in Figure 1. Energy-wise, solar radiative forcing in the visible, near-ultraviolet and infrared bands constitutes by far the dominant input into the Earth's environment, and represents over 99% of all the natural forcings. However, if we consider instead the variable contribution of each input, then a completely different picture emerges, in which several seemingly weak mechanisms end up having significant leverage too. For example, the amplitude of the variations observed in the ultraviolet band becomes comparable to that observed in the visible band. Likewise, the energy input associated with energetic electrons that precipitate intermittently into the upper atmosphere cannot be neglected anymore. However, the impact of each of these solar inputs on the terrestrial environment also strongly depend on what time-scale they are operating. As it turns out, these time scales cover a wide range, and our simple energy budget reveals one side only of the full story.

Most Sun-climate investigations proceed by comparing a solar driver at one particular time scale, and looking for a possible climate response at that same time scale, with the hope of using their possible correlation to disentangle the multiple mechanisms that could lead to such a response. Scientists have been hunting down for cycles, because these can be more easily identified in noisy observations. This systemic approach wherein the response of a system (the Earth's environment) to a small perturbation by an external driver (solar variability) is studied, is widely used in many disciplines. However, for a system that is as complex and nonlinear as the Sun-Earth system, unexpected results may show up.

---

[1] LPC2E, University of Orléans, France

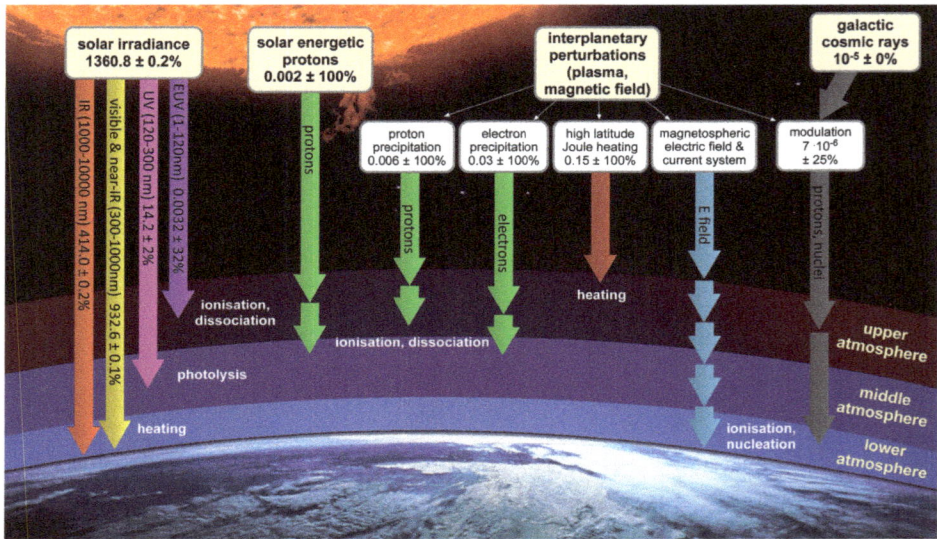

Fig. 1. Main solar inputs into the terrestrial environment. Numbers refer to their average energetic contribution in W m^{-2}, followed by their relative variability (when known) over an 11-year solar cycle.

2 Variability at all time scales

For climate change studies, the quantity of interest is the relative variability of the solar input, not its absolute power. The Sun varies on all time scales, ranging from fractions of a second to millennia (see Figure 2), and will eventually decay into a white dwarf, in about 6.5 billion years. The causes of this variability, and its manifestations are highly time scale-dependent, as are their impacts on the terrestrial environment (see Chapter 2.1, and 2.5). By analogy with terrestrial weather and climate, we are witnessing today the emergence of two new disciplines that are respectively called space weather and space climate. Clearly, both are two facets of the same issue.

The magnitude of the variability of the Sun-climate connection is sketched in Figure 3 by means of the power spectral density (or Fourier spectrum) of the total solar irradiance (TSI), which is the total solar radiative input, integrated over all spectral bands. Also shown is the power spectral density of the mean temperature of the Northern Hemisphere. While the two should not be considered as being representative of the full diversity of the Sun-climate connections, they do reveal several interesting properties. Here, the smallest period is arbitrarily set to one day, although much shorter variations can be observed.

The Sun, which has often been considered as a hallmark for stability, shows up in this figure as a highly dynamic object, which varies on all time scales. Note in particular how both power spectral densities gradually increase at low frequency, which indicates that long time scale variations, on average, have much larger amplitudes than short time scale ones.

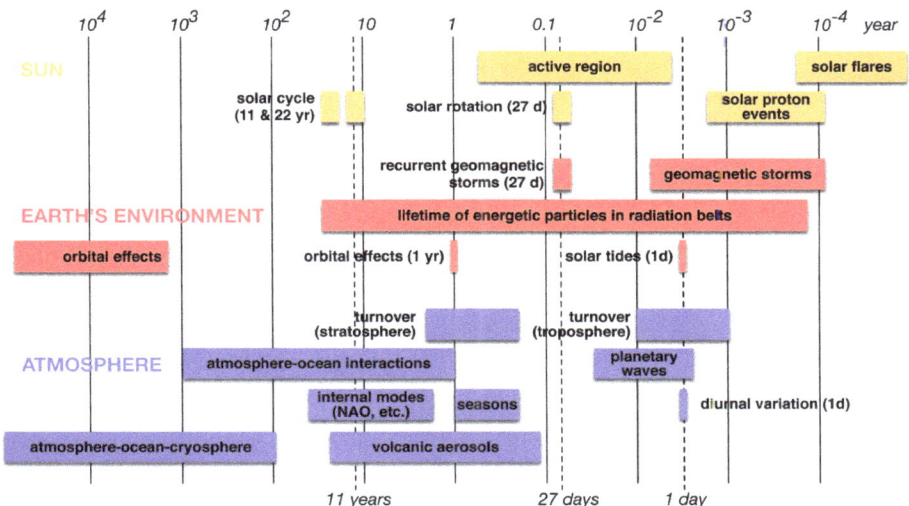

Fig. 2. Main characteristic time scales involved in the Sun-climate connection. Several (e.g., chemical reactions) occur on shorter time scales, and are not shown.

Two of the most conspicuous features in the power spectral density of the TSI are a broad peak near 27 days, and a much sharper one at about 11 years. The former corresponds to quasi-periodic modulation of about 27 days of the solar forcing, which is caused by solar rotation. Although the Sun does not rotate like a rigid body, many observables exhibit this 27-day quasi-periodicity, whose nature is geometric and not intrinsically solar.

Such periodic or quasi-periodic external forcings offer a unique opportunity for hunting for signatures in climate records by looking for variations that exhibit the same period, and are in phase with the forcing. Figure 3 does not show any in the terrestrial temperature, which might be erroneously taken as an indication for the lack of solar signature in climate records at those time scales.

Four important conditions need to be fulfilled in order for such a periodic signal to be detectable.

1. The driver should be highly coherent, in the sense that several consecutive oscillations of the same period should be observable. This is indeed the case with bright solar features, such as plages and faculae, whose lifetime of several months allows them to be repeatedly observed during several solar rotations. This repetition results in periodic enhancements of the solar UV flux, similar to a lighthouse beam striking the Earth, see Chapter 2.2. The same occurs for the alternation between fast and slow solar winds, which

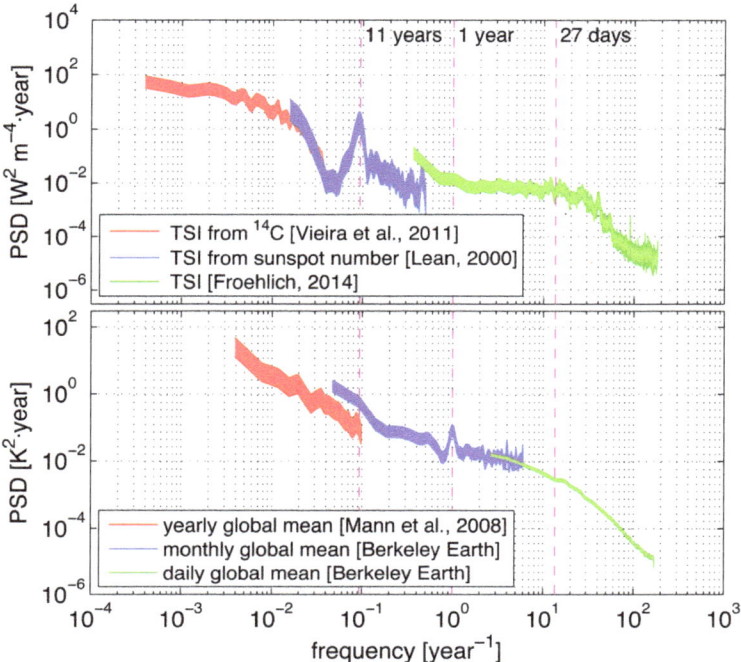

Fig. 3. Power spectral density of the one particular solar input (the TSI), and one particular climate response (average temperature over the Northern Hemisphere). The shaded regions represents a confidence interval of 95 %; Welch periodograms were used. The different colors correspond to different observations or reconstructions, and so no perfect match between them should be expected.

periodically strike the Earth and cause recurrent geomagnetic storms, see Chapter 2.4. Flare-emitting active regions on the Sun, on the contrary, have shorter lifetimes, and thus rarely have a strong periodic imprint.

2. The second condition for having a detectable signal deals with the characteristic time scale of the physical and chemical process(es) by which the solar input eventually affects climate. Variations in the UV flux immediately heat the upper atmosphere (which strongly absorbs it) and consequently lead to a sizable response of local parameters, such as neutral density and temperature, see Chapter 4.1. The solar impact on lower atmospheric layers in contrast involves processes that tend to have much longer response times, which results in a strong dampening of their response. Signatures of the 27-day modulation in the UV flux have indeed been well documented in the ionosphere, much less so in the mesosphere, and not below the stratosphere. This does not necessarily mean that long-term (e.g., centennial of millennial) changes in its amplitude has no impact on climate. However, they are

very unlikely to play a role, and this definitely cannot be diagnosed only by looking at their periodic signature. This is a typical situation wherein climate models are a precious ally for doing attribution, i.e., for ascertaining the mechanisms that are responsible for such changes.

3. A third condition for having a detectable signal is the absence of additional forcings that may disrupt the periodicity. A typical example is the irregular rate of volcanic eruptions, which considerably hinders the search for solar cycle periodicities in aerosol signals. Another example is the terrestrial rotation, whose alternation of sunlit and dark phases strongly modulates the ion content, and each night acts as a periodic reset of the the local plasma conditions.

4. Finally, the observable should properly capture the periodic cycle. Sometimes apparently innocent changes can bring in major differences. The temperature record used in Figure 3, for example, is heavily based on tree rings, which mostly reflect summer temperatures only. Ice core and speleothem records are more sensitive to winter temperatures, and show a stronger solar signal (Steinhilber et al., 2012). Likewise, winter temperatures from the North Atlantic exhibit a stronger solar signal than global temperatures, presumably because solar variability influences the blocking situations that affect winter temperatures in that region (Lockwood et al., 2010).

There is one additional type of coupling which is not properly revealed by Figure 3. Some solar inputs are impulsive and highly intermittent; however, their rate of occurrence is modulated by the Schwabe cycle. Solar flares are a typical example, whose impact on the middle and upper atmosphere is described in Chapter 4.1. Whether a modulation of their rate on multi-decadal time scales may affect climate, is still an open question.

From this, we can already conclude that a strong modulation in a solar signal does not necessarily imply an equivalent signature in climate records, and that the situation is likely to be much more complex.

The next most conspicuous cycle in Figure 3 is the yearly one. However, since it has no immediate connection to solar variability, we shall not consider it. Let us stress, however, that the Earth's environment exhibits a great variety of internal climate oscillations, or climate cycles, that have a priori no connection with solar variability. To name a few: the El Niño southern oscillation (2 to 7 years), the Atlantic multidecadal oscillation (50 to 70 years), and the Pacific decadal oscillation (8 to 12 years). These cycles are not strictly periodic, which explains the absence of a clear peak in the power spectral density of the global temperature.

The cycle that has attracted most attention is the Schwabe cycle, whose period is approximately 11 years, and which is associated with the solar magnetic cycle. However, let's proceed with longer time scales first. There has been considerable controversy over the existence of longer cycles in solar and climate signals. The Gleissberg (87 years), Suess (210 years), and Hallstatt (2300 years) cycles are well documented (Usoskin, 2008). These cycles have received much attention because of their potential impact on the long-term predictability of the Sun, and thus climate as well. The study by Eddy (1976) has demonstrated convincingly that irregular variations in solar surface activity could be connected with climate shifts, and has largely contributed to raise awareness for this connection. The physical explanation of such astronomical cycles has remained elusive: although some could be ascribed to planetary influences, a physical explanation for their impact on solar variability has yet to be found. Moreover, false claims for periodicities have been reported, and a more rigorous use of advanced statistical techniques, coupled with physical models, is definitely the way forward.

Even longer time scales bring us to the Milanković cycles, by which climate is strongly influenced through orbital forcing, with periodicities that are due to variations in the eccentricity of the Earth's orbit, and in the tilt and the precession of the Earth's rotation axis. These cycles have received overwhelming theoretical support, but there are also some open questions too. Their time scales, which are beyond 10 000 years, are beyond the scope of this chapter, but are further addressed in Infobox 2.1.

3 The 11-year solar cycle: more complex than it looks

The characteristic time scale that has by far received most attention is the 11-year Schwabe cycle, which is also the strongest one in Figure 3. The true period of the solar magnetic cycle is actually about 22 years, with two periods of alternate magnetic polarity. However, most (but not all) solar inputs to the Earth's environment are sensitive to the absolute value of the magnetic field only, hence their conspicuous 11-year modulation.

Many studies have investigated the Sun-climate connection by searching for signatures of this 11-year periodicity in climate records. Multiple examples of statistically significant correlations have been reported in the upper atmosphere (i.e., in the ionosphere/thermosphere), because most of the variations occurring at that altitude are directly driven by the strong modulation of the UV flux. Relative changes of 100% between periods of low/high solar magnetic activity are not uncommon, and so they can be easily detected.

This situation, however, becomes much more complex when moving down to the lower atmosphere (i.e., the troposphere), where less numerous convincing examples of significant correlation have been found (Pittock, 1983). Some of the reasons for this absence have already been given above. Energetically speaking, the prime solar input which directly heats the lowermost layers is the visible and near-infrared flux. Both vary by about 0.1% over a solar cycle. The resulting modulation of climate parameters is completely dominated by the internal variability of the climate system, thus making it extremely difficult to detect statistically meaningful correlations without having long records.

Correlations are frequently misinterpreted as an evidence for causation, whereas they actually give no physical insight on the mechanisms that are at play. In some occasions, additional information may come from the phase lag between the solar driver and the terrestrial response. Indeed, this may allow to discriminate between fast responding mechanisms (e.g., direct heating of the atmosphere) or more complex feedbacks, such as a bottom-up coupled ocean-atmosphere surface response.

By analogy with dynamical systems, one may also expect some of the internal modes of the climate system to couple with the weak solar modulation, and eventually synchronise with it. One of the internal modes that has come under close scrutiny is the North-Atlantic Oscillation (NAO), whose quasi-decadal periodicity is suspected of synchronising with the solar cycle when the latter is strong enough. Such a synchronisation has important and unexpected implications, to be considered next.

4 Detection and attribution of climate effects

Most correlation studies, whether based on periodic signals or not, rely on the idea of connecting the response of a perturbed system to that of its driver. As long as the perturbations are relatively small, one may reasonably assume that the climate system responds in a linear way too, even if the underlying physics is described by nonlinear equations. By linear we mean that a doubling of the small perturbation will produce a response that is about twice as large.

This linear perturbative approach is well founded in many disciplines. It also applies reasonably well to the climate system, and is often the best strategy for disentangling the impact of various forcings. But there are limitations to it, such as tipping points, wherein a system may respond linearly until it suddenly switches to another state. Such a bifurcation is precisely what may occur if indeed some internal mode couples to the solar cycle, such a synchronisation may be absent under quiet solar conditions (for example during the Maunder minimum, see Chapter 1.4), and then suddenly switch on.

In the case of a possible coupling with the North-Atlantic Oscillation, another unexpected consequence is the emergence of regional impacts. Indeed, the North-Atlantic Oscillation strongly influences winter conditions in Europe by affecting the westerly winds that bring warm and moist air. Recent results show that weaker westerly winds occur in winters with a less active Sun. Several studies corroborate this new paradigm wherein solar influences may be weak or undetectable at a global scale, and on the contrary significant at a regional scale (see also Chapter 4.3). Incidentally, this may also explain why the Maunder minimum had its strongest impact on Europe, regardless of its better documentation in that area.

The response to other forcings (e.g., greenhouse gases, and aerosols) is also largely controlled by regional feedback processes, rather than directly by the forcing itself. This loose connection between the spatial pattern of the climate response, and the forcing that has led to it, makes the attribution problem very challenging (see also Chapter 3.9).

5 To conclude

Perturbative experiments, wherein the response of a system (the Earth's climate) responds to an external driver (the Sun) provide a powerful means for shedding light on the possible mechanisms by which solar variability may affect climate. Signatures of the 11-year solar cycle have been reported in many climate observables. The response is generally strongest in the upper atmosphere, but signatures have been observed as well in the troposphere.

However, great caution is needed: extrapolating such correlations to other time scales, and drawing for example conclusions on what happened for example during the Maunder minimum, is very risky, because of the complexity of the processes involved. In addition, correlations between solar variations and climate observables may at best suggest the existence of some connection. They rarely give deeper insight because the same variation at the Sun often affects different solar forcing mechanisms, which thus cannot be disentangled. Furthermore, effects such as positive feedback may end up leading to the same signature for different forcings, which complicates their attribution even more. Smoking guns are hard to find.

And finally, there are roads for further investigation: the plausible impact of solar variability on internal modes of the climate system is likely to change in the future our interpretation of Sun-climate correlations. So will be the more rigorous application of statistical techniques that go beyond mere correlation tests.

Further reading

Feldman, T.: 2014, *Changing Sun, changing climate*, http://www.aip.org/history/climate/solar.htm (last visited 12/2014)

Haigh, J. D., M. Lockwood, M. S. Giampapa, I. Rüedi, M. Güdel, and W. Schmutz (eds.): 2005, Saas-Fee Advanced Course 34: *The Sun, Solar Analogs and the Climate*, Springer Verlag, Berlin.

Pap, J., P. Fox, C. Fröhlich, H. S. Hudson, J. Kuhn, J. McCormack, G. North, W. Sprigg, and S. T. Wu (eds.): 2004, *Solar Variability and its Effects on Climate*, vol. 141 of Geophysical Monograph Series, Washington DC, 2004, American Geophysical Union.

Pittock, A. B. 1983: *Solar variability, weather and climate: an update*, Quarterly Journal of the Royal Meteorological Society, 109, 23.

Usoskin, I. G.: 2013, *A History of Solar Activity over Millennia*, Living Reviews in Solar Physics 10, 1, http://solarphysics.livingreviews.org/Articles/lrsp-2008-3/

CHAPTER 1.4

THE ROLE OF THE SUN IN CLIMATE CHANGE: A BRIEF HISTORY

José M. Vaquero[1] and Ricardo M. Trigo[2]

1 Introduction

Since ancient times, the Sun has been recognized as the main source of light and heat for our planet. Therefore, it is not surprising that we can find numerous examples in history of people who were conscious that changes in the Sun had necessarily a major impact on the Earth and, in particular, on its climate. In this chapter, we provide a brief introduction to this topic, identifying the major steps over the centuries, including the fundamental ideas and basic concepts of the Sun-Earth relationship.

Making such a synthesis is not a simple task as there are many remarkable facts, innovative ideas and characters of interest. We opted to focus on the key dates that have changed the paradigm of the Sun's influence on terrestrial climate. Based on those major milestones, four distinct eras are distinguished here (Figure 1) that help to better understand the development of these ideas.

The first stage encompasses the initial ideas about the influence of the Sun on the Earth. However, these early speculations were later further investigated by the telescopic discovery of sunspots in 1610. Starting from this date, a new era of ideas and events unfolded, namely by considering the darker regions of the solar photosphere as essential observational elements. The culmination of the process initiated by telescopic observers of sunspots was the discovery by Schwabe of the solar cycle in 1844 (Schwabe, 1844). Since then the idea of finding the same periodicities in the Sun's variability and Earth's climate become attractive to a wide community of scientists, in particular to find the 11-year periodicity in solar radiation measurements from the Earth surface. However, despite the numerous efforts undertaken in the next century, it was found impracticable to make out

---

[1] Departamento de Física, Centro Universitario de Mérida, Universidad de Extremadura, Avda. Santa Teresa de Jornet 38, 06800 Mérida, Badajoz, Spain

[2] Instituto Dom Luiz – Faculdade de Ciências, Universidade de Lisboa, Campo Grande, Edf. C8, 1749-016 Lisboa, Portugal

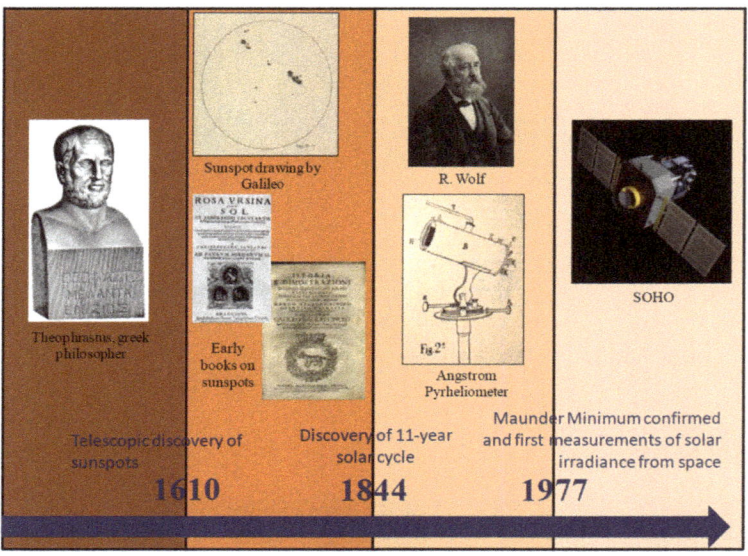

Fig. 1. Main historical milestones and corresponding eras that proved the way in our understanding of the role of the Sun on the terrestrial climatic system.

any solar periodicity based on measurements of solar radiation from the surface of the Earth. This incapacity led to the start of the fourth stage of this brief history, which was largely based on two distinct post-war technological revolutions: first the robust and precise measurements of solar irradiance from satellite platforms, and also the access to modern super-computers that allowed to probe and analyze the solar signal through sophisticated numerical models of climate.

2 Early intuitions and beliefs

As in so many other areas of science, if one is looking for the first theories it is appropriate to go back to the first centuries BC when classical Greek culture flourished. These early observers were watching sunrise and sunset to firmly establish the calendar. Although there are no clear evidence of naked-eye sunspot observations in Greek classical texts, we can infer that naked-eye sunspot observations were probably connected with changes in the weather. For example, the surviving fragments of "De Signis Tempestatum" (On weather signs) by Theophrastus (371–287 BC), contain rather explicit references to sunspots in three fragments. This work by Theophrastus consists of several chapters describing signs of different weather meteors. In particular, one of the sentences indicates the following: "if the Sun has a black mark when it rises, or if it rises out of clouds it is a sign of rain" (Vaquero and Vázquez, 2009).

This ability to report occasional examples of naked-eye observations of sunspots is not restricted to the Greek context, as it can be found in other ancient cultures. For example, a description of an naked-eye observation of a sunspot associated with heavy rainfall in the following days appears in the Arabic historical source "Al-Muqtabis-V": "At the end of this year, a strange and unknown prodigy occurred in the solar disc, being covered by a patent spot, visible by eye, situation that continued seven complete days (October 14–20, AD 939). At the end of the week that spot in the Sun disappeared because of a copious rain fallen down during the night [...]".

The Greek belief in weather harbingers persisted in the Western culture for centuries. Thus, the first books printed in Europe on weather forecasts contain "harbingers of the Sun" to make weather predictions. The book "Chronology or repertoire of times" (Sevilla, 1548) by Jerónimo de Chaves contains many examples of these prediction rules that were abundantly replicated in the popular astrological literature for centuries.

3 Sunspot and the first crude Sun-Earth studies

The first telescopic observations of sunspots were made by Galileo and others in 1610 and contributed to the scientific revolution of the 17th century. The detailed observations of dark areas at the Sun's surface, its alterations and rapid evolutions provide the first scientific evidence of the structural changes that occur often in our star. Hoyt and Schatten (1997) have shown that early on, some of these observers tried to associate the occurrence or absence of sunspots with weather or climate events. For instance, in his book "The eye of Enoch and Elijah" (1645), Rheita proposed that sunspots were the cause of a cold June in Colonia (Germany) in 1642. Likewise, in his book "New Almagest" (1651), Riccioli stated that low temperatures are associated, according to his observations, with the presence of more sunspots. It is still possible to find many other examples in the 18th century. Thus, after the deep solar minimum of 1784, the Mexican astronomer Alzate wondered: "Would the seasonal alterations experienced in Europe, and here, depend partly on that cause? Only experience will tell".

However, the most outstanding work on the relationship between the Sun and Earth weather was published at the beginning of the 19th century by Sir William Herschel. In 1801 he wrote two papers in the "Philosophical Transactions of the Royal Society of London" (Figure 2) in which he noted that the absence of sunspots coincided with severe weather seasons. Due to the lack of long-term meteorological series in that epoch, he used records of the price of wheat in England as a proxy for meteorological information. Herschel hypothesized an anticorrelation between the sunspot number and the market price of wheat. Dozens of works have revisited and updated Herschel's results along the years. However, modern studies have shown that we should treat this hypothesis by Herschel with some skepticism.

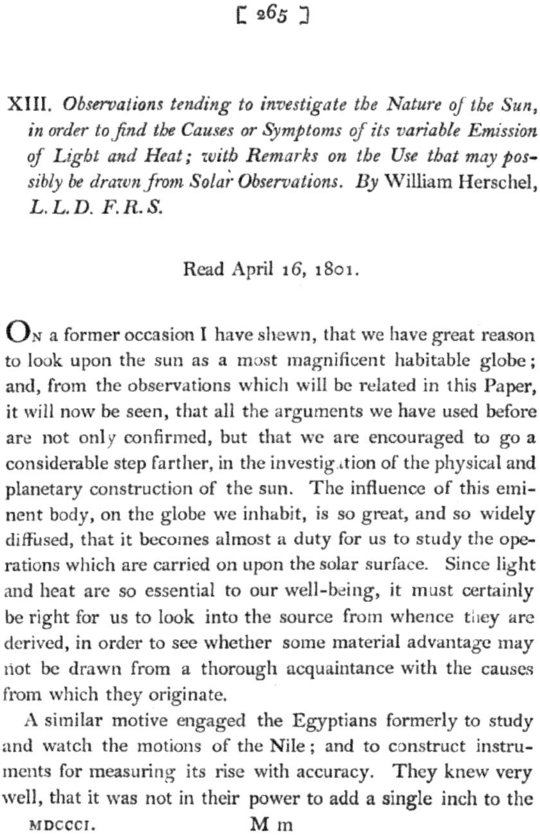

> [265]
>
> XIII. *Observations tending to investigate the Nature of the Sun, in order to find the Causes or Symptoms of its variable Emission of Light and Heat; with Remarks on the Use that may possibly be drawn from Solar Observations.* By William Herschel, L.L.D. F.R.S.
>
> Read April 16, 1801.
>
> On a former occasion I have shewn, that we have great reason to look upon the sun as a most magnificent habitable globe; and, from the observations which will be related in this Paper, it will now be seen, that all the arguments we have used before are not only confirmed, but that we are encouraged to go a considerable step farther, in the investigation of the physical and planetary construction of the sun. The influence of this eminent body, on the globe we inhabit, is so great, and so widely diffused, that it becomes almost a duty for us to study the operations which are carried on upon the solar surface. Since light and heat are so essential to our well-being, it must certainly be right for us to look into the source from whence they are derived, in order to see whether some material advantage may not be drawn from a thorough acquaintance with the causes from which they originate.
>
> A similar motive engaged the Egyptians formerly to study and watch the motions of the Nile; and to construct instruments for measuring its rise with accuracy. They knew very well, that it was not in their power to add a single inch to the
>
> MDCCCI.   M m

Fig. 2. First page from famous article by Herschel (1801), investigating the relationship between sunspots and price of crops (Courtesy Historical Archive of Sunspot Observations).

4 Laying the foundations; solar cycle and irradiance measurements

The next key event in this brief chronology took place in 1844. During this year, H. Schwabe, an amateur German astronomer, published in the "Astronomische Nachrichten" a brief note announcing the discovery of a 11-year cycle in the solar activity. This discovery had a low impact in the astronomical community at that time, but with hindsight, we know that this discovery was the trigger for two revolutionary developments in the study of solar-terrestrial physics over the subsequent decades. The first one was that development by R. Wolf of the Sunspot Number series (Clette et al., 2014). Thus, for the first time, a long and standardized time series was now available for the scientific community for systematic comparisons with meteorological records. The second advance resulted from the new evidence also found in the 19th century for a similar cycle in geomagnetic observations. Moreover, the first observation of a solar flare on September 2, 1859 by Carrington and Hodgson (independently) and the related large geomagnetic storm was the confirmation of the close relationship between the Sun and the Earth's magnetic field (Vaquero and Vázquez, 2009).

Thus, a period began with studies of the relationship between the Sun and climate being dominated by the search for common cycles (or extreme events) in both the Sun and climate systems. In this context, scientists needed to measure the variations of the solar brightness to associate it with the sunspot number. In many observatories around the Earth, systematic observations of solar radiation began. However, the first measurements with actinometers revealed the huge difficulty of separating the contribution due to the Sun from the contribution due to other local factors. This difficulty led to the development of increasingly precise (and complex) instruments, such as the pyrheliometer. Since the energy per unit area received at the Earth did not seem to vary with time, this amount was named solar constant (1361 W m^{-2}).

Among the various attempts to improve the solar constant, we should highlight the long-term effort led by the Smithsonian Astrophysical Observatory Solar Constant Program (Hoyt, 1979). This program derived the solar constant from 1902 to 1962 using measurements from many locations on the Earth's surface, such as Mt. Montezuma (Chile), Mt. Wilson (USA) and Bassour (Algeria). Although data from Smithsonian Astrophysical Observatory show a positive correlation between solar constant and the sunspot number, it is not statistically significant. This low correlation indicates the extreme difficulty of detecting a solar signal at ground level due to atmospheric absorption and scattering. This led to the conclusion that the solar constant should be measured at very high altitudes, above the atmosphere, i.e with dedicated satellite platforms. In this regard, the Smithsonian program was not a complete failure, as it allowed to establish the required level of accuracy and reproducibility that space instruments should reach for measuring accurately the solar irradiance.

During the first half of the 20th century, some pioneering studies on the influence of the Sun on biological systems were initiated. In particular, we would like to emphasize the ground-breaking work of A. Chizhevsky (Kornblueh, 1965). The search for common cycles between sunspot number (or other indices of solar activity) and meteorological parameters continues even nowadays. During this third stage, other long-term cycles (or quasi-cycles) were found in the series of solar activity indices, particularly the Gleissberg cycle (about 88 years) and the Vries or Suess cycle (approximately 210 years). More relevant within the context of this book, the 11-year cycle was found to be related to various terrestrial phenomena, such as the classic examples of tropical cyclones and the levels of Lake Victoria (Hoyt and Schatten, 1997).

In the 1870s, C. Meldrum and A. Poey pioneered the idea of a solar connection with tropical cyclones in the Indian and Atlantic Oceans, respectively. Later, in the 1920s, a strong association between the levels of Lake Victoria (Africa) and the sunspot number was reported (Figure 3). This strong positive correlation spanned the period from 1896 to 1922. However, around 1927, it weakened and reversed sign, prompting the publication of several papers investigating the causes of these discontinuities in the relationship. Recent works, relying on more complex analyses, show that maxima of the 11-year sunspot cycle were matched by Victoria level maxima during the 20th century. This association is due to the occurrence of positive rainfall anomalies roughly one year before solar maxima (Stager et al., 2007).

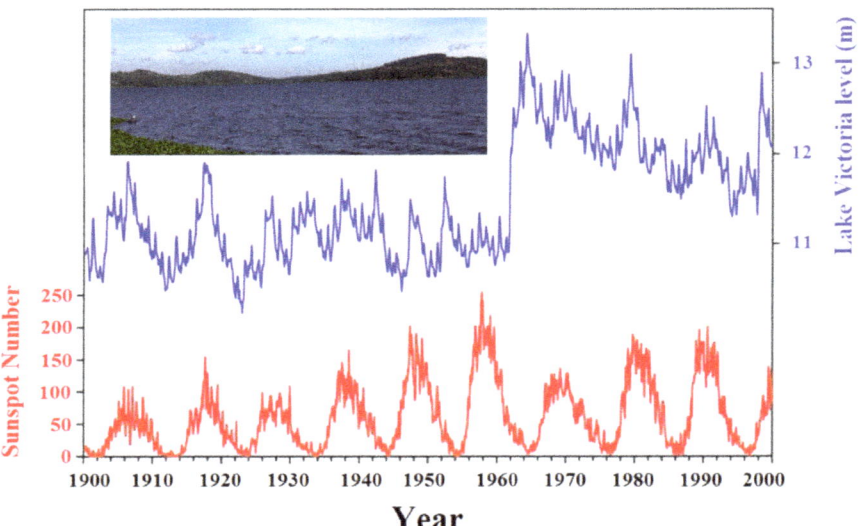

Fig. 3. A comparison between the Sunspot number (red line) and Lake Victoria level (blue line).

5 Birth of Sun-Earth science: space data and computer simulations

Our last key date is 1977, the year when J. Eddy confirmed that a period in the 17th century virtually without sunspots occurred, the so-called Maunder minimum (c. 1645–1715). The association of this episode of grand solar minimum with the most intense phase of the Little Ice Age fueled the idea of a clear relationship between solar activity and Earth's climate (Soon and Yaskell, 2003). Retrospectively, historians and natural scientists have established the links between a number of historical documents that recorded the coldest years of the little ice age with the Maunder minimum, including paintings of the frozen river Thames in London or the numerous canals in Holland (Figure 4). Finally, systematic measurements of the solar constant (irradiance) from space also started around 1978. This new space technology allowed the in-situ measurements of the solar wind (density, composition, magnetic field), improving the physical models of the heliosphere that shed new lights on the mechanisms of Sun-Earth relations.

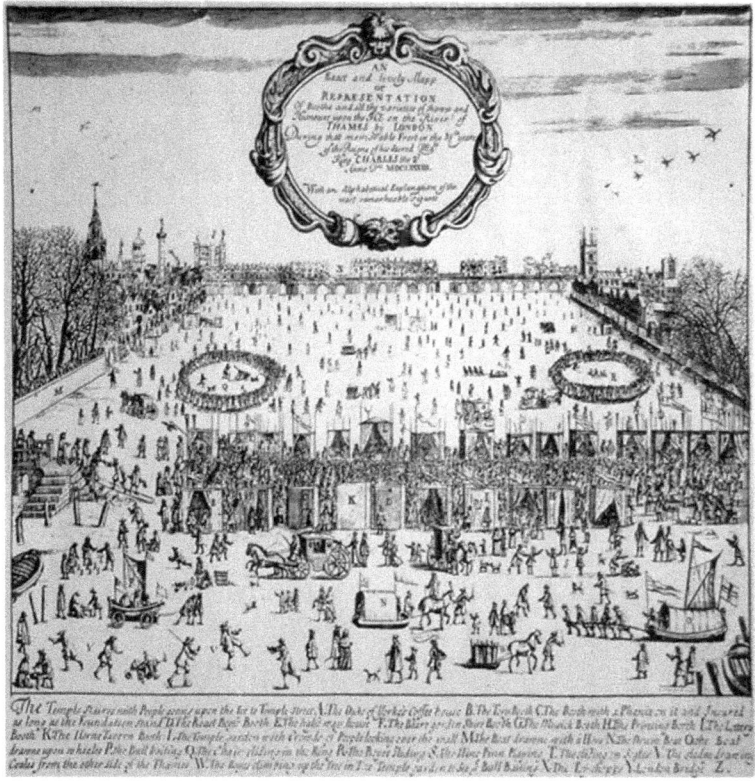

Fig. 4. Ice on the River of Thames (London) during the memorable frost of 1683. (Courtesy Historical Archive of Sunspot Observations).

Of course, in recent decades, an increasing number of studies have found other statistically significant correlations between solar variability and various changes in the climate system. In 1980s, K. Labitzke and co-workers suggested that an influence of solar activity on temperatures in the stratosphere (approximately 10–50 km) appears when the winters are grouped according to the phase of the quasi-biennial oscillation, an approximately 2-year oscillation of easterly and westerly zonal winds in the equatorial lower stratosphere (Labitzke and van Loon, 1988). In 1990s, E. Friis-Christensen and K. Lassen reported a correlation between Northern Hemisphere land air temperature and the solar cycle length (Friis-Christensen and Lassen, 1991). Although other authors had previously proposed similar relationships, this work had a widespread impact on the scientific community, which strongly criticized these results. More recently, in the 2000s, N. Marsh and H. Svensmark presented a strong positive correlation between the low cloud amount and the galactic cosmic rays over the period 1983–2005, triggering tens of papers on this matter during the last decade (Marsh and Svensmark, 2000). However, these last years in the history of the ideas of the solar-terrestrial relationships were also characterised by the use of modern technology to improve our understanding of the problem.

Total solar irradiance (TSI) has been measured consecutively and accurately from space over the past 35-40 years. It shows a clear solar cycle variation of about 0.1%, positively correlated with sunspot numbers. TSI values show similarities during the last three solar cycles (labeled 21, 22 and 23), but also important differences as the extremely low TSI value during the most recent solar cycle minimum. TSI values during solar cycle minima also show a long-term change correlating with the open magnetic field of the Sun (Fröhlich, 2013).

The increase in computing power over recent decades has also been crucial for the formulation and testing of fresh ideas about the influence of solar variability on the Earth. We now have access to numerical climate models that allow performing thousands of computational experiments. So, we can better understand and test the mechanisms of solar forcing on climate. In addition, we have much better access to large weather datasets, both real observations and model simulations, allowing studies that were previously impossible. Moreover, a great effort have been made in the collection of parameter related with the middle and higher atmosphere, including stratospheric soundings, chemistry of the mesosphere-thermosphere, total electron content, FoF2 index, etc. Thus, scientists have proposed a number of different solar forcing mechanisms, nevertheless a hot debate has surrounded this topic and a consensus view is still lacking (Gray et al., 2010). As a final remark, it is worth mentioning that in recent decades, astronomy and climatology have been characterised (as most other branches of science) by an increasing speed in the rate of new results and publications. In this perspective, we are quite sure that the next big milestone in the fascinating quest for the elusive Sun-climate link is likely to occur sooner than we might currently expect.

Further reading

Benestad, R.E.: 2006, *Solar activity and Earth's Climate*, 2nd edition, Springer.

Hoyt, D.V. and Schatten, K.H.: 1997, *The role of the sun in climate change*, New York, Oxford University Press.

Soon, W.-H. and Yaskell, S.H.: 2003, *The Maunder Minimum and the Variable Sun-Earth Connection*, Singapore, World Scientific.

Vaquero, J.M. and Vázquez, M.: 2009, *The Sun Recorded Through History*, Berlin, Springer.

CHAPTER 1.5

THE ROLE OF THE SUN IN CLIMATE CHANGE: A SOCIETAL VIEWPOINT

Christian Muller[1]

1 Introduction

The early man was fascinated by the Sun, its daily sunrise and sunset, its role as a purveyor of energy and light. Its cult in Egypt was even a first step towards monotheism. A scientific view of the part played by the Sun in human health appears in the Greek medicine as it was practiced in the Asclepion of Hippocrates on the island of Kos: exposure to sunshine was seen as a part of a curative environment as well as a healthy regime.

The scientific renaissance began with the Copernicus heliocentric system followed by the systematic observation of sunspots by Galileo beginning in 1610. The relation between solar radiation and climate got wide publicity at the end of the 18th century when Benjamin Franklin (1784) made the link between the June 1783 eruption of the Laki volcano in Iceland, the dry haze which followed and the abnormally cold winter of 1783-1784 (Franklin, 1784). These conditions lead to a shortage in cereals and price increases which can be related to the revolutions of the end of the decade. Presently, a much more detailed meteorological analysis is possible and some authors (D'Arrigo et al., 2011) indicate that the volcanic eruption might have only amplified a global perturbation caused by an El-Niño event.

The next step was taken by Herschel (1801) when he compared the sunspot number to the first economical statistical report: the "Essay on the wealth of the nations" by Adam Smith. Herschel hypothesised a relation between crop yield and the price statistics, he noticed then that the price values were anti-correlated with the sunspot number. He did not discover the 11-year cycle in this analysis as his series still included the Maunder minimum of the 17th century, which dominated the 11-year cycle.

---

[1] B.USOC, Avenue Circulaire 3, B-1180 Brussels, Belgium

Schwabe (1844) was the first to put it in evidence after three textbook solar cycles. The Hershel anti-correlation was met with criticism since its first publication and revisited in an appendix of the analysis of sunspots made by Carrington (1863). Besides being a superior solar observer, Richard Carrington was also a brewer and had his income tied to the cereals price. He concluded that in the 19th century, wheat prices, and especially the import duties, were dependent on tax laws passed by the British parliament. These taxes were amounting to import subsidies in order to have British farmworkers leave their job to participate in the new industrial expansion in the cities and be sufficiently fed with minimal wages. However the work of Herschel after more than 200 years is still open for discussion in both the study of the influence of geophysical parameters on economy and in the use of statistics to determine causality (see for example, (Love, 2013)).

More recently, scholars studying the Maya classical periods found that the successive collapses of the Maya civilisations were periodical with a period close to 200 years and corresponding to a possible solar forcing (Hodell et al., 2001). Here again, after 30 years of abundant publications, the issue is still open (Turner and Sabloff, 2012) because it involves the governance and agricultural practices of a closed society which could have built its own instability and repeated the same errors at each recovery. The Maya cycles constitute in any way an important case study for the validation of regional climate models including the interaction with human activities (Cook et al., 2012).

The best known historical example is of course the Maunder minimum where a coincidence between the absence of sunspots and the "little ice age" in Northern Europe was observed with a peak in the 17th century. The "little ice age" corresponds to real historical observations, such as the growth of glaciers, population extinctions in Greenland and retreat of crops towards the South (Le Roy-Ladurie, 1967). Possible mechanisms are developed in the review paper by Gray et al. (2010), associating solar UV forcing with the North Atlantic Oscillation, this last article concludes that further research is required both on the reconstruction of past spectral solar irradiances and on the possible triggers of this cooling.

2 UV climate and its biological and societal effects

Until the middle of the 20th century, medicine followed the Hippocrates hypothesis that exposition to solar radiation was beneficial. However, these effects could only be quantified when physicists developed both lamps and UV sensors, and used them in a hospital environment. Subsequent studies showed damages comparable to the effects of ionising radiation on plants and test animals and ended up establishing a relation between UV and the genesis of human skin cancer (Latarjet, 1959). The progress of UV sensors led not only to this new development on the biological effects of solar radiation but also to studies of what was called at the time "the UV limit of the solar spectrum", these were precursors of the determination of the UV effects on atmospheric chemistry and its climate effects.

From a biological point of view, UV radiation is divided between UV-A (400–315 nm), UV-B (315–280 nm) and UV-C (280–100 nm). It is commonly admitted that UV-A is necessary in moderate doses to initiate metabolic processes, such as for example the production of vitamin D in the human skin.

Stratospheric ozone filters entirely UV-C and the largest part of UV-B; ozone is not the only effective UV filter, as carbon dioxide filters all UV below 220 nm, but a reduction in ozone would expose the biosphere to UV-B with consequences on both plants and animals.

The decrease of the ozone layer observed since the 1970s and especially since the Antarctic spring ozone hole has led to the 1987 Montréal protocol on ozone depleting species. Solar variations play a role in the equilibrium of the ozone layer but this effect is buffered by the fact that atomic oxygen produced by the dissociation of molecular oxygen more than compensates the destruction of ozone by UV radiation of longer wavelengths. Thus, the solar climate has very little effect on the UV radiation received at the Earth's surface, but there again scientific research should never overlook the effects of perturbations and never consider a part of the Earth's system as definitively auto-stabilising.

3 Present: assessment of anthropogenic influence on climate

The Intergovernmental Panel on Climate Change (IPCC) was established by the United Nations Environment Programme (UNEP) and the World Meteorological Organisation (WMO) in 1988 to provide the world with a clear scientific view on the current state of knowledge in climate change and its potential environmental and socio-economic impacts. In the same year, the UN General Assembly endorsed the action by WMO and UNEP in jointly establishing the IPCC.

The IPCC has up to now generated five global reports each divided in three parts: 1) physical scientific basis, 2) vulnerability of socio-economic and natural systems, and 3) mitigating climate change through limiting or preventing greenhouse gas emissions and enhancing activities that remove them from the atmosphere.

The successive IPCC reports evolved in both complexity and assertiveness together with the collection of new evidence of the surface temperature increase, the first four reports were more and more precise in comparing the observed warming with anthropogenic forcings until reaching an encyclopaedic size in 2007 with the Assessment Report 4 (AR4). AR4 was criticised in detail points, as ignoring the growth of some high altitudes Himalayas glaciers and largely dismissing fluctuations in natural forcings. It is in this respect that the 5th Assessment Report (AR5) increased its coverage to include and discuss the solar forcings.

4 Natural and extra-terrestrial forcings in AR5 of the IPCC

The Sun is the main driving force of our climate, but in general its changes appear in cycles or on a relatively long scale. Short term fluctuations as in Natural forcings and especially extra-terrestrial forcings are detailed in chapter 8 of the first report. The IPCC documents being consensus reports, the solar variations are considered from the point of view of the total solar irradiance in the modelling chapters, as non-linearity is not yet present in the models. The conclusions of IPCC WG-1 concerning the balance of forcings are shown in Figure 1: since 1750, solar irradiance has been a negligible cause of warming compared to the anthropogenic warming by the greenhouse gases.

Chapter 7 (Clouds and Aerosols) of AR5 dedicates a whole subsection on the impact of cosmic rays on aerosols and clouds, coming to the following synthesis: cosmic rays enhance new particle formation in the free troposphere, but the effect on the concentration of cloud condensation nuclei is too weak to have any detectable climatic influence during a solar cycle or over the last century.

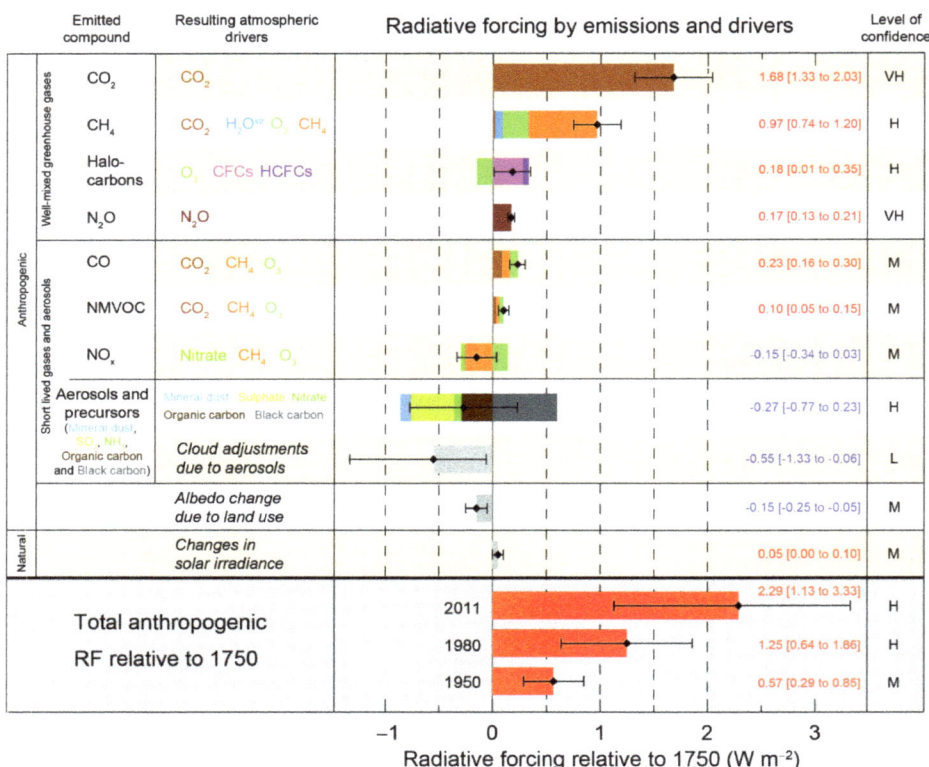

Fig. 1. Radiative forcing estimates in 2011 relative to 1750 and aggregated uncertainties for the main drivers of climate change. Volcanic forcing is not included as its episodic nature makes is difficult to compare to other forcing mechanisms. Total anthropogenic radiative forcing is provided for three different years relative to 1750. The change in solar irradiance forcing correspond to the "grand maximum" of solar activity of the 20th century. (Figure from (IPCC, 2014)).

4.1 Anthropogenic forcing and acceptance of AR5

From the summary for policy makers of (IPCC, 2014), we learn that the WG 1 contribution to AR5 considers new evidence of climate change based on many independent scientific analyses from observations of the climate system, paleoclimate archives, theoretical studies of climate processes and simulations using climate models. Warming of the climate system is unequivocal, and since the 1950s, many of the observed changes are unprecedented over decades to millennia. The atmosphere and ocean have warmed, the amounts of snow and ice have diminished, sea level has risen, and the concentrations of greenhouse gases have increased.

Human influence on the climate system is clear. This is evident from the increasing greenhouse gas concentrations in the atmosphere, positive radiative forcing, observed warming, and understanding of the climate system.

AR5 has been until now much less criticised by climate skeptics than the previous reports as it contains almost all the scientific arguments previously put in evidence to dismiss the values of the conclusions. Governments have now come to accept that the 2015 ministerial conference planned in Paris will lead to a binding agreement on the emission of anthropogenic greenhouse gases and begin the preparation of corresponding legislation. Businesses representing more than 92 trillions of dollars have federated since 2000 in the carbon disclosure process[2] in the United States, China, Japan, and the entire American continent to consider the risks and opportunity aspects of both global warming and regulation. Similar studies are more discrete in Europe where the adaptation information comes mostly from documents of the European Commission. [3] However, a business-based process similar to the carbon disclosure process does not seem to exist. It is clear that European companies are now considering global warming and urban heat islands in their extension plans and planning of their air conditioning needs.

5 Evolution of solar-societal aspects non-related to climate: UV and visible solar radiation

The IPCC and climate research in general do not consider the human need for natural visible light and moderate UV-A, most of the legislations do not go further than the European Council directive : "Workplaces must as far as possible receive sufficient natural light" (Council Directive of 30 November 1989 concerning the minimum safety and health requirements for the workplace). National regulations even recommend the obstruction of natural light sources to prevent the glare on computer screens. However, more and more architects incorporate in buildings light wells and atriums to achieve a feeling of comfort and spaciousness but it is far from being a general practice in housing projects and workplaces. For the UV part, the dangers of exposure to UV-B are better known than they were before

---

[2] https://www.cdp.net/en-US/Pages/HomePage.aspx
[3] See for example http://ec.europa.eu/clima/policies/adaptation/what/docs/com_2013_216_en.pdf

the ozone layer was endangered by the release of atomic chlorine and nitric oxide in the stratosphere, excessive exposure is however not yet recognised as a health hazard related to profession. It can be said that both the information campaigns and fashion evolution have mitigated the increase in human skin cancer. On the other side of the spectrum, morbidity associated to deprivation of UV-A is rarely studied and not considered at all in work legislation.

6 Future research related to climate

The IPCC did not yet set a schedule for a sixth report but many improvements are possible in the domain of solar forcing, the first being to use a forecast of the solar cycles in the models instead of repeating a typical solar cycle (in AR5: cycle 23). Future IPCC reports will certainly consider regional variations and will thus increase their societal impacts as they would influence insurance rates. These future reports will also be supported by improved data: solar total and spectral irradiance observations continue with better instruments on both ends of the frequency spectrum: from the short term variations associated with space weather to several 11 years cycles, these provide already better inputs to weather and climate models. Climate and General Circulation models are also rapidly progressing in spatial and time scales. Moreover, more complex earth system models including the ocean, land and cryosphere are appearing. These new developments lead to hope for a better description of the influence of all manifestations of solar activity on climate.

Further reading

The societal effects of climate change generated a large number of articles. The most complete and certainly the most verified is the IPCC synthesis report (IPCC, 2014), which summarises conclusions from all three working groups and which can be directly downloaded from the website of the IPCC http://www.ipcc.ch/pdf/assessment-report/ar5/syr/SYR_AR5_SPMcorr1.pdf

CHAPTER 1.6

THE DEBATE ABOUT SOLAR ACTIVITY AND CLIMATE CHANGE

Rasmus E. Benestad[1]

1 Introduction

The question whether Earth's climate is influenced by sunspots was posed already in the 17th century, and is one of the oldest scientific hypotheses that is still debated. Harriot discovered the sunspots in 1610 and Galileo, Scheiner, and Riccioli suggested in 1651 that sunspots may influence temperature on Earth (Benestad, 2002). These speculations were in a sense premature due to an extremely limited sample of data, and perhaps were a result of the need for humans to see the universe as ordered (Kahneman, 2012). An illustration of this point is that the stars, which are known to be arbitrarily spread across the skies, have been grouped into constellations and fed superstitions, such as astrology. This common mental bias of the human brain ("obsession" with order) may, according to Kahneman, explain some potential pitfalls and some other aspects of the on-going debate about the Sun's role in climate change, as well as perhaps implicit assumptions and the way statistics has been used (see Chapter 1.5).

In recent years, the global warming has been blamed on solar activities (Scafetta and West, 2007; Friis-Christensen and Lassen, 1991: Solheim et al., 2011; Svensmark and Friis-Christensen, 1997), despite the fact that no solar indicator exhibits any long-term trend nor is there a clear ∼11-year cycle in the global mean temperature record (Figure 1). There is a number of non-linear processes in Earth's climate responsible for the varying weather and weather regimes, which perhaps may swamp any connection to solar activity or even generate variations that resemble the solar variations. The variations in the Sun have only a weak effect on its energy output (∼0.1%), which means that solar variations must be amplified to have a substantial impact on Earth's climate. While such amplifying processes play a key role, they are often not acknowledged in the debate (see Chapter 4.1). Correlations between a small set of variables combined with an absence of the comprehensive physics picture and other relevant information in this debate

---

[1] The Norwegian Meteorological Institute, PO Box 43 Blindern, 0313 Oslo, Norway

has enabled the *amplification paradox* to exist, the counter-factual position that solar activity is amplified by climatic processes and plays the most important role for the ongoing global warming in the presence of a more pronounced greenhouse gas forcing subject to the same amplification.

A debate about the role of the Sun, however, is both valid and a natural scientific discourse, but the solar link to climate change is still debated exactly because the evidence for a link is elusive (see Chapters 1.3 and 1.4). There are some key points in this debate that can bring us further, such as the question whether conclusions are based on proper and well-suited statistical methods, valid assumptions, and whether the hypotheses are properly physics-based or ill-conceived.

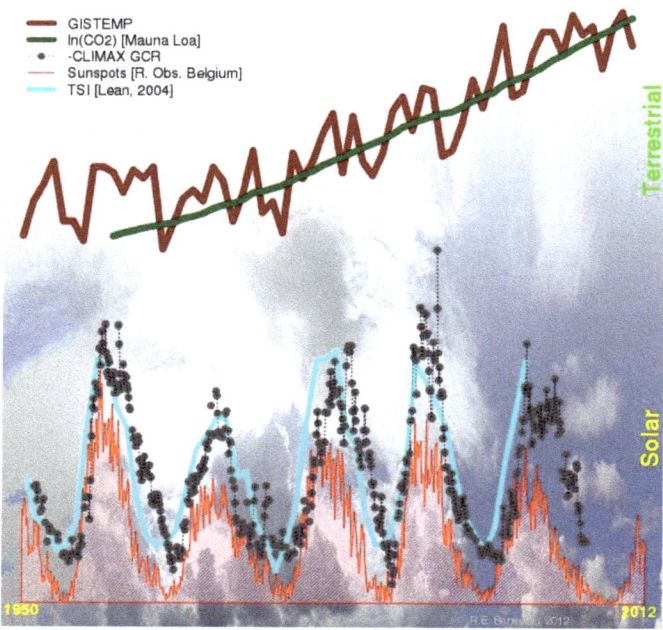

Fig. 1. Global mean temperature (GISTEMP), the logarithm of CO_2 concentration, and various solar indices: galactic cosmic rays (GCR) multiplied by -1, total solar irradiance (TSI), and sunspots. Only $\ln(CO_2)$ shows a trend that matches the global warming since 1950. There are no clear 11-year cycles in the global mean temperature. All the curves are scaled to have a standard deviation of 1, but the CO_2-curve is placed on top of the temperature whereas the solar data are shown below.

2 The empirical evidence

In empirical evidence-based studies, such as statistical analyses, it is important to make use of all available relevant information in order to arrive at convincing conclusions. This information includes limitations about methods, model evaluation, robustness, physics, and comprehensive consistency (Chapters 3.1, 3.5 and 3.9). It is furthermore important that analyses based on empirical evidence and statistics can be replicated independently by others. Otherwise, such results are unconvincing and any debate about their findings gets dogmatic. Open-source code and transparency let others see how such analysis is carried out, assumption made, and show exactly which methods are used. Studies based on empirical evidence are often not computation intensive, involve limited data and methods, and lend themselves to transparency and replication. The methods used can be tested further in a synthetic setting, for instance with test data for which the answer is known *a priori* (e.g., true and false). In some cases, methods can be tested in a Monte-Carlo type setting or compared with analytical results (Benestad et al., 2013).

2.1 Why some statistical methods are unsuited

In some cases, the statistical evidence presented is based on inappropriate choice of statistical methods, and it may be possible to replicate the analysis and demonstrate that the method chosen was predisposed to give a certain result, such as inflated significance or a false confirmation (Benestad et al., 2013). Another subtle mistake made in empirical studies is inappropriate sampling or making invalid assumptions (Benestad and Schmidt, 2009). Statistical theory assumes that a sample is a representative selection of a larger universe, which can be achieved if it is made up of randomly drawn observations. Hence, any true hypothesis should not only be valid for just one sample, but should also apply to out-of-sample data. This point is relevant for an analysis which discarded part of the data that did not fit the set of cycles that was claimed to be externally forced (Humlum et al., 2011). This analysis both failed to take the additional information associated with independent data into account for model validation, and illustrated potential pitfalls associated with a narrow view on spectral cycles. Spectral analysis methods will find some frequencies real or imagined as a result of non-linear dynamics. Another typical mistake is when conclusions are drawn on the basis of a small number of events or periods (Humlum et al., 2011). Figure 2 illustrates all these points.

2.2 Coincidental resemblance

If a scholar looks for a solar signal in a large selection of different records, then (s)he will find some that have a high correlation just from pure chance. This is expected from the null hypothesis itself. Let's say (s)he has a set of 100 random data series with no connection to the sunspots (the size of the selection of tests is 100), then 5 of these records are expected to have a correlation that is statistically significant at the 5-percentage level, just because this is how the statistical significance is defined. The size of the selection of statistical tests results is increased if one examines not just synchronised variations, but extends the search to lagged relationships. The total number of searches and tests must be reported in order to get an unbiased estimate of the significance, and there are ways to

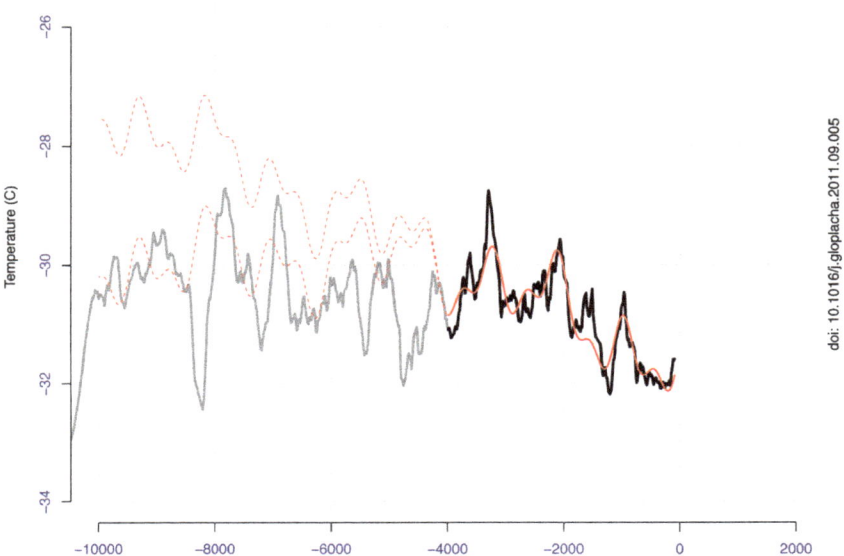

Fig. 2. Replication of the analysis of Humlum et al. including out-of-sample data. The black curve shows the part of the data they considered and the solid red line is their curve fit. The grey line shows the data they discarded and the dashed red lines show two attempts to extend their model: one taking the trend into account and one setting the trend (artificially) to zero. This example shows that the data selection in their analysis was not representative for the Greenland temperature from GIST2.

account for multiple tests (Wilks, 2006, 2011). A failure to account for all tests (searches) combined with the human tendency to find patterns, just like the stellar constellations and astrology, has a similar effect as *cherry picking*.

One typical issue may be regional climate variations that have a high correlation with some index of solar activity (Lockwood et al., 2010). While this correlation may be statistically significant it may not necessarily imply a real physical link. However, this kind of empirical evidence may also suggest that there is a real link, but one for which we may not yet know the physics behind. Lockwood et al. suggest a number of possible mechanisms, such as enhanced cooling through the increased maritime clouds connected to cosmic ray flux increase or an influence of solar forcing on stratospheric temperatures, the stratospheric polar vortex, and the tropospheric jet streams involving a downward propagation or a refraction of tropospheric eddies. These physical mechanisms are plausible, but a causal link has yet to be demonstrated and there are many other non-linear processes present that interfere (Chapters 3.5 and 3.7). Hence, we can neither dismiss such a link from the absence of scientific evidence, nor can we say with 100% confidence that solar activity has an effect on regional climate variability. Such cases call for more research. One way to establish whether they are real can be done by comparing predictions with out-of-sample data, independent data which have not been used in training the statistical model or motivating a hypothesis. Out-of-sample data is typically corresponding observations made in the future or from the past, but a general lack of observations often makes such comparisons difficult.

Another point is that data processing further increases the likelihood of a good match between variations in the solar activity and climatic records, whether real or absent. Filtering can highlight variations on similar time scales as solar data, and hence we must expect to see a number of cases with higher correlation if applied to a range of local or regional climate variables (Coughlin and Tung, 2006). The size of the test selection is usually increased with the introduction data processing, such as filtering. Filtering solar cycle lengths has been done in the past (Friis-Christensen and Lassen, 1991), but comparing such analyses to climate records on a decade-scale is physically meaningless because the former would imply a kind of time warp, where 10-year mean temperature is compared with the solar cycle length for the epoch about 10 years preceding and the epoch for the subsequent 10 years that have not yet happened. Furthermore, there is no physics involved in the connection, and there is no similar correlation for the sunspots. Indeed, there is little physics explaining how solar cycle length can influence Earth's climate. Hence, it is important that there is not just statistical evidence for a connection, but also that these are backed up with physics and a demonstration of the chain of causation. Statements about solar influence on climate change based on correlation or regression are for all intents and purposes prediction based on resemblance, so finding a physics mechanism and using models based on this physics to mimic the behaviour are the best way to ensure that similar behaviour really are related. This has rarely been done for solar-terrestrial hypotheses, however, there are some exceptions (Meehl et al., 2009). In other words, there is a difference between predictions based on resemblance and predictions based on established theory such as physics.

2.3 The need of physical consistency

We must be able to explain causation if an explanation is considered to be convincing, in addition to empirical evidence, statistics, and resemblance. There are three main physics-based hypotheses on how solar activity may influence Earth's climate: (i) changes in the total solar irradiance, (ii) changes in the stratospheric UV, and (iii) changes in galactic cosmic rays (Chapters 4.5 – 4.7). One common trait for many of these hypotheses is that the solar signal is weak and needs to be amplified by some mechanisms internal to the Earth's climate. Such mechanisms may involve clouds modulating the energy balance as for (iii) and feedback mechanisms that respond to changes in the temperature. The latter must apply for all types of forcings, and involve increased atmospheric moisture, melting of snow and ice (albedo feedback), and changes in the atmospheric lapse rate. These mechanisms play a role for the amplification paradox introduced by some proponents of a dominant solar influence on the global warming (Scafetta and West, 2007; Svensmark and Friis-Christensen, 1997; Friis-Christensen and Lassen, 1991; Solheim et al., 2011), which will be addressed later on. In our modern age, we can probably measure any physical condition that affects the Earth, and we can assume that any link between changes in the Sun and conditions on Earth must be detectable by modern instruments.

One example illustrating some of the shortcomings connected to size of the selection of tests, lack of physics, coincidental match, and the amplification paradox is found in a paper that made predictions for the winter temperature at Svalbard (Solheim et al., 2011). It neglected known relevant factors, such as greenhouse gases, argued that the solar cycle length affects the (regional) Arctic winter temperature when the latter lagged the former by one solar cycle, and then predicted Svalbard winter temperature to be $-17.2\,°C$ for 2009-2019 (95% confidence interval of $-20.5\,°C$ to $-14\,°C$) while the temperature has kept rising. This prediction is already likely to fail with wide margins unless the current trend reverses abruptly and the mean temperature of the next 5 winters are exceptionally cold (Figure 3).

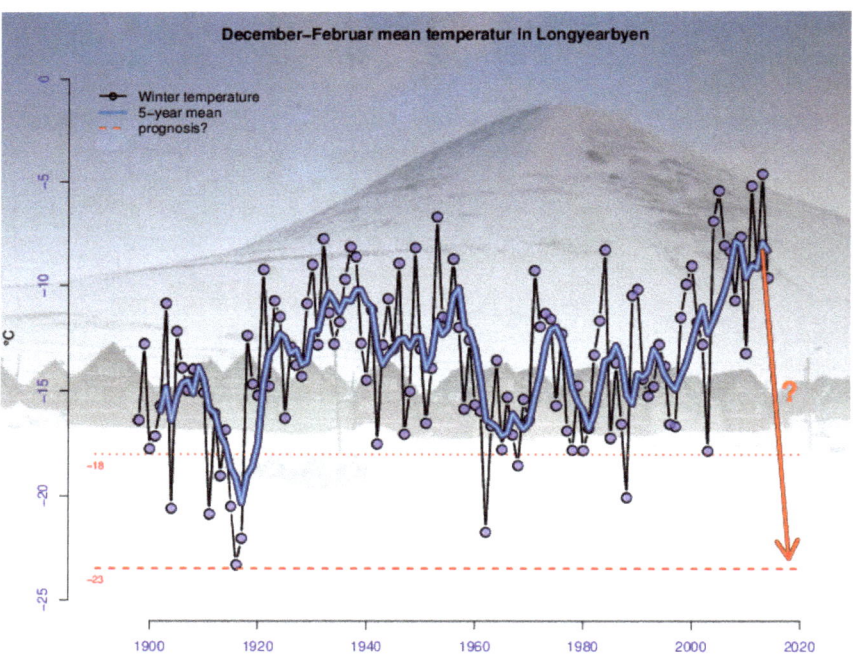

Fig. 3. The winter temperature at Svalbard and its 5-year moving average. The long red dashed line indicates $-23.5\,°C$, the mean winter temperature for the next 5 years (thick blue curve) needed for a successful prediction of Solheim et al. The short-dashed red line shows the lower limit for Solheim et al.'s 95% confidence estimate: $-18\,°C$.

There could of course be some unknown mechanism, and then we may look to statistics and empirical-evidence studies for clues. Does the statistics indicate any clear resemblance to the solar variations that can be explained in terms of physics? We cannot rule out new types of connections, but so far the evidence for such a connection has not been very persuasive, hence the on-going debate. However, there are science debates in other fields too, and there are probably more factors involved in the controversies as well (Sherwood, 2011).

A contender in the debate has involved galactic cosmic rays and their hypothesised effect on clouds (Svensmark and Friis-Christensen, 1997), which subsequently may affect the global temperature through their reflecting capacity. According to this hypothesis, clouds are a modulating mechanism rather than a feedback mechanism which is a function of the global temperature $f(\langle T \rangle)$. They respond almost instantly to changing conditions, and variations will almost be in sync with potential forcings. Furthermore, changes in the cloudiness are expected to prompt a response on a shorter or similar time scale as volcanoes, and hence, if the galactic cosmic rays had an effect on the clouds and climate, we would see an ~11-year cycle in the global mean temperature. There is no such cycle in global temperature (Figure 1). Several scholars reviewed past studies on galactic cosmic rays and climate and concluded that the alleged effect probably is overrated (Laken et al., 2012; Damon and Laut, 2004; Kristjánsson et al., 2002), and Farrar revisited a study on the connection between galactic cosmic rays and climate that highlighted the resemblance between cloudiness and galactic cosmic rays (Svensmark and Friis-Christensen, 1997), but found that these were most probably coincidentally similar due to El Niño (Farrar, 2000).

Many such studies have contributed to polarise the debate further, but it could be less contentious if the authors of the controversial papers (Friis-Christensen and Lassen, 1991; Svensmark and Friis-Christensen, 1997; Scafetta and West, 2007; Solheim et al., 2011; Humlum et al., 2011) had provided open access to both data and method, and hence made the analysis more transparent (Le Page, 2009). Transparent evidence and science are more compelling than elusive analyses, and one cannot force anybody to be convinced without actually convincing. It is possible that some of the controversies are connected to intuition biases (Kahneman, 2012), but such controversial questions are also the most interesting ones in science as they are specific and highlight where there is a need for further study. It is crucial to pinpoint what exactly are the contested or uncertain information and then find ways to shed further light on the issue. The picture needs to be both comprehensive and consistent if it is to be credible – and that does not just involve temperature, but other aspects as well, such as physical mechanisms. For instance, the amplification paradox concerns feedback processes associated with higher or lower temperatures and can be expressed as a function of the global mean temperature $f(\langle T \rangle)$ that should act on all factors causing a change in the temperature, unless there are physical reasons for being limited to only one type. It is therefore incorrect to argue that weak solar forcing is amplified and explains most of the global warming, when we also know that greenhouse gases are increasing (Figure 1) and they too invoke the same kind of amplification.

3 Conclusion

The amplification paradox has not been properly addressed, partly due to a misguided idea about scientific progress and a debate that has clouded an otherwise consistent and comprehensive picture. It is manifested in the naive and misconceived idea that a strong role of greenhouse gases diminishes the role of solar forcing or vice versa. The effect of changes in the Sun and changes in the greenhouse gases are both subject to an amplification that is connected to changing temperature, such as a mutual connection between diminishing ice/snow and a darker surface that absorbs more heat from the Sun. The influence of changes in greenhouse gases is well-known but that of changes in the Sun is weaker and more obscure. Any amplification must act on the more prominent forcings as well as the weak ones. Moreover, it would therefore be a paradox if Earth's climate were sensitive to solar forcing though reinforcing processes in Earth's climate, but insensitive to more pronounced forcing, such as greenhouse gases subject to the same type of temperature-dependent feedback mechanisms.

Further reading

P. Yiou, *et al.*, *Climate of the Past* **6**, 461 (2010).

R. Proctor, L. Schiebinger, eds., *Agnotology: The Making and Unmaking of Ignorance* (Stanford University Press, 2008).

D. Bedford, *Journal of Geography* **109**, 159 (2010).

R. G. Harrison, K. P. Shine, A review of recent studies of the influence of solar changes on the earth's climate, *Tech. Rep. HCTN6*, Hadley Centre for Climate Prediction & Research, UK Met Office, Bracknell, RG12 2SY, UK (1999).

R. E. Benestad, H. O. Hygen, R. van Dorland, J. Cook, D. Nuccitelli, *Earth System Dynamics Discussions* **4**, 451 (2013). Revised version on Benestad, Rasmus (2014): Replication of a number of contrarian climate change studies. figshare. http://dx.doi.org/10.6084/m9.figshare.951971.

References of Part I

Benestad, R. E. Solar Activity and Earth's Climate. Praxis-Springer, Berlin, 2002.

Benestad, R. E., and G. A. Schmidt. Solar trends and global warming. *Journal of Geophysical Research (Atmospheres)*, **114**, 14101, 2009. 10.1029/2008JD011639.

Benestad, R. E., H. O. Hygen, R. van Dorland, J. Cook, and D. Nuccitelli. Agnotology: learning from mistakes. *Earth System Dynamics Discussion*, **4**, 451–505, 2013. 10.5194/esdd-4-451-2013.

Carrington, R. C. Observations of the Spots on the Sun. Williams and Norgate, London, 1863.

Clette, F., L. Svalgaard, J. M. Vaquero, and E. W. Cliver. Revisiting the Sunspot Number. A 400-Year Perspective on the Solar Cycle. *Space Science Reviews*, **186**, 35–103, 2014. 10.1007/s11214-014-0074-2.

Cook, B. I., K. J. Anchukaitis, J. O. Kaplan, M. J. Puma, M. Kelley, and D. Gueyffier. Pre-Columbian deforestation as an amplifier of drought in Mesoamerica. *Geophysical Research Letters*, **39**(L16706), 2012.

Coughlin, K. T., and K. K. Tung. Misleading patterns in correlation maps. *Journal of Geophysical Research (Atmospheres)*, **111**, 24102, 2006. 10.1029/2006JD007452.

D'Arrigo, R., R. Seager, J. E. Smerdon, A. N. LeGrande, and E. R. Cook. The anomalous winter of 1783–1784: Was the Laki eruption or an analog of the 2009–2010 winter to blame? *Geophysical Research Letters*, **38**(L05706), 2011.

Damon, P. E., and P. Laut. Pattern of strange errors plagues solar activity and terrestrial climate data. *EOS, Transactions American Geophysical Union*, **85**(39), 370–374, 2004.

Eddy, J. A. The Maunder Minimum. *Science*, **192**, 1189–1202, 1976. 10.1126/science.192.4245.1189.

Farrar, P. D. Are cosmic rays influencing oceanic cloud coverage–or is it only El Nino? *Climatic Change*, **47**(1-2), 7–15, 2000.

Franklin, B. Meteorological imaginations and conjectures. In Manchester Literary and Philosophical Society Memoirs and Proceedings, vol. 2, 1784, 1784.

Friis-Christensen, E., and K. Lassen. Length of the solar cycle: an indicator of solar activity closely associated with climate. *Science*, **254**, 698–700, 1991. 10.1126/science.254.5032.698.

Fröhlich, C. Total Solar Irradiance: What Have We Learned from the Last Three Cycles and the Recent Minimum? *Space Science Reviews*, **176**(1–4), 237–252, 2013. 10.1007/s11214-011-9780-1.

Gray, L. J., J. Beer, M. Geller, J. D. Haigh, M. Lockwood, et al. Solar Influences on Climate. *Reviews of Geophysics*, **48**, 4001, 2010. 10.1029/2009RG000282.

Herschel, W. Observations tending to investigate the nature of the Sun, in order to find the causes or symptoms of its variable emission of light and heat; with remarks on the use that may possibly be drawn from solar observations. *Philosophical Transactions of the Royal Society of London*, **95**, 261–318, 1801.

Hodell, D. A., M. Brenner, J. H. Curtisa, and T. Guilderson. Solar forcing of drought frequency in the Maya Lowlands. *Science*, **292**, 1367–1370, 2001.

Hoyt, D. V. The Smithsonian Astrophysical Observatory Solar Constant Program. *Reviews of Geophysics*, **17**(3), 427–458, 1979. 10.1029/RG017i003p00427, URL http://dx.doi.org/10.1029/RG017i003p00427.

Hoyt, D. V., and K. H. Schatten. The Role of the Sun in Climate Change. Oxford University Press, Oxford, 1997.

Humlum, O., J.-E. Solheim, and K. Stordahl. Identifying natural contributions to late Holocene climate change. *Global and Planetary Change*, **79**, 145–156, 2011. 10.1016/j.gloplacha.2011.09.005.

Kahneman, D. Thinking, Fast and Slow. Penguin, 2012.

Kornblueh, I. H. In memoriam Alexander Leonidovich Tchijevsky. *International Journal of Biometeorology*, **9**(1), 99–99, 1965. 10.1007/BF02187321.

Kristjánsson, J. E., A. Staple, J. Kristiansen, and E. Kaas. A new look at possible connections between solar activity, clouds and climate. *Geophysical Research Letters*, **29**, 22–1, 2002.

Labitzke, K., and H. van Loon. Associations between the 11-year solar cycle, the QBO (quasi-biennial-oscillation) and the atmosphere. Part I: the troposphere and stratosphere in the northern hemisphere in winter. *Journal of Atmospheric and Terrestrial Physics*, **50**, 197–206, 1988.

Laken, B. A., E. Pallé, J. Čalogović, and E. M. Dunne. A cosmic ray-climate link and cloud observations. *Journal of Space Weather and Space Climate*, **2**, A18, 2012.

Latarjet, R. Radiations in relation to carcinogenesis and mutation. In Genetics and Cancer. University of Texas Press, 1959.

Le Page, M. Sceptical climate researcher won't divulge key program. *New Scientist*, **18 December**, 2009.

Le Roy-Ladurie, E. Histoire du climat depuis l'an mil. Flammarion, Paris, 1967.

Lockwood, M., R. G. Harrison, T. Woollings, and S. K. Solanki. Are cold winters in Europe associated with low solar activity? *Environmental Research Letters*, **5**(2), 024001, 2010. 10.1088/1748-9326/5/2/024001.

Love, J. J. On the insignificance of Herschel's sunspot correlation. *Geophysical Research Letters*, **40**, 4171–4176, 2013.

Marsh, N. D., and H. Svensmark. Low cloud properties influenced by cosmic rays. *Physical Review Letters*, **85**(23), 5004, 2000.

Meehl, G. A., J. M. Arblaster, K. Matthes, F. Sassi, and H. van Loon. Amplifying the Pacific Climate System Response to a Small 11-Year Solar Cycle Forcing. *Science*, **325**, 1114–, 2009. 10.1126/science.1172872.

Pittock, A. B. Solar variability, weather and climate: an update. *Quarterly Journal of the Royal Meteorological Society*, **109**, 23–55, 1983. 10.1256/smsqj.45902.

Scafetta, N., and B. J. West. Phenomenological reconstructions of the solar signature in the Northern Hemisphere surface temperature records since 1600. *Journal of Geophysical Research (Atmospheres)*, **112**, 24–+, 2007. 10.1029/2007JD008437.

Schwabe, H. Sonnen-Beobachtungen im Jahre 1843. *Astronomische Nachrichten*, **21**(15), 234–235, 1844.

Sherwood, S. Science controversies past and present. *Physics Today*, **64**(10), 39–44, 2011. 10.1063/PT.3.1295.

Solheim, J.-E., K. Stordahl, and O. Humlum. Solar Activity and Svalbard Temperatures. *Advances in Meteorology*, **2011**, 8, 2011. URL http://dx.doi.org/10.1155/2011/543146%]543146.

Soon, W. W.-H., and S. H. Yaskell. The Maunder Minimum : the variable sun-earth connection. World Scientific Publishing Co, Singapore, 2003.

Stager, J. C., A. Ruzmaikin, D. Conway, P. Verburg, and P. J. Mason. Sunspots, El Niño, and the levels of Lake Victoria, East Africa. *Journal of Geophysical Research (Atmospheres)*, **112**, 15106, 2007. 10.1029/2006JD008362.

Steinhilber, F., J. A. Abreu, J. Beer, I. Brunner, M. Christl, et al. 9,400 years of cosmic radiation and solar activity from ice cores and tree rings. *Proceedings of the National Academy of Sciences*, **109**(16), 5967–5971, 2012. 10.1073/pnas.1118965109, http://www.pnas.org/content/109/16/5967.full.pdf+html, URL http://www.pnas.org/content/109/16/5967.abstract.

Stocker, T., and D. Qin, eds. Climate Change 2013 - The Physical Science Basis. Working Group I Contribution to the Fifth Assessment Report of the IPCC. Cambridge University Press, Cambridge, 2014.

Svensmark, H., and E. Friis-Christensen. Variation of cosmic ray flux and global cloud coverage—a missing link in solar-climate relationships. *Journal of Atmospheric and Solar-Terrestrial Physics*, **59**(11), 1225–1232, 1997.

Turner, B. L., and J. L. Sabloff. Classic Period collapse of the Central Maya Lowlands: Insights about human–environment relationships for sustainability. *Proceedings of the National Academy of Sciences of the USA*, **109**, 13908–13914, 2012.

Usoskin, I. G. A History of Solar Activity over Millennia. *Living Reviews in Solar Physics*, **5**, 3–67, 2008. 0810.3972, URL http://solarphysics.livingreviews.org/Articles/lrsp-2008-3/.

Vaquero, J. M., and M. Vázquez. The Sun Recorded Through History: Scientific Data Extracted from Historical Documents, vol. 361 of *Astrophysics and Space Science Library*. Springer Verlag, 2009.

Wilks, D. S. On "Field Significance" and the False Discovery Rate. *Journal of Applied Meteorology and Climatology*, **45**, 1181–1189, 2006. 10.1175/JAM2404.1.

Wilks, D. S. Statistical Methods in the Atmospheric Sciences: an Introduction. International Geophysics. Academic Press, San Diego, 3rd edition, 2011.

Part II

SOLAR AND SPACE FORCING

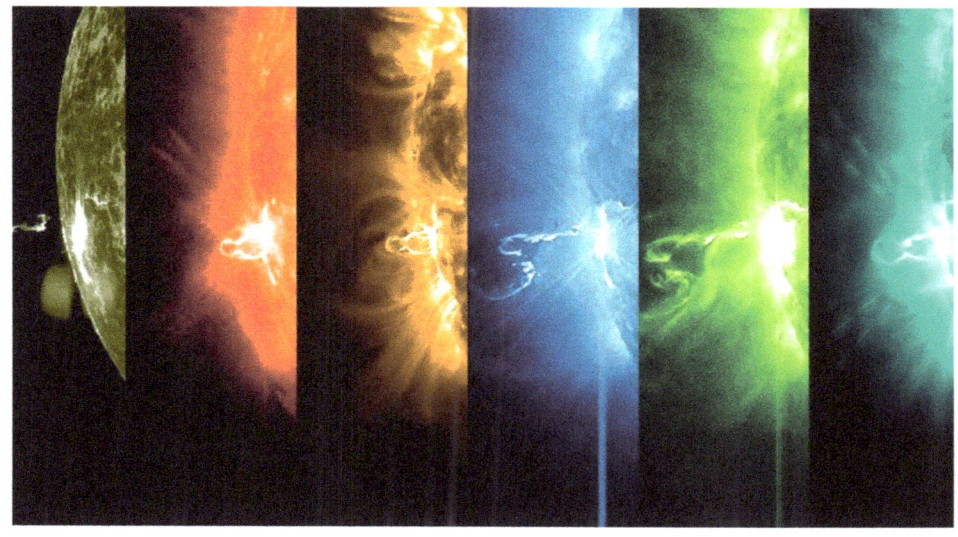

CHAPTER 2.1

BASICS OF SOLAR AND HELIOSPHERIC MODULATION

Astrid M. Veronig[1]

The Sun is the nearest star to Earth, and the central body governing our solar system by its gravity, radiation, magnetic field and emanating particles. The continuous flow of charged particles emanating from the Sun, the so-called solar wind, builds up the interplanetary magnetic field and the heliosphere in which our solar system is embedded. Changes on the Sun are thus directly affecting the state of our heliosphere. These changes may occur impulsively on time scales of minutes to hours, such as in solar eruptions but also on longer periods, such as the 11-year solar cycle during which the Sun's magnetic field and its radiation output is varying. This chapter gives an overview on the basic structure and physics of the Sun, and how changes on the Sun modulate the conditions in our heliosphere.

1 The Sun

The Sun is a hot gaseous ball, an almost perfect sphere with a radius of 695.66 thousand km (Haberreiter et al., 2008) located at a mean distance of about 150 million km, from Earth. In the solar interior, the temperature is extremely high (up to 15 million Kelvin in its center), implying that the gas is ionised, i.e., the electrons have been separated from their atomic nuclei. Thus, the Sun consists predominantly of free charged particles, i.e., ions and electrons, which form a special type of gas that we call plasma. The most frequent elements on the Sun are hydrogen (about 70 % of its total mass or 91 % by particle) and helium (28 % by mass or 8.9 % by particle). Only 2 % of the mass resides in heavier elements.

The Sun and our solar system have been formed about 4.6 billion years ago, when a cloud of interstellar dust and gas has collapsed under its own gravity. Most of the mass in the cloud was accreted to form the central star, i.e., the Sun, while the matter in the orbiting disk formed the planets. When such a cloud collapses, the matter at its center may heat up so strongly that thermonuclear fusion sets in – a star is born. Since its formation, the radius and the luminosity

---

[1] Institute of Physics/Kanzelhöhe Observatory, University of Graz, Austria

of the Sun have gradually increased due to the continuously changing chemical composition in the solar interior caused by the nuclear reactions. Early in its life, the Sun's luminosity was only about 70 % of its current value. The hydrogen reservoir in the Sun's core provides that it will shine for another 5 billion years before it will expand to a red giant star and then finally collapse to form a white dwarf star of about the size of the Earth, fading away over billions of years.

2 Basic structure of the Sun and the heliosphere

The Sun is composed of physically distinct layers. The solar interior consists of the core, in which the Sun's energy is produced, followed by the radiation zone and the convection zone. The solar atmosphere defines those layers from where the radiation is released. It consists of the photosphere, which is the "surface" of the Sun, the chromosphere, transition region and corona. The corona seamlessly extends as the solar wind to form the heliosphere, inside of which our solar system lies.

The energy that is radiated by the Sun is constantly produced in its central core, where the pressure and temperature are so high that nuclear fusion processes occur. In a chain of reactions, four hydrogen nuclei fuse to form a helium nucleus. The mass difference (Δm) between the newly formed helium nucleus and the sum of the four hydrogen nuclei is released as energy (ΔE) in the form of gamma-ray radiation, according to Einstein's famous formula $\Delta E = \Delta m \cdot c^2$, where c is the speed of light. This energy is transported outward from the core, first through radiation of high-energy gamma-ray photons that are efficiently scattered in the very dense interior, and further out through convection motion of hot plasma parcels (Figure 1). Once reaching the solar surface, the energy is radiated to space, in each second $3.8 \cdot 10^{26}$ J. This energy has to be continuously replenished by the thermonuclear reactions in the Sun's core.

In contrast to the solar interior, which is opaque to the radiation produced, the solar atmosphere refers to those layers from where the radiation can escape into space, i.e., the layers that we can actually observe. Over 99 % of the Sun's radiation comes from the photosphere, which is the innermost atmosphere layer located at the top of the convection zone. The photosphere has a thickness of only 500 km, i.e., less than 0.1 % of the solar radius. This is the reason, why the Sun appears to have a very sharp boundary although it is a gaseous body. The energy spectrum that is emitted from the Sun closely follows the continuum spectrum of a black body with an effective temperature of 5772 K. The major part of the solar radiation resides in the visual and the infrared range of the spectrum (about 90 %). The remaining about 10 % are mostly emitted in the ultraviolet, with only minor contributions (less than 0.1%) from the longer (radio) and shorter (X-ray, gamma-ray) wavelengths domains. The wavelength-integrated flux observed at the

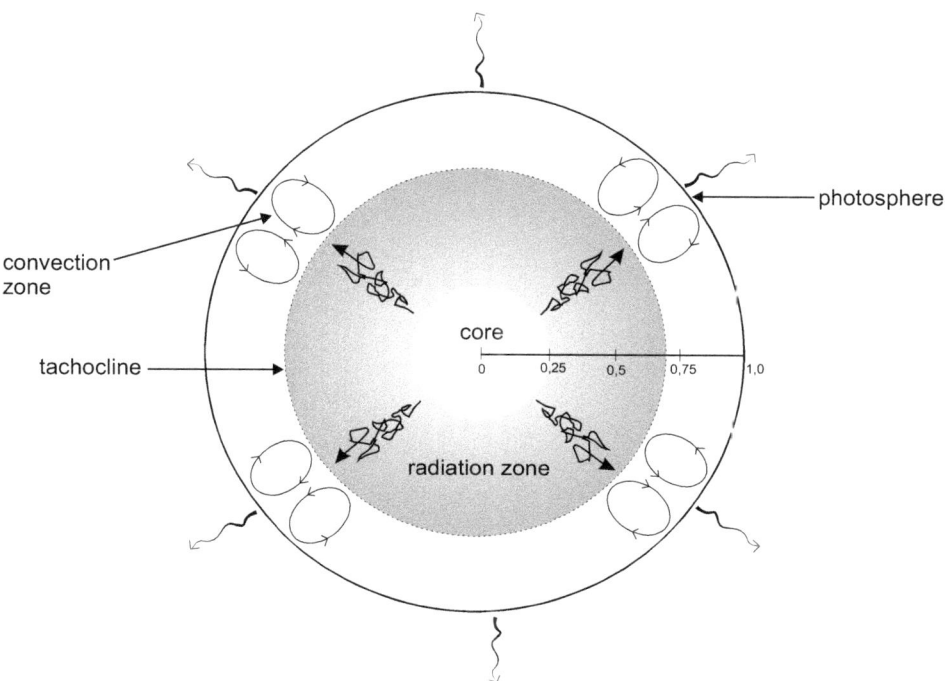

Fig. 1. Schematic view of the solar interior. In the core, the energy is produced by thermonuclear fusion of hydrogen to helium. The energy is transported first through high-energy photons (radiation zone) and then through convective motions of hot plasma parcels (convection zone) to the solar surface (photosphere), where it is radiated to space.

top of the Earth's atmosphere amounts to 1361 W m^{-2} (Kopp and Lean (2011); Schmutz et al. (2013))[2].

Watching the Sun during a total solar eclipse, when the moon blocks out the predominant emission from the solar disk, we can see the faint atmospheric layers that lie above the photosphere. First we see the chromosphere which appears reddish due to the strong emission in the Hα spectral line of neutral hydrogen in the red part of the spectrum. At the moment of total eclipse, we can see the solar corona which extends many solar radii into interplanetary space. In the solar photosphere, the temperature decreases from about 6 500 K at its base to about 4 100 K at its top. This is a natural consequence of the energy loss by the Sun's radiation. Surprisingly, in the adjacent chromosphere, the temperature

---

[2] For the date of the Picard/PREMOS first light on July 27, 2010 the exact total solar irradiance values has been independently confirmed to agree between 1360.9 ± 0.4 W m^{-2} as measured by Picard/PREMOS and 1361.3 ± 0.5 W m^{-2} as observed with SORCE/TIM (Schmutz et al., 2013). The TSI value for the 2008 solar minimum condition is 1360.8 ± 0.5 W m^{-2} as observed with SORCE/TIM (Kopp and Lean, 2011).

starts to slowly rise again to some 10 000 K. This is followed by a steep temperature increase in the so-called transition region that separates the chromosphere from the corona, which has typical temperatures in the range of 1 to 2 million Kelvin. It is important to note that the density falls off sharply in the solar atmosphere. In the chromosphere and corona, the densities are so low that the number of collisions between the particles is too small to provide a thermodynamic equilibrium in these layers. Thus, the chromospheric and coronal temperatures of the order of 10^4 and 10^6 K, respectively, are not truly thermodynamic temperatures but only relate the mean kinetic energy of the particles to the temperature value.

Due to its high temperature, the solar corona emits predominantly at soft X-ray (SXR; 0.01–10 nm) and Extreme UltraViolet (EUV; 10–120 nm) wavelengths (Figure 2d). Since the emission at these short wavelengths is absorbed by the Earth's atmosphere, the EUV and SXR telescopes observing the corona have to be deployed on satellites in space. Coronagraph instruments placed on satellites allow observations of the coronal structures over many solar radii at optical wavelengths. Basically, a coronagraph produces an artificial solar eclipse by blocking out the radiation from the solar photosphere in the optical path of the telescope. In contrast, the solar chromosphere can be well observed by Earth-based observatories, especially in strong absorption lines, so-called Fraunhofer lines, that intersperse the solar continuum spectrum in the visual range. Examples include the Hα line and the Ca II H and K lines of ionised calcium (Figure 2c). From space, the chromosphere can be also observed in emission lines in the UV part of the spectrum that are formed due to the rising temperature gradient in the chromosphere. The solar photosphere can be most easily observed from ground-based observatories in the visual and in the near infrared part of the spectrum, both in the continuum as well as in Fraunhofer absorption lines (Figure 2a). Thus, observing then Sun in different wavelengths, allows us to observe different heights and structures in the solar atmosphere.

The photosphere appears as the "surface" of the Sun, as it emits most of the solar radiation. The photosphere is structured by a network of granulation cells, which are the result of overshooting mass motions from the convection zone below. Each granulation cell has a diameter of about 1000 km and a typical lifetime of the order of 10 min. In the center of the granulation cells, hot plasma parcels move upwards. When reaching the surface, the plasma cools due to the radiation loss and sinks back at the boundaries of the granulation cells. Due to the lower temperature, these intergranular lanes appear darker than the granulation cells (Figure 3).

The Sun's atmosphere has actually no sharp outer boundary. Instead, the solar corona continuously extends to the interplanetary space in the form of the solar wind, a stream of charged particles released from the Sun with speeds in the range of 300 to 800 km s^{-1}. The fast solar wind streams originate in coronal holes, which are *open-field* regions, i.e., they connect the solar surface to interplanetary space. The slow solar wind originates in regions that are predominantly closed low in the corona but eventually open up to interplanetary space. The heating of the outer atmospheric layers (*coronal heating problem*) and the acceleration of the

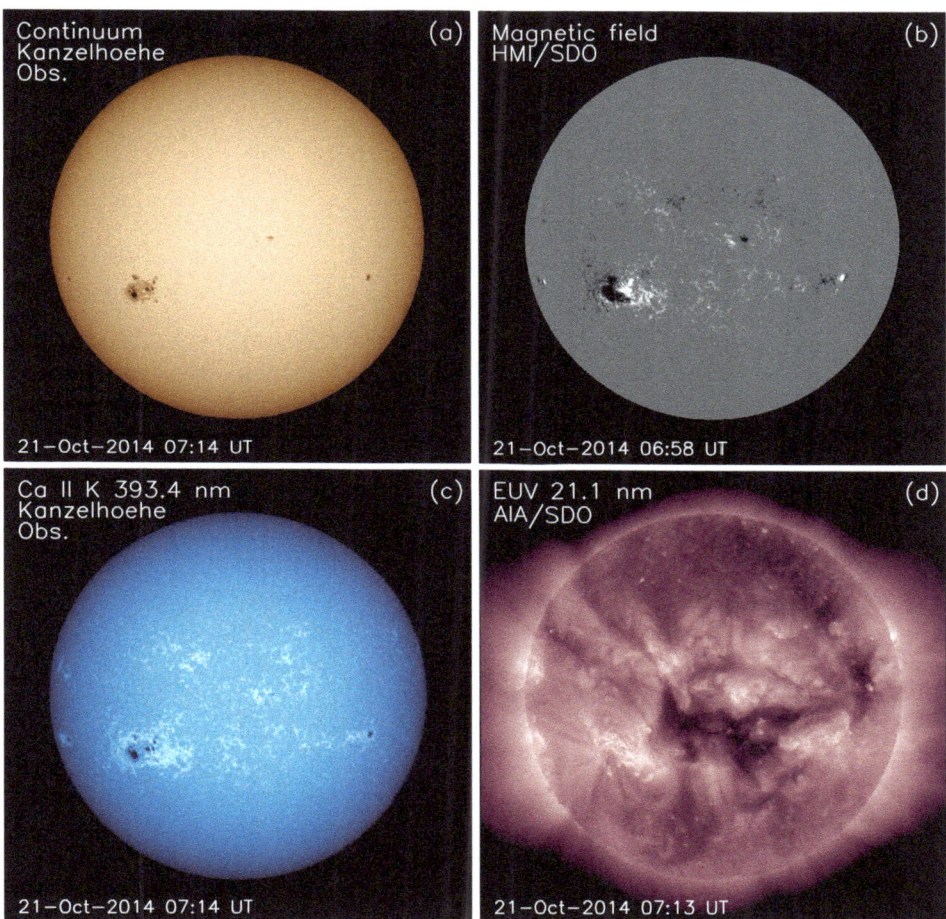

Fig. 2. Full-disk images of the Sun from October 21, 2014 obtained in different wavelengths: a) optical continuum (showing the photosphere), b) line-of-sight magnetic field map (photosphere), c) Ca II K image (chromosphere), d) EUV 21.1 nm image (corona). Data sources: a, c) Kanzelhöhe Observatory (University of Graz, Austria), b) Helioseismic and Magnetic Imager (HMI) onboard NASA's Solar Dynamics Observatory (SDO), d) Atmospheric Imaging Assembly (AIA) onboard SDO.

solar wind particles are intimately coupled, as both require energy transport from the lower layers of the Sun to the corona, in addition to the escaping radiation. The ultimate source is assumed to lie in the kinetic energy of the turbulent plasma motions in the solar convection zone.

A plasma has very special properties compared to a normal gas. One of these properties is that due to the large number of free charged particles, the electrical conductivity is extremely high. As a consequence, in a plasma, the magnetic field

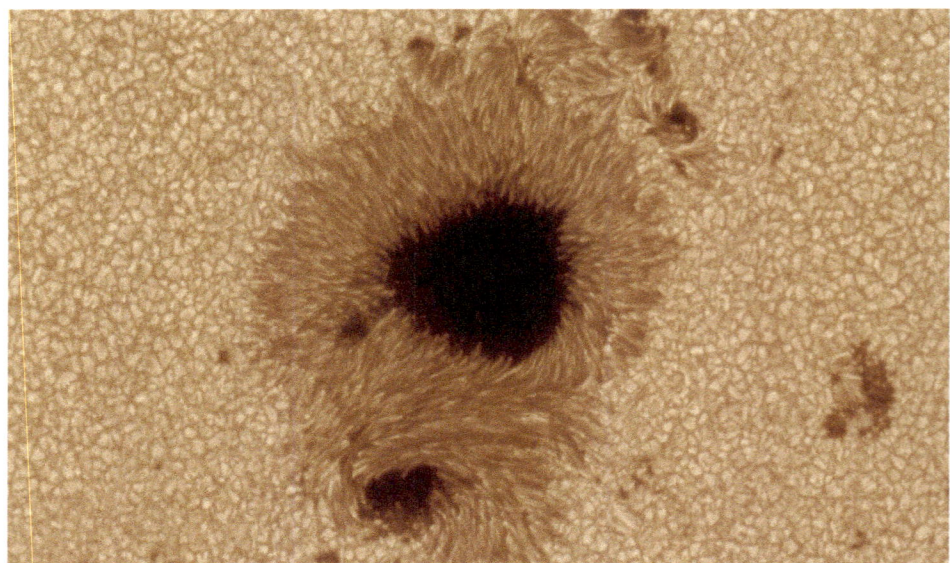

Fig. 3. High-resolution image of the Sun recorded in the optical continuum showing a complex sunspot group and the granulation pattern in the surrounding quiet-Sun regions. The plotted field of view has a size of 107 000 km in the horizontal and 57 000 km in the vertical direction. Data source: Solar Optical Telescope (SOT) onboard the Japanese Hinode satellite.

and the matter cannot evolve independently from each other, but the magnetic field and plasma are said to be *frozen-in*, i.e., the field lines are tied to the plasma and vice versa. One consequence of the frozen-in condition is that the outflowing solar wind particles carry the magnetic field from the Sun far into interplanetary space, where it forms the Interplanetary Magnetic Field (IMF). Since the Sun rotates (once in about 27 days as observed from Earth), the IMF has a spiral shape. The region in interstellar space where the solar wind pressure dominates over the pressure exerted by the stellar wind, defines the heliosphere, a tenuous bubble in which our solar system is embedded. In 2012, Voyager 1 was the first spacecraft to leave our heliosphere, when it traversed its outer boundary, the heliopause, at a distance of about 120 AU.

3 Disturbances on the Sun and the heliosphere

Sunspots are the most striking appearance of strong magnetic field concentrations on the Sun. They appear much darker than their surrounding because they are less hot (Figures 2a and 3). According to Stefan-Boltzmann's law, the total power radiated by a black body across all wavelengths is proportional to the fourth power of its thermodynamic temperature. The temperatures in a sunspot are "only" about 4000 to 5000 K, in contrast to 5800 K in regions that are not populated by strong fields (quiet-Sun regions). The reason for the reduced temperature is the strong magnetic field in sunspots (up to 0.3 Tesla) that hinders the efficiency of the energy transport by the convective motions from below. Figure 2b shows the magnetic field measured in the solar photosphere. The white and black patches indicate regions of strong magnetic fields of opposite polarities, i.e., regions where bundles of magnetic field lines enter from the convection zone to the solar surface (white; positive polarity) and where they re-enter again to the convection zone (black; negative polarity). They coincide with the dark sunspot regions that can be seen in the corresponding continuum image in Figure 2a. Due to the solar rotation, sunspots appear to move across the solar disk when we observe them over consecutive days.

Due to their lower temperature, sunspots are regions of a radiation deficit compared to quiet-Sun regions. In the surrounding of sunspots, however, we often observe regions that are somewhat brighter than the quiet Sun, i.e., they reveal a radiation excess, so-called faculae. Like sunspots, faculae are also related to magnetic field concentrations, but on much smaller scales. The chromospheric counterparts of the photospheric faculae are called plages. Plages appear very prominent in the temperature sensitive Ca II K filtergrams (Figure 2c). In addition to the plage regions surrounding the dark sunspots, the Ca II K filtergrams also show a large-scale network with bright cell boundaries covering the whole chromosphere. This chromospheric network is a consequence of the magnetic network that is formed by the large-scale horizontal plasma flows in the solar photosphere called *supergranulation*.

The solar corona observed in EUV and SXR wavelengths appears structured into a multitude of thin strands, so-called coronal loops, most pronounced in regions associated with sunspots (Figure 2d). These loops are signatures of the extending magnetic fields that are rooted in the solar surface. Due to the frozen-in condition, hot emitting plasma is concentrated along the magnetic field and quasi "illuminates" the field structure. The total set of phenomena in the different atmospheric layers that occur in association with sunspots (faculae, plages, coronal loops, etc.) is called an active region. The second striking class of features in the solar corona are coronal holes. They appear as large-scale dark regions in EUV and SXR images, due to the lower density and temperature compared to the surrounding corona (an example can be seen in the center of Figure 2d). Coronal holes are associated with "open" field lines connected to interplanetary space, along which particles can escape.

The most energetic solar events are Coronal Mass Ejections (CMEs) and flares. CMEs are magnetised plasma clouds expelled from the Sun with speeds of hundreds up to $3\,000\,\mathrm{km\,s^{-1}}$ (Figure 4). Flares cause a sudden enhancement of the solar radiation, most prominently at short (EUV, X-rays) and long (radio) wavelengths resulting from the heated flare plasma and accelerated particles. CMEs and flares are related processes and often occur together, as they are both driven by magnetic instabilities and reconnection of coronal fields in active regions. Magnetic reconnection is a fundamental process in magnetised plasmas,

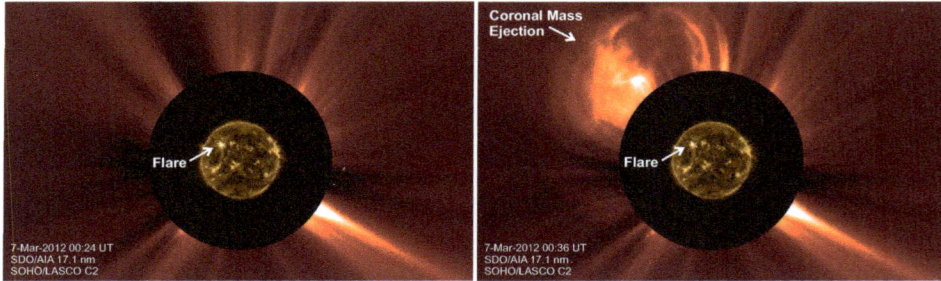

Fig. 4. Composite images showing the corona in EUV 17.1 nm (inner part) and in scattered white light above the coronagraph occulter (outer part) on March 7, 2012. The image sequence shows the evolution of a coronal mass ejection and the associated flare event. Data source: SDO/AIA and Large Angle and Spectrometric Coronagraph (LASCO) C2 onboard the ESA/NASA Solar and Heliospheric Observatory (SOHO).

by which the field connectivity is reconfigured and part of the magnetic energy is impulsively released to heat and accelerate the plasma. The energy that is released in flares and CMEs is originally stored in the strong magnetic fields associated with sunspots. The turbulent plasma motions, driven in the convection zone and photosphere, shear and twist the coronal magnetic field. In this way, electric currents are built up in the corona and the "free" energy stored in these currents may eventually be released. Whereas the energy build-up occurs gradually over days to weeks, the energy release via flares and CMEs occurs impulsively, i.e., on the order of minutes to hours.

The Earth has a shielding magnetic field, our magnetosphere, at which CMEs and solar wind particles are generally deflected. However, when the magnetic field of an impacting CME is anti-parallel to the Earth magnetic field, magnetic reconnection will occur at the day-side magnetosphere. In this process, energy and particles from the solar plasma can couple to the Earth magnetosphere, thus causing a geomagnetic storm, i.e., a temporary disturbance of the Earth's magnetosphere, and also polar lights. The major difficulty in predicting the geo-effectivity of CMEs lies in the strength and orientation of the CME magnetic field on which we have very little information before it arrives at satellites close to Earth. CMEs and flares are both capable of accelerating Solar Energetic Particles (SEPs). SEPs pose a severe hazard to astronauts in space due to their ionising radiation, and may also damage the electronics onboard satellites.

The promptest effects of solar disturbances at Earth are actually due to the flare, as the radiation reaches Earth about eight minutes after the outburst happened on the Sun, followed by the effect of SEPs. Moving at relativistic speed, SEPs may reach Earth some ten minutes after their release from the Sun. The geo-effects of CMEs show a longer delay, with the fastest CMEs traversing the interplanetary space from the Sun to Earth in less than one day (slow CMEs may take up to about 5 days). However, fast CMEs are the most severe solar events

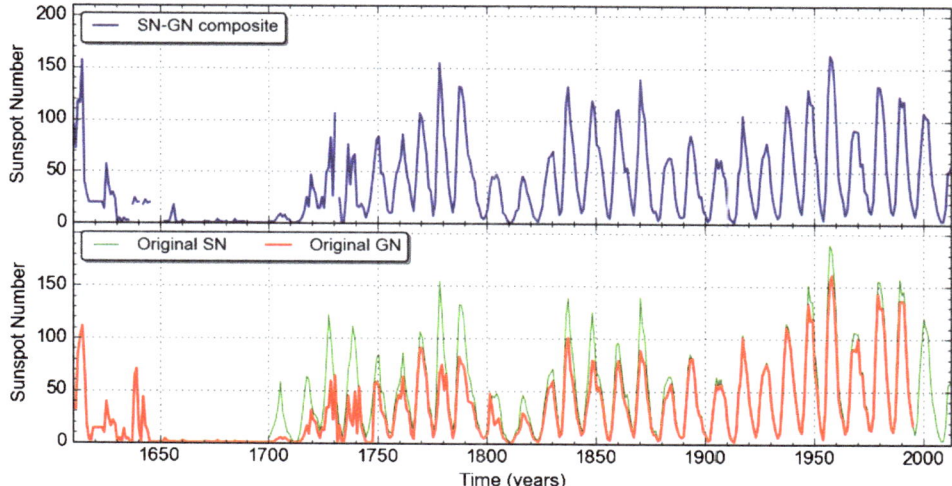

Fig. 5. Evolution of the yearly mean sunspot number from 1610 to 2014. Top panel: Composite of the Sunpot Number (SN) and Group Number (GN) based on the recalibrated data sets following Clette et al. (2014), using the group number before 1843 and the sunspot number after 1843. The individual datasets closely match after applying the independent corrections identified in each series. Bottom panel: the red curve shows the original Group Number (GN) and the green curve shows the original Sunspot Number (SN). Data credit: WDC-SILSO, Royal Observatory of Belgium (http://www.sidc.be/silso/).

as they are the cause of the strongest geomagnetic storms. At times of low solar activity, Earth is mostly embedded in the slow and fast solar wind streams, which impact Earth about 3 to 6 days after their release from the Sun. High-speed wind streams are then dominating the geo-effects, as they may cause moderate geomagnetic storms.

4 The 11-year solar cycle

The number of sunspots varies with an average period of 11 years, and so does the occurrence rate of flares and CMEs that are powered by the sunspots' magnetic fields. This periodic change is called the solar activity cycle, or Schwabe cycle (Figure 5). The primary quantity describing the solar activity cycle is the relative sunspot number, defined as $R = k(f + 10g)$, where f is the number of individual sunspots, g the number of sunspot groups and k a correction factor that serves for the inter-calibration of the data collected from different observatories. The maximum and minimum of a solar cycle are actually defined on the basis of the sunspot number.

At the beginning of a solar cycle, sunspots form at heliographic latitudes of about ∼40°, whereas in the course of the cycle, the sunspot belts migrate closer to the solar equator. George Ellery Hale was the first to establish in 1908 that sunspots are indeed magnetic phenomena, by measuring the Zeeman splitting in spectral lines. He also showed that sunspot groups reveal a predominant east-west alignment and that their magnetic polarity in the northern and the southern hemisphere is opposite to each other (Figures 2a and 2b). After one 11-year solar cycle, the sunspot polarities in the two hemispheres switch. This means that there is actually a fundamental 22-year magnetic cycle underlying the 11-year sunspot cycle. In addition to the magnetic polarity of sunspots that changes during solar minimum, it was also found that the Sun's polar fields cyclically reverse, with the field reversal occurring around the time of the solar cycle maximum.

These observational findings imply that the Sun's large-scale magnetic field undergoes a cyclic regeneration. The most important ingredients for such a *solar dynamo* are the turbulent convective plasma flows and the differential rotation of the Sun, i.e., the fact that the Sun rotates faster at the solar equator than at higher latitudes. Starting from an originally weak poloidal magnetic field, the differential rotation stretches the field lines into an azimuthal direction and coils them like a rope (Ω-effect). This Ω-effect is presumed to take place in the tachocline, which is the transition layer between the rigidly rotating radiation zone and the differentially rotating convection zone (Figure 1). To fulfil one magnetic cycle, this predominantly azimuthal field has to be reformed again to a poloidal field but of opposite polarity. This is done by the so-called α-*effect*, which describes the interaction of solar rotation and convection, where Coriolis forces exert a twist on the rising magnetic flux tubes in the rotating solar plasma.

The variation of the number of sunspots is the most obvious sign of the solar cycle (Clette et al., 2014). However, the solar cycle causes a variety of modulations in the appearance of the different layers of the solar atmosphere (sunspots, plages, coronal holes, etc.) as well as the heliosphere, and modulates the occurrence frequency of energetic solar eruptions. CMEs and flares are most frequent at solar maximum, whereas high speed solar wind streams at Earth orbit are most frequent at the descending phase of a solar cycle. The solar luminosity also varies over the solar cycle by about 0.1%, being higher during times of solar maximum than minimum, due to the predominance of the excess radiation in solar faculae over the radiation deficit in sunspots. This predominance arises from the fact that faculae cover larger areas and have longer lifetimes than sunspots.

The sunspot number is the longest available indicator of the solar cycle that is directly derived from solar observations, and is available since the year 1610 when the Sun was first observed with the newly invented telescope. Figure 5 shows the sunspot number evolution from 1610 to present, where the basic 11-year cycle is evident as well as modulations on longer time-scales of about 80–90 years. Note also the sunspot deficit in the period 1645 to 1715, the so-called *Maunder minimum*, which coincided with the coldest part of the Little Ice Age affecting Europe and North America. More recent measurements of the solar cycle (available since 1946) include the 10.7 cm solar radio flux that is predominantly emitted from plasma trapped in the magnetic field of active regions. Dating further back in the history of solar activity, we have to rely on indirect, i.e., non-solar, measurements, so-called *proxies*. The most common proxy of solar activity are data from cosmogenic radionuclides, such as ^{14}C and ^{10}Be, which reside in tree rings and polar ice cores, respectively. Cosmogenic radionuclides are produced by cosmic rays in the Earth's atmosphere. The changing magnetic field of the Sun modulates the flux of galactic cosmic ray particles entering our heliosphere, which is smaller at times of high solar activity when the Sun's field is strong. Thus, the concentrations of cosmogenic radionuclides in tree rings and ice cores vary in anti-phase with the solar activity cycle.

Further reading

Hathaway, D., The Solar Cycle, Living Reviews in Solar Physics, 7:1, 2010, DOI:10.12942/lrsp-2010-1

Haigh, Joanna D., The Sun and the Earth's Climate, Living Reviews in Solar Physics, 4:2, 2013, DOI: 10.12942/lrsp-2007-2

Foukal, Peter V., Solar Astrophysics. 3rd, revised edition, Wiley, 2013

Kamide, Y., A. C.-L. Chian (Eds.), Handbook of the Solar-Terrestrial Environment. Springer, 2007

Owens, Mathew J., Robert J. Forsyth, The Heliospheric Magnetic Field, Living Reviews in Solar Physics, 10:5, 2013, DOI: 10.12942/lrsp-2013-5

Pulkkinen, Tuija, Space Weather: The Terrestrial Perspective, Living Reviews in Solar Physics, 4:1, 2006, DOI: 10.12942/lrsp-2007-1

Schwenn, Rainer, Space Weather: The Solar Perspective, Living Reviews in Solar Physics, 3:2, 2006, DOI: 10.12942/lrsp-2006-2

Stix, Michael, The Sun: An Introduction. 2nd edition, Springer, Astronomy and Astrophysics Library, 2004.

Usoskin, Ilia, A History of Solar Activity over Millenia, Living Reviews in Solar Physics, 10:1, 2013, DOI: 10.12942/lrsp-2013-1

CHAPTER 2.2

SOLAR RADIATIVE FORCING

Natalie A. Krivova[1] and Ilaria Ermolli[2]

1 Radiative forcing

Radiative forcing (RF) quantifies the impact of various factors on Earth's surface temperature and climate. In equilibrium, the global average temperature of the Earth is determined by a balance between the incoming energy absorbed in the atmosphere and by the surface and the energy of the thermal infrared radiation emitted back into space, so that the net radiative flux (globally and annually averaged) at the top of the atmosphere[3] is zero. When the amount of the incoming or outcoming radiation changes, then before a new equilibrium is established the net flux at the top of the atmosphere is not zero. Thus, radiative forcing is the instantaneous change in the value of the net downward radiative flux. Radiative forcing is defined positive when the amount of energy retained by the Earth-atmosphere system increases. This could be due to an increase in the incoming or a decrease in the outcoming energy and leads to an increase in the Earth's temperature.

2 Solar irradiance

The dominant energy source to Earth's climate system is the Sun. Solar energy is produced in the Sun's core, where hydrogen atoms fuse to form, via a chain reaction, helium. Photons released in this process travel inside the Sun until, tens or even hundreds of thousands years later, they reach up to the so-called solar visible surface (or the photosphere), above which the solar atmosphere is transparent for the photons. They escape as visible light and part of them reaches the Earth just minutes later.

---

[1] Max-Planck-Institut für Sonnensystemforschung, 37077 Göttingen, Germany
[2] INAF, Osservatorio Astronomico di Roma, Monte Porzio Catone, Italy
[3] Radiative forcing can also be defined at the tropopause.

The total solar electromagnetic energy flux received at the top of the Earth's atmosphere and reduced to the mean distance from the Sun to the Earth (one astronomical unit, AU) is called the total solar irradiance (TSI). This definition takes care to differentiate between changes in the solar radiative output itself and changes in the Earth-Sun distance (see Infobox 2.1) or the interaction between the solar radiation and the Earth's atmosphere.

Solar electromagnetic energy is emitted over essentially entire spectrum. Almost half of this energy originates in the range visible to the human eye (roughly 400 to 800 nm), more than 40% come from the infrared, IR, range (wavelengths above 800 nm), and only less than 10% are contributed by the ultraviolet, UV, radiation (wavelengths below 400 nm). Solar irradiance measured at a certain wavelength or within a wavelength interval (in units of W m^{-2} per unit wavelength interval, e.g. W m^{-2}nm^{-1}) is called spectral solar irradiance (SSI).

Total solar irradiance is one of the more recently discovered variables of the Sun, generally being referred to as the solar constant in the older literature.

3 Measurements of solar irradiance

3.1 TSI

Measurements of TSI are done with absolute radiometers. They include a cavity with a blackened inner side, operating as a black body to efficiently absorb the incident radiation, which heats the cavity. A built-in electrical calibration heater maintains the heat flux constant by adjusting the supplied heating power. The electrical power required to maintain the equilibrium is proportional to the incoming radiative power absorbed by the cavity.

Due to atmospheric extinction, measurements of total solar irradiance sufficiently precise to reveal its variability are only possible from space and have been regularly carried out since 1978 by over a dozen of different radiometers (Figure 1). On an absolute scale, results from individual experiments differ significantly. Recent tests and experiments have helped to identify main sources of the offsets. The TSI value of 1360.8 ± 0.5 Wm^{-2} recorded by SORCE/TIM in 2008 is currently believed to best represent the solar minimum conditions.

The relative TSI changes measured by space-borne radiometers agree quite good with each other on time scales of days to years. Thus, TSI varies on all observable time scales, from minutes to decades. Most prominent are the short-term fluctuations on time scales of days to roughly a week and the modulation in phase with the 11-year solar activity cycle (see Figure 1).

But there are also differences between the data coming from individual experiments, most importantly in the longer-term trends. This is because changes in the sensitivity (i.e., the calibration) of the instruments due to their aging and degradation once they are launched and exposed to the solar light, are quite individual and often tricky to assess. Owing to the offsets in the absolute values, combining the data from individual instruments into a single composite record is non-trivial.

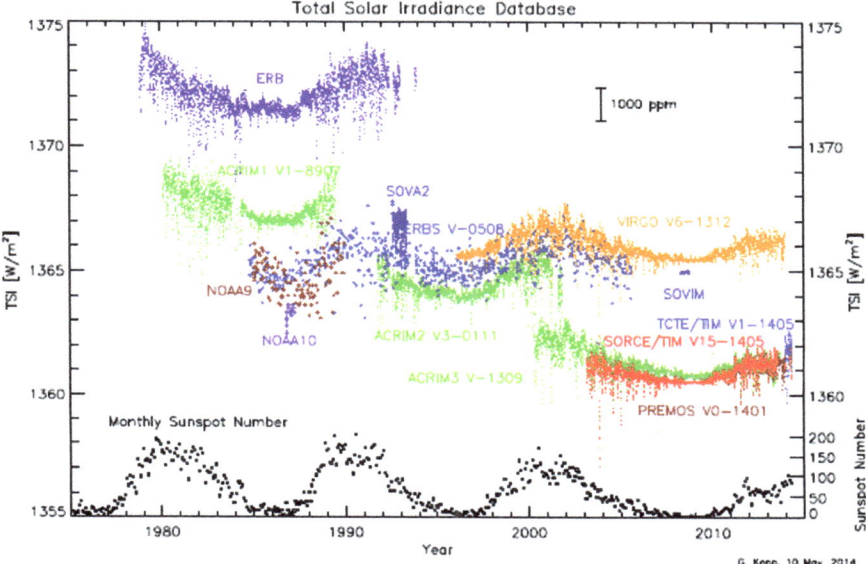

Fig. 1. Space-borne measurements of the total solar irradiance (TSI) covering the period from 1978 to 2014. Individual records are shown in different colors, as labeled in the plot. The bottom part of the plot shows the monthly mean sunspot number. Courtesy of G. Kopp (http://spot.colorado.edu/koppg/TSI/). Abbreviations: Active Cavity Radiometer Irradiance Monitor (ACRIM) instruments on the Solar Maximum Mission (SMM), Upper Atmosphere Research Satellite (UARS), and ACRIMSat; Earth Radiation Budget (ERB) instrument on Nimbus-7; Earth Radiation Budget Satellite (ERBS), the Earth Radiation Budget Experiment (ERBE) on the Earth Radiation Budget Satellite (ERBS) National Oceanic and Atmospheric Administration (NOAA), Precision Monitor Sensor (PREMOS) on PICARD, Solar Variability Experiment (SOVA) on EUropean Retrievable Carrier of ESA (EURECA), Solar Variability and Irradiance Monitor (SOVIM) on the International Space Station (ISS), Total Irradiance Monitor (TIM) on Solar Radiation and Climate Experiment (SORCE), Variability of Solar Irradiance and Gravity Oscillations (VIRGO) on SoHO.

Currently, three different composites exist (see Figure 2). The most critical discrepancy concerns their conflicting secular trends, best seen as the difference in the TSI levels during activity minima in 1986, 1996 and 2008. Models of solar irradiance (see Section 5) return an either downward trend (Figure 2), as in the PMOD (Physikalisch-Meteorologisches Observatorium Davos) composite, or in case of most proxy models, no trend. None of the models returns an upward or alterating trend.

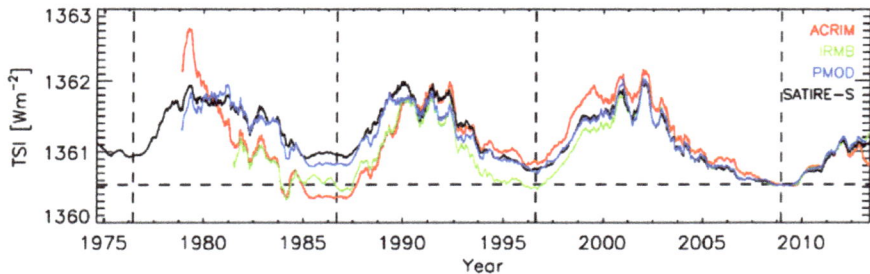

Fig. 2. The ACRIM (red), IRMB (green) and PMOD (blue) TSI composite records smoothed over 181 days. Also plotted is the SATIRE-S model reconstruction of TSI. The vertical dashed lines mark solar minima. From Yeo et al. (2014).

3.2 SSI

Measurement of the solar spectral irradiance is more sophisticated. Very low amplitudes of the variability in the visible and IR ranges, a rapid increase in the variability in combination with significant decrease of the amount of the radiation with decreasing wavelength (Figure 3) and the need for a meaningful spectral resolution are only some of the complicating factors. Thus in addition to the high radiometric accuracy and precision, a spectrometer has to maintain a sufficient wavelength accuracy and account for the wavelength-dependent long-term degradation effects.

Regular monitoring of the solar spectrum in the UV (below 400 nm) also started in 1978. Most of our knowledge is based on the results of two experiments, SOLSTICE (Solar Stellar Irradiance Comparison Experiment) and SUSIM (Solar Ultraviolet Spectral Irradiance Monitor), onboard UARS (the Upper Atmosphere Research Satellite), in operation from 1991 to 2001 and 2005, respectively. SSI monitoring over a broad range from the UV to the IR (up to about 2 400 nm) with absolute radiometric calibration has been carried out by SORCE/SIM since 2003.

Whereas most of the solar radiant energy is emitted at visible and infrared wavelengths, the variability is significantly stronger in the UV spectral range (Figure 3). The amplitude of the variation grows from about 0.1% in the visible and even less in the IR to almost 100% around the Ly-α at 121.6 nm, which is the strongest line in the solar spectrum. The contribution of the UV (below 400 nm) part of the spectrum to the total irradiance variability most probably exceeds 50%, although the exact number is still a matter of debate (Figure 4). Thus, the more recent observations by SORCE/SIM suggest stronger variability in the UV than had been measured before SORCE and predicted by the models. This excessive variability would need to be partly compensated by a presumably anti-phase (with the solar cycle) variation in the visible (400–700 nm). This pattern of SSI variability is, however, in contradiction not only with other, partly contemporary, observations, but also with the up-to-date models, which are at the same time

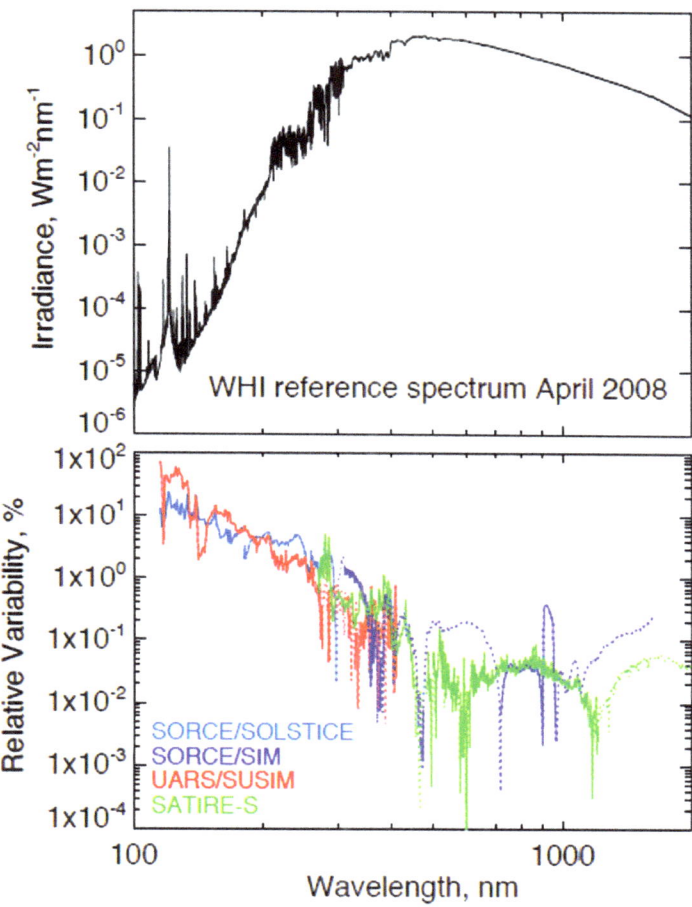

Fig. 3. Top: Reference solar spectrum recorded in April 2008 within the Whole Heliosphere Interval (WHI) project. Bottom: Relative SSI variability as observed by UARS/SUSIM (red curve) between the maximum of cycle 23 (March 2000) and the preceding minimum (May 1996), as well as by SORCE/SOLSTICE (light blue) and SORCE/SIM (dark blue) (Harder et al., 2009) between April 2004 and December 2008. Also shown is the variability between 2000 and 1996 predicted by the SATIRE-S model (green). For each period, averages over one month are used. Negative values are indicated by dotted segments. Following Solanki et al. (2013). Abbreviations: UARS - Upper Atmosphere Research Satellite; SUSIM - Solar Ultraviolet Spectral Irradiance Monitor; SORCE - Solar Radiation and Climate Experiment; SOLSTICE - Solar Stellar Irradiance Comparison Experiment; SIM - Spectral IrradianceMonitor; SATIRE-S - Spectral and Total Irradiance Reconstruction for the Satellite era.

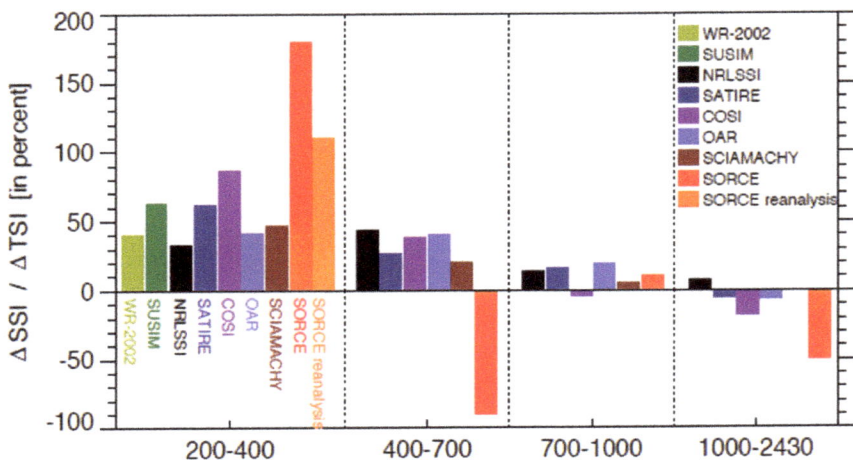

Fig. 4. Relative contribution of the UV (200–400 nm), visible (400–700 nm), near-IR (700–1000 nm) and IR (1000–2430 nm) ranges to the TSI change over the same period. For SORCE/SIM, only the period between 2004 and 2008 can be considered. For other data and models, the plotted relative differences are between solar maximum an minimum. Within each wavelength bin, from left to right: UARS/SUSIM (green), NRLSSI (black), SATIRE-S (blue), COSI (purple), OAR (light blue), SCHIAMACHY/ Mg index empirical model (orange), SORCE/SIM (red). Following Ermolli et al. (2013). Abbreviations: UARS - Upper Atmosphere Research Satellite; SUSIM - Solar Ultraviolet Spectral Irradiance Monitor; NRLSSI - Naval Research Laboratory Solar Spectral Irradiance; SATIRE-S - Spectral and Total Irradiance Reconstruction for the Satellite era; COSI - COde for Solar Irradiance; OAR - Osservatorio Astronomico di Roma; SCIAMACHY - SCanning Imaging Absorption spectroMeter for Atmospheric CHartographY); SORCE - Solar Radiation and Climate Experiment; SIM - Spectral IrradianceMonitor.

very successful in reproducing all other available TSI and SSI observations (see Figure 4). A growing body of evidence has been found that SIM/SORCE measurements might still suffer from unaccounted instrumental effects but the resolution of the debate is still pending.

4 Origin of solar irradiance variations

Variations of solar irradiance on very short (minutes to hours) time scales are irrelevant to Earth's climate and are not discussed here. Variations on time scales longer than one day are driven by changes in the amount and distribution of the solar surface magnetic field. The magnetic field emerging on the surface manifests itself in form of different brightness features, such as sunspots, faculae, plage or network.

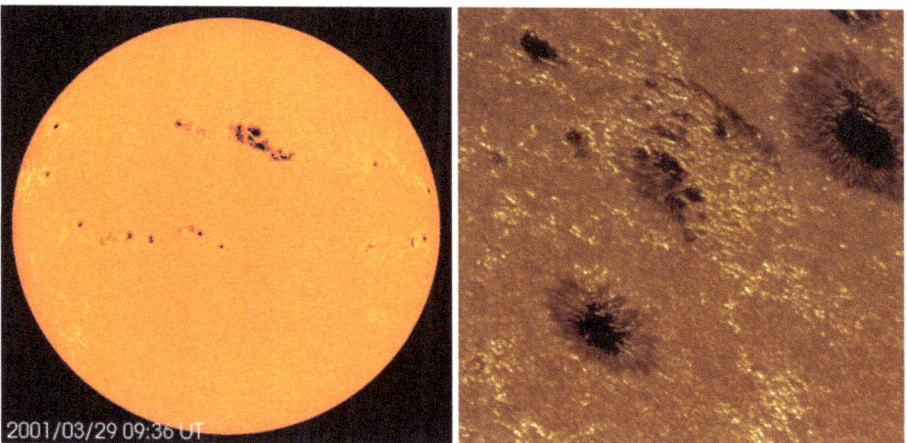

Fig. 5. Left: Continuum image of the Sun taken by the SoHO/MDI instrument on 30 March 2001. Courtesy of SoHO/MDI consortium. SOHO is a project of international cooperation between ESA and NASA (http://sohowww.nascom.nasa.gov/gallery/bestofsoho.html). Right: Active regions with sunspots and faculae near solar limb. Image by Dan Kiselman and Mats Löfdahl in G-band (430.5 nm) taken on June 29, 2003 with the Swedish 1-m Solar Telescope (SST). The SST is operated on the island of La Palma by the Institute for Solar Physics of the Royal Swedish Academy of Sciences in the Spanish Observatorio del Roque de los Muchachos of the Instituto de Astrofsica de Canarias (http://www.solarphysics.kva.se/gallery/images/2003/).

Sunspots are strong concentrations of the magnetic field, in which the heat transport from the interior is inhibited. Therefore they are cooler than the surrounding photosphere and appear dark (Figure 5). Sunspots crossing the visible solar hemisphere cause a darkening of the Sun by up to 0.3% on time scales of hours to about a week (Figure 1). The magnitude of the brightness dip is roughly proportional to the projected area of the spot. Smaller magnetic field concentrations cause brightenings on the solar surface termed faculae, plage and the network. Faculae and plage are usually observed in active regions, in the vicinity of sunspots (Figure 5). The weaker network elements are found in the quiet Sun all over the disc. Faculae and the network are responsible for the overall brightening by about 0.1% (for recent cycles) from activity minimum to maximum (Figure 1).

5 Models and reconstructions of solar irradiance variations

Records of TSI and SSI observations, covering less than four decades, are too short to allow reliable linking of climate change on Earth with solar irradiance variability. It is therefore necessary to extend the irradiance time series back in time with the help of suitable models. The aim of such models is two-fold: (1) to understand and reproduce the directly measured irradiance variations and (2) to reconstruct past (and ideally, also predict future) irradiance changes.

A number of models have been constructed that can be divided into two classes: the so-called proxy and semi-empirical models. Proxy models regress different indices of solar magnetic activity (two or more to describe sunspot darkening and facular brightening) to the measured irradiance variation. The need to have a record of direct irradiance observations that is stable over a sufficient period of time makes in particular reconstructions of SSI with this technique circuitous. The other type of models follows a more physics-based approach. The brightness spectra of individual surface components (sunspots, faculae, plage, network and the quiet Sun) are calculated using radiative transfer codes from corresponding semi-empirical model atmospheres. The latter describe how the temperature and density of the solar atmosphere change with height. The corresponding features are then identified on the solar visible surface and their fractional area disc coverage is quantified. Typically high resolution full-disc solar images are employed for this, but disc-integrated indices are also often used, e.g. when going back in time to periods when such maps were not available. The irradiance, as a function of time, is then calculated by weighting the intensity spectra of each surface component with the corresponding area coverage and summing up all contributions. State-of-the-art models reproduce TSI variations in great detail (see, e.g., Figure 2).

Reconstructions of solar irradiance into the past require suitable proxies of solar activity. Most of the reconstructions on time scales of centuries rely on the historical record of the sunspot number, which is the longest record of direct solar observations. It goes back to 1610 and describes the cyclic evolution of the active regions fairly accurately. However, the emergence rate of the weaker ephemeral magnetic regions does not exactly follow the evolution of active regions. The activity cycles of these weak elements are significantly weaker in amplitude (if any at all) and are most probably somewhat shifted and stretched in time (i.e., their activity cycles are longer than the corresponding sunspot cycles, which have an average period of roughly 11 years). It has been proposed that long-term changes in the emergence rate of such weak features is the main source of the secular change in irradiance. As no direct long-term proxy of this component exists, the magnitude of the secular variation remains heavily debated (see Figure 6 for different TSI reconstructions since 1610). The situation is further complicated by the fact that the secular change over the period of spaceborne irradiance observations was weak and rather uncertain (Section 3). Thus most models agree with the direct measurements within the measurement uncertainty, while diverging on longer scales. The estimated magnitude of the change in TSI since 1700 ranges between 0.6 and 3 W m^{-2}.

Estimates of the irradiance on yet longer time scales rely on the records of the cosmogenic isotopes ^{10}Be and ^{14}C (Chapter 2.5).

6 Summary

Measurements of solar irradiance are now available for around 3.5 solar cycles. These data are invaluable. Nevertheless, the debate about the presence or absence of a secular trend is ongoing. To understand the role of the Sun on climate, longer

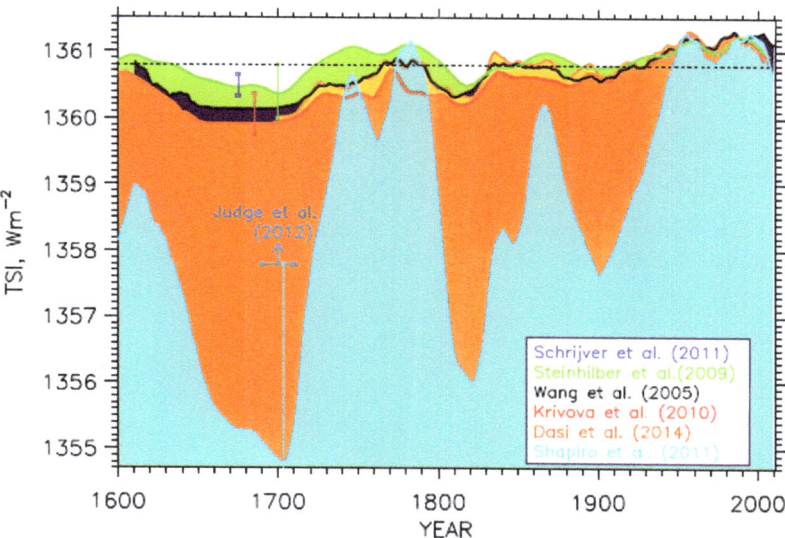

Fig. 6. Various reconstructions of TSI since 1610. Vertical bars denote uncertainties of the models plotted in the same colours and identified in the plot. The dark blue bar shows the estimate of the possible TSI change following Schrijver et al. (2011); no reconstruction done. The uncertainty of the model by Shapiro et al. (2011), ± 3 W m^{-2}, extends downward outside the plot. The revised estimate for this model by Judge et al. (2012) is marked by the blue bar and arrow. The black dotted line marks the TSI value at the solar minimum in 2008 according to SORCE/TIM measurements. Other references are (Steinhilber et al. (2009); Krivova et al. (2010); Dasi-Espuig et al. (2014), priv. comm.)

time series are imperative. These can only be obtained using suitable models. To ensure the correct interpretation of the physical processes, successful models have, in the first step, to reproduce the observed changes. State-of-the-art models have progressed to a high level of accuracy on time scales up to the solar cycle, although further progress is still needed and possible.

On longer (and for climate studies more important) time scales, a number of reconstructions going back to 1610 or even longer exist. But the magnitude of any secular trend remains a matter of intense debate. Recent estimates converge to the value of about 1 to 1.5 W m^{-2}, but range from 0.6 to 3 W m^{-2}, thus differing by a factor of five. Further work in this direction is of great importance.

Further reading

Stocker, T., and Qin, D., eds., 2014. Climate Change 2013 - The Physical Science Basis. Working Group I Contribution to the Fifth Assessment Report of the IPCC. Cambridge University Press, Cambridge.

Fröhlich, C., 2010. Solar radiometry, In: Observing Photons in Space. Eds.: M.C.E. Huber, A. Pauluhn, J.L. Culhane, J.G. Timothy, K. Wilhelm, and A. Zehnder. ISSI Scientific Reports Series 9, 525-540

Kopp G., 2014. An assessment of the solar irradiance record for climate studies, J. Space Weather Space Clim. 4, A14, doi:http://dx.doi.org/10.1051/swsc/2014012

Solanki, S.K., N. A. Krivova and J. D. Haigh 2013. Solar Irradiance Variability and Climate. ARA&A 51, 311-351. doi:10.1146/annurev-astro-082812-141007

CHAPTER 2.3

VARIABILITY OF SOLAR AND GALACTIC COSMIC RAYS

Galina A. Bazilevskaya[1] and Irina A. Mironova[2]

1 History

Cosmic rays were discovered in 1912 by Austrian physicist Victor Hess who in 1936 was awarded the Nobel Prize in Physics for his discovery. The term "rays" does not mean that this radiation has purely electromagnetic nature, as sunlight, radio waves or X-rays. This term was introduced after the discovery of the phenomenon of cosmic rays whose nature was not known at that time. Later, however, it became clear that the main component of cosmic rays is related to energetic particles, mostly protons.

2 Origins of cosmic rays

Cosmic rays may be of galactic or solar origin.

2.1 Galactic cosmic rays

The main source of galactic cosmic rays (GCR) in the Galaxy is believed to be explosions of supernova. The observed range of GCR energy extends by nearly 15 orders of magnitude, from 1 MeV ($1.6 \; 10^{-13}$ J), to an enormous value of $3 \; 10^{14}$ MeV (48 J). The flux of particles with these ultra-high energies is very low, about 1 particle per 10 km^2 per year, while in the lower part of energy spectrum, say around 100 MeV, intensity of GCR protons is about 1 per cm^2 per second. More than 90% of GCR are protons, helium comprises about 8%, less than about 2% being the other elements. Electrons are a minor constituent of GCR (about 1% or lower).

2.2 Solar cosmic rays

Strong increases of the intensity of energetic particles can sporadically occur near Earth, due to arrival of solar energetic particles (SEP) also known as solar cosmic

---

[1] Lebedev Physical Institute, Russian Academy of Sciences, Moscow, Russia.
[2] Institute of Physics, St.Petersburg State University, St.Petersburg, Russia

rays (SCR), or solar proton events (SPE), since solar protons are far more abundant than other elements. SEP is associated with fast powerful energy release on the Sun, such as solar flares and/or coronal mass ejections (CME). The energy of SEP usually extends from 1 MeV to several GeV/nucleon, but, rarely, can reach several tens of GeV/nucleon. The energy spectrum of SEP covers more than four orders of magnitude in energy and more than 8 orders of magnitude in intensity. However SEP, existing only sporadically during periods of high solar activity, may occasionally enhance the flux of particles with energy tens-hundreds MeV by several orders of magnitude for hours-days, and in rare events called ground level enhancement (GLE) up to several GeV, for minutes-hours. Although the elemental composition of SEP changes from one event to another, it is dominated (larger than 90%) by protons.

3 Cosmic rays in the Earth's vicinity

The flux of primary cosmic rays (CR) outside the Earth's atmosphere depends on the level of solar activity and conditions in magnetosphere. Also, the CR flux is modified while propagating through the atmosphere.

3.1 Cosmic rays in the Earth's geomagnetic field

The Earth possesses its own magnetic field, which can, at a zero approximation, be considered a dipole. This geomagnetic field deflects charged particles and acts as a spectrometer separating the arriving CR particles according to their rigidity which is defined as $R = \frac{cP}{Ze}$ (here c is the speed of light, P is the particle's momentum, Z is the particle's charge number, and e is the unit charge). For protons, R is connected to kinetic energy as $R = \sqrt{E^2 + 2E_o E}$, where $E_o = 0.938$ GeV is the proton's rest mass, R is given in GV and E in GeV. Roughly speaking, the effect of the geomagnetic shielding can be characterised by a cut-off rigidity, Rc, so that particles with lower rigidity cannot reach the given location. The geomagnetic cut-off rigidity Rc varies over the globe from $Rc = 0$ in polar regions (although the atmospheric shielding cut-off of about 1 GV is always present) to $Rc = 13$–17 GV in the equatorial region.

3.2 Cosmic rays in the atmosphere

The Earth's atmosphere also acts as a particle energy spectrometer, separating particles according to their energy. Primary CR particles with energies below several hundreds of MeV/nucleon are simply stopped and absorbed in the atmosphere due to ionisation losses. However, if the energy of a primary particle is sufficiently high, it can collide with a nucleus of an atmospheric gas atom. This results in the generation of secondary cosmic rays via development of a nuclear-muon-electromagnetic cascade in the atmosphere, which involves different species, such as electrons, X-rays, muons, pions, kaons and nucleons. Due to development of the cascade, the flux of ionising particles first increases downwards in the atmosphere reaching maximum at the altitude of 17–27 km depending on the cut-off rigidity Rc and the level of a solar activity, but then decreases with atmospheric altitude due to prevailing absorption (Dorman (2004)). Solar protons with energy below 100 MeV lose their energy in the atmosphere at altitudes above 30 km, mostly due to ionisation of the ambient air (Bazilevskaya et al. (2008)). Moreover, such particles can penetrate only in the polar cap region where there is no geomagnetic shielding. Since such low-energy particles are much more abundant in the SEP spectrum, the strongest effect of SEP events is ionisation produced in the upper polar stratosphere and mesosphere at about 40–90 km altitude (Quack et al. (2001)).

4 Variability of cosmic ray fluxes

Cosmic rays are subject to temporal variations on the time-scales from hours to millennia. Solar activity affects the CR fluxes in the lower-energy part of the energy spectrum (up to about 100 GeV).

4.1 11-year solar activity modulation

The 11-year solar cycle causes the most prominent CR modulation. The flux of GCR changes in the opposite phase with 11-year cycle of solar activity. During the solar maximum activity, GCR fluxes with energies above 100 MeV are lower by 35% than during solar minimum, while for GCR with energies above 15 GeV, this value is only 5%, see Figure 1. On the contrary, SEP events occur much more frequently during periods of high solar activity. Number of SEP events affecting the upper Earth's atmosphere (SEP energy above 10 MeV) around solar maximum is about 30 per year, SEP events with relativistic protons (E larger than 1000 MeV) observed on the ground level as ground level enhancements (GLE) occur near the maximum of solar activity at a rate of approximately 3 per year, see Figure 2.

4.2 Forbush decreases and 27-day recurrent variations

A sudden reduction of the CR flux usually associated with a geomagnetic storm was discovered by Forbush (Forbush (1937)) and later named after him. The effect is a result of CR modulation by interplanetary disturbances caused by coronal mass ejections (CME) or fast streams of solar wind. It lasts usually from several hours to 1–2 days and is followed by a longer phase of recovery. The amplitude of the majority of events is about several percent.

Coronal holes or active regions on the Sun live sometimes during several solar rotations and cause long-lived disturbances in the interplanetary space. In this case, Forbush effects may have a recurrent pattern associated with the 27-day synodic rotation of the Sun. Also, CR fluxes may demonstrate rather smooth 27-day recurrent modulation not explicitly due to Forbush effect. The 27-day variation is

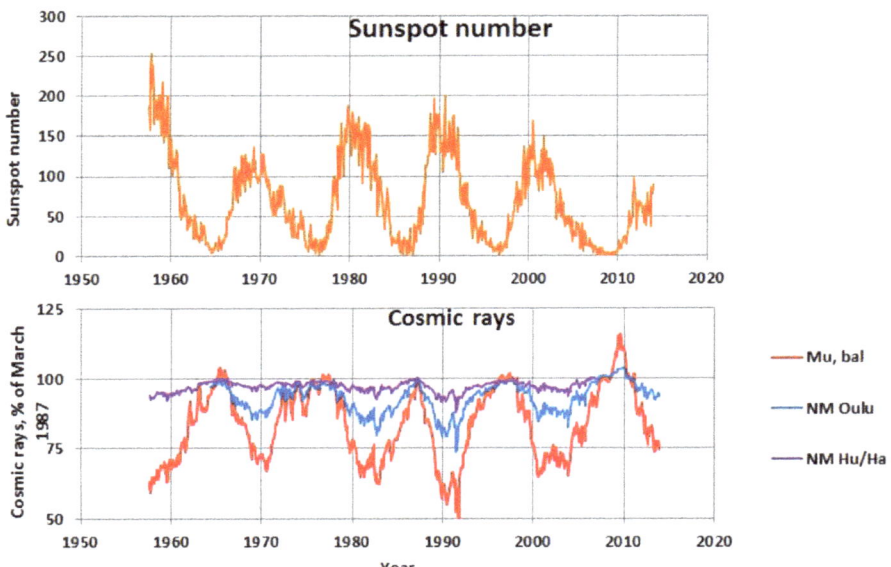

Fig. 1. 11-year modulation of GCR fluxes. Upper panel: monthly data on sunspot number. Lower panel: monthly data on GCR fluxes as observed by balloon-borne counter ($Rc = 0.6$ GV); polar NM Ouly ($Rc = 0.8$ GV); equatorial NM Huancayo/Haleakala ($Rc = 13$ GV). The data series are normalised to 100% at March 1987.

usually more distinctive and persistent during the descending or minimum phases of the solar activity cycle.

5 Measurements of cosmic rays

Since both energy and intensity ranges of GCR and SCR cover many orders of magnitude, a wide number of instruments is needed for the CR observation. In order to study relations between CR and atmospheric processes, the long-term homogeneous sets of CR data are necessary. CR with energy below several hundreds MeV are well measured onboard numerous spacecraft, while balloon-borne detectors are more suitable to detect CR with energy between 100 and 500 MeV. The most energetic CR (above 1000 MeV) which generate the nucleonic-electromagnetic cascade in the atmosphere are recorded by ground-based detectors, such as neutron monitors or muon telescopes.

5.1 Ground-based instruments

A ground-based neutron monitor (NM) was proposed by J.A. Simpson in 1948 as a standard instrument to measure cosmic ray intensities at Earth. He initiated the worldwide network of standard installations which has operated since the 1950s,

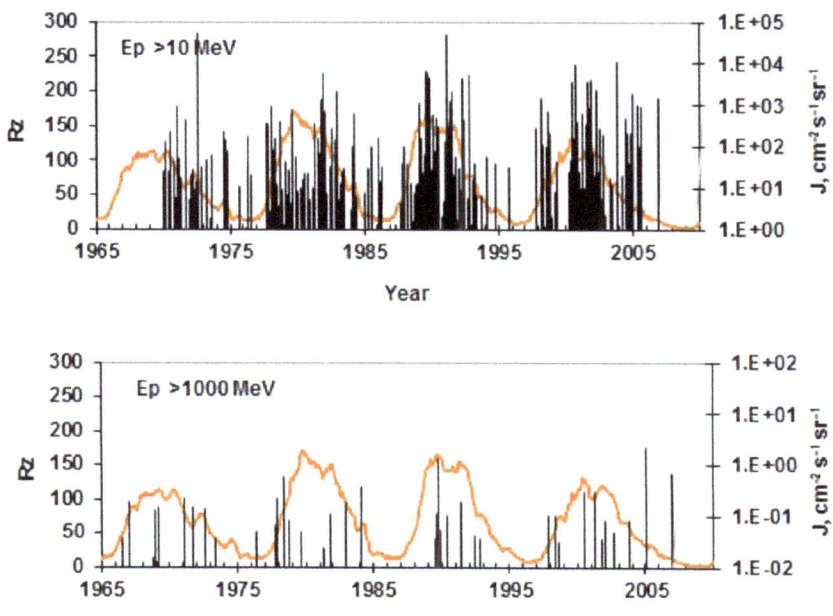

Fig. 2. Solar energetic particle events alongside with sunspot number Rz (7-months smoothed) vs. time. Each vertical bar indicates the time of a SEP event occurrence and represents maximum SEP flux during this event J. Upper panel: SEP events with J(> 10 MeV)> 1 cm^{-2}s^{-1}sr^{-1} (http://www.wdcb.ru/stp/index.en.html). Lower panel: ground level enhancements (GLE), recorded by neutron monitors (http://www.nmdb.eu/).

about 50 stations operating nowadays. The NM is an energy-integrated device which measures all cosmic rays above the detection threshold. The effective energy of CR measured by NM is about 6 to 20 GeV for the high-latitude stations and 20 to 40 GeV for equatorial stations. Another common type of ground-based cosmic ray detectors is a muon telescope, which measures the muon component of the cosmic ray induced atmospheric cascade. Because of the high penetration ability of muons, such detectors are often located at shallow underground depth to study higher energies of primary cosmic rays. The effective energy of muon detectors varies from 50 to 70 GeV for a ground-based instrument to TeV energy for underground.

5.2 Balloon-borne monitoring

Balloon-borne devices allow to measure radiation at various levels of the atmosphere. First series of regular balloon measurements of the ion production rate in the atmosphere were conducted by H.V. Neher from 1930 up to 1969. In 1957, A.N. Charckhchyan initiated regular monitoring of charged particle fluxes in the atmosphere with meteorological balloons that is being performed up to now. The balloons are launched several times a week at several geographic sites including polar latitudes. The device returns the flux of secondary cosmic rays at the atmospheric depths from the ground level up to altitude of about 35 km.

5.3 Satellite-borne monitoring

Satellites are indispensable for monitoring of SEP since the majority of SEP have energy below 100 MeV and do not penetrate into the atmosphere below 30 km. However, the low-energy SEP play a significant role in chemistry of the upper stratosphere and mesosphere. Regular measurement of SEP has been implemented by spacecraft since 1967. Geostationary GOES satellites have performed SEP monitoring from mid 1980s [http://spidr.ngdc.noaa.gov/spidr/]. It should be noted that satellite-borne detectors are being used for the GCR monitoring only recently with the advent of PAMELA and AMS instruments because earlier they were not suitable for measurements of fluxes of high-energy particles.

6 Modeling of cosmic rays induced ionisation of the Earth atmosphere

It is known that ions are involved in many atmospheric processes. CRs are the most important contributor to ion-pair production in the atmosphere from about 3–4 km up to about 50 km. When energetic particles penetrate the atmosphere, they lose energy for ionisation. The process can be modeled by different approaches depending on the particle's energy. In all models, the ionisation energy losses of particles (either primary or secondary of different nature) are converted into the production of ion pairs, assuming that one ion–electron pair is produced, on average, per each 35 eV of deposited energy (Porter et al. (1976)). This process is well understood but cannot be modeled analytically and requires a full Monte-Carlo approach. It is worth noting that cosmic ray induced ionisation (CRII) rate will be dependent on three main variables: altitude, geographical location (via the local geomagnetic rigidity cut-off Rc), and time (via the variability of the integrand energy spectrum), see tables (http://cosmicrays.oulu.fi/CRII/CRII.html). Presently there are two basic numerical approaches to the CRII simulation: ATMOCOSMICS/ PLANETOCOSMICS model developed on the basis of the GEANT- (Desorgher et al. (2005)) and the CRAC (Cosmic Ray Atmospheric Cascade) model developed on the basis of CORSIKA+FLUKA packages (Usoskin and Kovaltsov (2006); Usoskin et al. (2010)). Figure 3 shows production over geomagnetic pole (Rc smaller than 1 GV) by GCR during maxima and minima solar activity as well as by SEP during a moderate and a GLE events. The maximum ionisation by galactic cosmic rays appears at about 15 km altitude in the polar region, where the ionisation rate is several orders of magnitude greater than that at sea level. The difference in the ionisation rate between the equatorial and polar regions is of the order of a factor 3–5. Variability of CRII over a solar

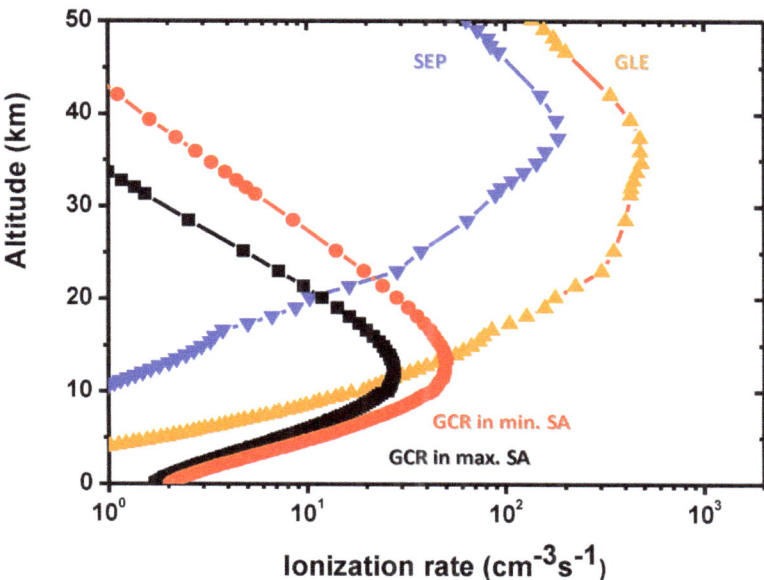

Fig. 3. Ionisation rates of galactic and solar cosmic rays. Cosmic ray induced ionisation (CRII) rates calculated by CRAC:CRII model. *Red circles* – Galactic CRII during minima solar activity (SA). *Black squares* – Galactic CRII during maxima solar activity (SA). *Upward pointing orange triangle* – Solar CRII like GLE of 20th January 2005. *Downward pointing blue triangle* – Solar CRII as lower energy SEP of 17th January 2005.

cycle is from 10–20% in the troposphere up to a factor of 2 in the polar stratosphere. The maximum ionisation by SEP appears at about 40 km altitude in the polar region, while the ionisation rate may be negligible at sea level.

7 Abbreviations used in Part 2.3

- CME - coronal mass ejection
- CR - cosmic rays
- CRII - cosmic ray induced ionisation
- GLE - ground level enhancement
- GCR - galactic cosmic rays
- NM - neutron monitor

- SCR - solar cosmic rays
- SEP - solar energetic particles
- SPE - solar proton events
- (SEP event ≡ SPE ≡ SCR)

Further reading:

Mironova, I. et al. 2015, Energetic Particle Influence on the Earth's Atmosphere, Space Science Reviews, in press, doi: 10.1007/s11214-015-0185-4

CHAPTER 2.4

VARIABILITY AND EFFECTS BY SOLAR WIND

Kalevi Mursula[1] and Eija I. Tanskanen[2]

1 Average structure of solar wind

A stream of charged particles, called the solar wind, continuously flows from the Sun into the interplanetary space. Solar wind carries with it the magnetic field of the Sun, which is called the interplanetary magnetic field (IMF) or the heliospheric magnetic field, reflecting the fact that the region of space dominated by the Sun via the solar wind and IMF is called the heliosphere (*helios* = Sun in Greek). While the solar wind is flowing radially away from the Sun, the magnetic field is turned to a spiral structure due to the rotation of the Sun, much in the same way as the water running out from a rotating garden hose.

The time of the solar wind to reach the Earth at its typical speed of about 400 km s^{-1} takes about 4 days. While expanding into open space, the solar wind gets diluted, and at the Earth, solar wind is already a very tenuous gas, containing only some of 5–10 particles per cubic centimeter. During this expansion, the solar wind cools down roughly by a factor of ten from the initial temperature of a couple of million degrees of the solar corona. The strength of the IMF also weakens from the Sun to the Earth to about 5 nanoTesla, which is only one in ten thousand when compared to the Earth's magnetic field on the ground. Most of solar wind energy is in the form of kinetic energy related to its anti-solar motion, with smaller contributions in thermal and magnetic energy.

The properties of the solar wind and IMF vary significantly, reflecting the nature of their coronal source. The solar wind can be roughly classified into two groups: the slow solar wind and the fast solar wind. The fast solar wind (faster than about 500 km s^{-1}) originates from large regions of solar corona that are seen as dark when viewed in normal light. Darkness is due to the low density of these regions. These rather empty regions of solar corona are called coronal holes. The low density results from the specific magnetic structure of these regions, which opens directly into space, having no magnetic loops that can contain high densities

---

[1] Department of Physics, University of Oulu, 90014 Oulu, Finland
[2] Finnish Meteorological Institute, 00101 Helsinki, Finland

of plasma particles. Obviously, solar wind can better be accelerated to high speeds within the open field lines of large coronal holes. However, the reason to this preference is not yet very well understood and remains a topic of intense research. On the other hand, the slow solar wind originates from the proximity of solar active regions. Since these regions are fairly dense, slow solar wind is also denser than fast solar wind. The magnetic structure of those regions of solar corona emitting slow solar wind is rather complicated, and the field tends to experience non-radial expansion. The speed difference also affects the winding of the IMF spiral, which is more tight for slow solar wind.

The properties of the solar wind and IMF are continuously changing, from very short time scales below one second to intermediate scales of several hours to one solar rotation (about 27 days), and to long time scales of a solar cycle (about 10–11 years) up to a century and even beyond. The short time scale variations mainly develop during the interplanetary space, while the intermediate scales mainly reflect the momentary distribution of solar active regions on solar surface, and the longer time scales reflect the changes in solar dynamo during the solar cycle and longer. The daily averaged values of solar wind speed vary roughly by a factor of five from about 200 to 1000 km s^{-1}. All other solar wind parameters vary even more, especially the solar wind density and the IMF strength, which can vary by two orders of magnitude, reaching their highest values in interplanetary shocks.

2 Solar wind transients

On top of the average solar wind, various temporary phenomena and processes can significantly modify the properties of the solar wind. One can divide these phenomena in two groups: those that have their origin in the Sun and those that develop during the solar wind flow in the interplanetary space. Of the latter, the most important phenomena are the corotating interaction regions (CIR), which are interplanetary shocks that form as a result of the collision of fast and slow solar wind streams. When the fast solar wind stream attains the preceding slow solar wind, it cannot overtake it because the magnetic fields of these two regions strongly oppose mutual mixing. So, instead of a smooth change of solar wind parameters, the two regions form sharp boundaries over which the solar wind parameters vary dramatically. Since the two different solar wind streams often have opposite magnetic polarities, the CIR also typically includes an IMF sector boundary.

Since the source regions of IMF of opposite polarity are typically located rather far from each other on solar surface, the time difference between the fast and slow streams is several days, and the CIR is formed only rather far away from the Sun. Indeed, most CIRs develop beyond the Earth's orbit. The term corotating refers to the fact that the CIRs tend to appear repeatedly, once per solar rotation, as if the CIRs would rotate with the Sun. This repeating pattern takes place because the global solar magnetic field structure, which is produced by

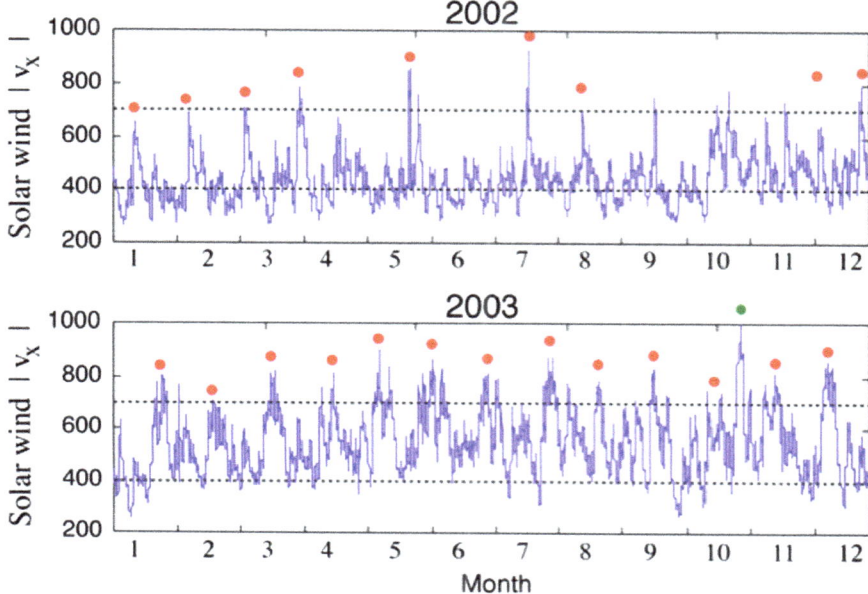

Fig. 1. Periodic high-speed streams from the solar coronal holes in 2002 and 2003. The same solar source region emits fast solar wind towards the Earth repeatedly at the 27-day rotation period of the Sun.

coronal holes and active regions, tends to remain roughly similar during several solar rotations. Figure 1 shows the repetition of high-speed solar wind streams at 27-day intervals during most of the years 2002 and 2003. CIRs are typically found to hit the Earth in the declining phase of the solar cycle, when the polar coronal holes are expanding and have an asymmetric structure in solar longitude. They often have an extension from the pole towards the equator, which can emit fast solar wind at low solar latitudes, thus reaching the Earth. When the production of new flux stops at the end of the cycle, the polar coronal holes become more symmetric and the high-speed streams become again more rare at the Earth. During sunspot maxima, there are active regions all over the solar surface, thus no large coronal holes exist.

The solar originated transients include, in particular, interplanetary coronal mass ejections (ICME) and solar flares. Solar flares can accelerate a fair amount of solar particles to very high energies, forming a burst of solar energetic particles (SEP), also called solar cosmic rays. However, the number of SEPs is rather small and, because of their high energy, they do not behave similarly as the solar wind particles. Therefore flares do not contribute much to the properties of the solar wind, and we will not discuss flares or SEPs in more details here.

Coronal mass ejections are large coronal loops with a huge amount of solar material, which burst into space typically during a few hours. These particles have roughly the same energy as solar wind particles, so they can become part of solar wind in the interplanetary space. Moreover, the ICMEs include so many particles that they can dominate over the background solar wind and thus determine its properties. ICMEs can be faster or slower than the ambient solar wind but, during the interplanetary travel, the ICME speed tends to approach closer to the speed of the backgound wind. Very fast and strong ICMEs, however, do exist and can reach the Earth even in less than one day, as during the famous Carrington storm in 1859.

Since the ICMEs are large loops of magnetic field, which can pertain their structure even in the interplanetary space, the magnetic field observed at the Earth's orbit during ICME passage can be very differently oriented than the background IMF structure. A typical core of a ICME is a magnetic cloud, a dense magnetic flux tube, where the field lines are twisted and tied to the Sun on one or both legs. Note also that fast ICMEs produce a leading shock ahead of them, which is followed by a sheath region of very turbulent field and solar wind until the ICME core arrives. Since ICMEs are related to sunspots and the appearance of new flux tubes on solar surface, they tend to maximise around the sunspot maximum.

3 Solar wind and the Earth

Solar wind affects the Earth's magnetic field, compressing the field on solar side and forming a comet-like tail in the nightside. Solar wind sustains a complicated and extremely variable system of electric fields and currents in this magnetic cavity, the Earth's magnetosphere, and in the ionosphere. The electric fields also accelerate magnetospheric particles (which partly come from the solar wind) and make some of them precipitate into the atmosphere. Overall, there is 10^{13} W power in the solar wind, of which about 10–20% is used to maintain the shape and basic convection of the magnetosphere. Accordingly, the solar wind power is much smaller than, e.g., the power of solar electromagnetic radiation, whereby its possible climatic effects were originally assumed to be minor.

The most important factor controlling the rate of energy input from the solar wind to the near-Earth space is the IMF orientation. Energy input increases as the IMF becomes increasingly antiparallel (southward IMF) to the equatorial geomagnetic field. Then, large scale merging or reconnection of magnetic field lines can take place, producing important electric fields and accelerating particles effectively. The energy input is enhanced by fast speed and high density of the solar wind, as well as by strong IMF. Moreover, the role of the ultra-low frequency (ULF) waves, also called Alfvén waves, in possibly enhancing the magnetic coupling, is under active study. Alfvén waves, which are more often found in the high-speed stream, may amplify the north-south IMF component and thereby the dayside reconnection electric field.

The charged particles precipitating into the atmosphere collide with the ambient neutral air at heights depending on their energy, the auroral particles at around 100 km and the more energetic particles down to about 50 km. The collisions ionise the atmosphere, thus contributing to the formation of the ionised layer called the ionosphere (which is mainly produced by solar EUV radiation). Most of the kinetic energy of the precipitating particles is converted to the thermal energy of the neutral air. Joule heating dissipates some 10^{15} J of energy during substorms, the majority of the energy available from the solar wind. In addition, the precipitating particles turn on the auroral lights. Most of this energy to all the three forms is dissipated by electrons while ions contribute less.

High-speed solar wind streams and CIRs are known to be particularly effective in accelerating magnetospheric particles. Therefore the declining phase of the solar cycle seems to be the most effective time interval for solar wind-related atmospheric effects, both to the ionised and to neutral air. Interestingly, there is increasing evidence that the high-speed solar wind streams, by accelerating and precipitating charged particles into the atmosphere, can produce significant chemical and dynamical changes in the atmosphere. In particular, they can produce NO_x and HO_x molecules which can descend into the middle atmosphere where they cause massive ozone destruction. This may further cause enhanced meridional circulation, creating a stronger polar vortex and a positive phase of the North Atlantic Oscillation. Indeed, recent studies show that the positive phase of the NAO prevails in the declining phase of the solar cycle and affects the arctic Winter temperatures strongly. These relations and effects are currently under active study.

CHAPTER 2.5

VARIATIONS OF SOLAR ACTIVITY

Sami Solanki[1], Ilya Usoskin[2] and Maarit Käpylä[3]

1 Introduction

Solar activity is the sum of all phenomena driven by the Sun's magnetic field. These cover a very wide range of features and events and include dark sunspots seen on the solar surface, as well as powerful flares, often emitting energetic particles and high-energy radiation, or coronal mass ejections, eruptions of hot matter (10^4-10^6 K plasma) and magnetic field.

Solar activity varies on all observed timescales: from short term (e.g. when an active region, with its sunspots and associated flares, appears on the solar disk) to long-term, most prominently over the 11-year activity cycle. We are lucky to have had many generations of assiduous solar observers who followed the solar cycle with great care and dedication for over 400 years. This has given us unique insight into how solar activity varies over time. In recent decades, techniques to follow solar activity even further back into the past have been developed that make use of the fact that the ups and downs of the Sun's activity can leave traces on Earth. At the same time, our understanding of what causes the Sun's activity to vary has been steadily improving, although we are still far from gaining a full understanding. Thus, there are still many open questions regarding how well we can predict solar activity.

2 The solar activity cycle and its variations

Roughly every 11 years the strength of solar activity increases, reaching a peak and then decreasing again until a minimum in activity is reached. After this, the cycle starts again. Over such a solar activity cycle (often called a solar cycle, or

---

[1] Max-Planck-Institut für Sonnensystemforschung, 37077 Göttingen, Germany

[2] Sodankylä Geophysical Observatory, ReSoLVE Centre of Excellence, University of Oulu, 90014 Oulu, Finland

[3] ReSoLVE Centre of Excellence, Computer Science Department, Aalto University, FI-00076 Aalto, Finland

a Schwabe cycle, after its discoverer, Heinrich Schwabe), the atmosphere of the Sun changes dramatically. For example, between activity minimum and maximum the number of sunspots, dark structures on the solar surface with a size of typically a few 10 000 km characterised by a strong magnetic field reaching multiple 1000 Gauss (compared to the Earth's dipole field smaller than 1 Gauss), vastly increases, as does the coverage by plage regions (bright areas associated with a strong magnetic field, which, however, is weaker than that of sunspots), the number of solar flares, which are outbursts of energetic radiation (X-rays, gamma-rays and sometimes high-energy charged particles), and the number of coronal mass ejections. Even the brightness and shape of the whole solar corona changes. Thus, the corona at activity maximum is brighter and hotter and surrounds the whole Sun, just like a crown, from where it gets its name. Importantly, also the total radiative output of the Sun, the so-called solar irradiance, changes over the solar activity cycle, with the Sun being slightly brighter during activity maximum (see Chapter 2.2 of this handbook).

Solar activity in general is driven by the evolution of the Sun's magnetic field. This magnetic field is structured on small scales, but also displays large-scale patterns. The amount of magnetic flux correlates with the level of solar activity, but consecutive cycles display a reversal of the magnetic polarities. For example, if in one cycle bipolar active regions in the northern hemisphere tend to have positive polarity leading the negative polarity, then in the next cycle, the negative polarity leads the positive. The situation is antisymmetric with respect to the solar equator (Figure 1). Thus it takes two activity cycles, i.e., roughly 22 years, for a full magnetic cycle to complete, so that the magnetic polarities return to the original configuration. This $\approx$22-year cycle is often called the Hale cycle, after George Ellery Hale, who was one of its discoverers.

Many manifestations of solar activity have been studied during the space age (i.e., since the 1960s) thanks to modern instruments, in particular those flying in space, which also record radiation at wavelengths that do not reach the ground (such as ultraviolet and X-ray radiation). If we go increasingly further back in time, fewer such records are available, until we are left with just the number of sunspots visible on a given day, scrupulously counted by professional and amateur astronomers starting with Harriot, Galilei, Fabricius and Scheiner around 1610 AD, i.e., basically ever since the first telescopes were available. This so-called sunspot number record reveals (Figure 2):

1. Although the sunspot cycle is remarkably stable, solar activity is not periodic, but cyclic, i.e., the individual cycles have somewhat different lengths (ranging roughly between 9 and 13 years) and often very different amplitudes. Solar cycles are numbered, starting in the middle of the 18th century.

2. The last 400 years have been dominated by cyclic activity, but the activity level has varied a lot with time, implying the additional presence of secular variability. For example, the period 1645–1700 was characterised by an almost total lack of sunspots. This so-called Maunder minimum (named after

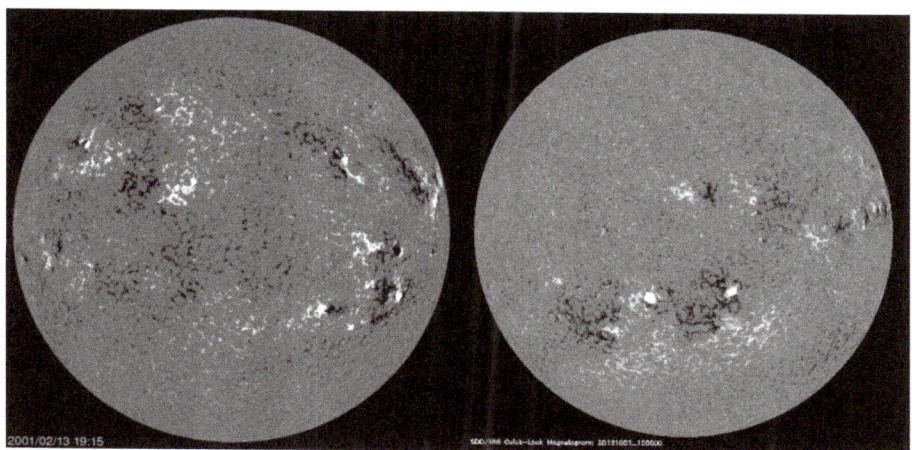

Fig. 1. HALE'S POLARITY LAW. Magnetograms showing the strength and distribution of the photospheric magnetic field (bright shading represent increasingly strong magnetic fields pointing out of the solar surface, dark shading increasingly strong magnetic fields pointing into the solar surface), measured with the SOHO/MDI and SDO/HMI instruments. The magnetogram on the left was measured in February 2001, near the maximum of cycle 23, the magnetogram on the right in October 2012, near the maximum of the current cycle 24. The leading and trailing spots of the nearly east-west (left-right) aligned bipolar structures are of opposite polarities, the pair showing different polarity on each hemisphere at any instant. Between two consecutive sunspot cycle maxima, the polarities on both hemispheres are reversed. Data credit: the SOHO data archive, ESA/NASA.

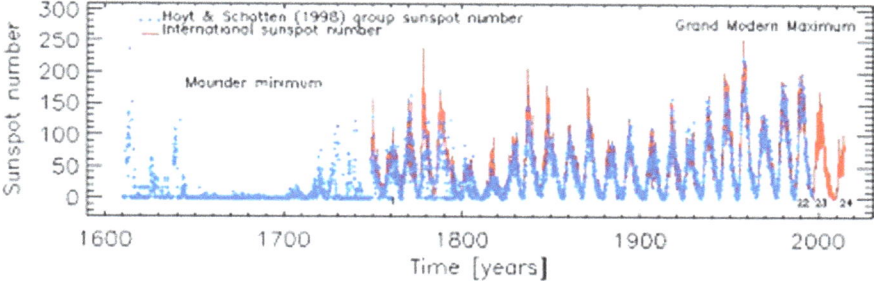

Fig. 2. SUNSPOT CYCLE. Monthly averaged group sunspot number (blue symbols) and the international sunspot number (red line) as function of time during the telescope era showing the 11-year sunspot cycle and its variations. The most notable epochs are the low activity state, Maunder minimum, with hardly any spots, and the current high state, Grand Modern Maximum. The current cycle 24 is the weakest for a century. Data credit: D.V. Hoyt and K.H. Schatten (blue crosses) and WDC-SILSO, Royal Observatory of Belgium, Brussels (red line).

Edward Walter Maunder) was not just an artefact caused by a lack of observations. Many observers looked hard, but could not find spots on the Sun. This indicated the extreme quietness of our star at that time. The other extreme in solar activity was reached in the second half of the 20th century, the modern Grand maximum. Solar cycle 19 was the strongest observed so far. The current cycle with the number 24 is to date the weakest cycle for nearly a century, indicating that the modern Grand maximum is coming to an end.

3 Solar activity on longer time scales

For many purposes, even the 400-year long dataset of telescopically measured sunspot numbers is too short, leaving open, e.g., the question of how typical are the changes in the solar activity level witnessed during the era of the telescope.

Unfortunately, quantitative and regular direct observations of the Sun do not exist for times prior to the 17th century. But luckily there is another method, based on cosmogenic radionuclide proxies, which makes it possible to reconstruct solar activity in the past, before the telescope era. Cosmogenic radionuclides (the most famous being ^{14}C or ^{10}Be) are produced by cosmic rays in the Earth's atmosphere.

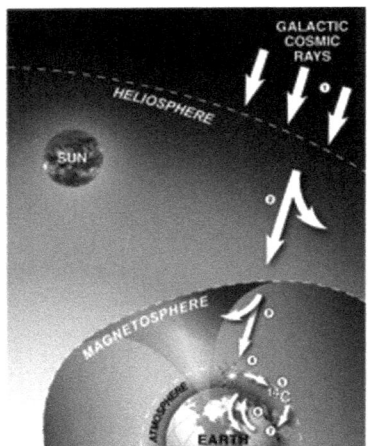

Fig. 3. A SKETCH OF THE MODULATION PROCESS. Artistic view of the link between ^{14}C record and solar activity, not to scale (credit to Dyon, Hulot and Gallet, IPGP). Galactic cosmic rays with roughly constant flux (item 1) enter the heliosphere and are modulated by solar magnetic activity (2) and by the geomagnetic field (3). Cosmic rays cause nuclear collisions in the Earth's atmosphere producing, in particular, radiocarbon ^{14}C (4), which takes part in the global carbon cycle (5) and finally is stored in a tree (6). By measuring ^{14}C in a tree sample dated to the past, one can assess the atmospheric ^{14}C production rate Q at that time. Using a physical model of ^{14}C production combined with an independent model of the past geomagnetic field, the production rate Q can then be converted into information about past solar activity.

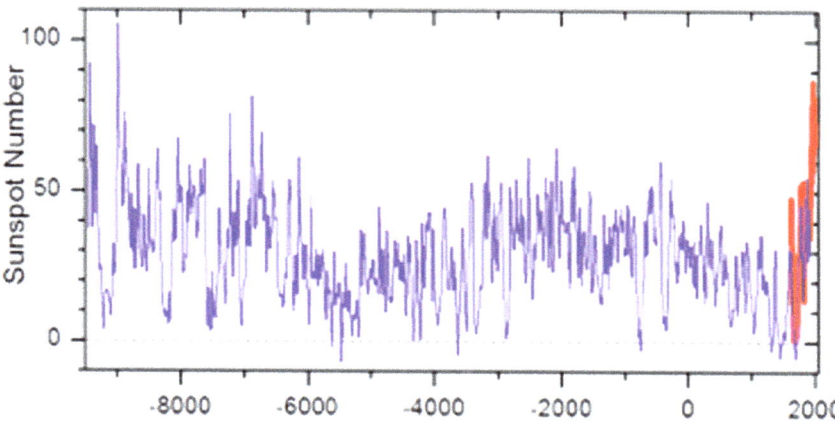

Fig. 4. RECONSTRUCTED SOLAR ACTIVITY. Sunspot number (10-year averages) reconstructed from ^{14}C data since 9500 BC (blue curve) and 10-year averaged group sunspot number (GSN) obtained from telescopic observations since 1610 (red curve). The horizontal dotted line marks the threshold above which we consider the Sun to be exceptionally active.

After production they are stored in natural stratified dateable archives (such as trees, polar ice, corals or lake/marine sediments) and their concentrations can be measured in specialized laboratories. Cosmic rays that produce the nuclides are modulated by solar activity, so that stronger solar activity leads to weaker cosmic ray flux and vice-versa. Thus, by measuring, in a modern laboratory, the concentration of a cosmogenic nuclide in an archive sample. one can reconstruct solar activity at the time when this sample was formed. For example, determining the amount of ^{14}C (a heavier radioactive isotope than normal carbon ^{12}C) in the trunk of a dead tree tells us how strong the cosmic ray flux was at the time that this tree lived (this time can be determined by dendrochronology from the pattern of tree rings). This in turn is related to the strength of solar activity at that time (for a sketch of the modulation process (Figure 3).

Solar activity reconstructed in this way from cosmogenic nuclides for the last eleven millennia of the Holocene period (Figure 4) demonstrates strong variability, with Grand minima and Grand maxima appearing every now and then. Grand minima (periods of very low activity, like the Maunder minimum in the 17th century) correspond to a very special state of solar activity that occupies about 1/6 of the time. Grand maxima correspond to periods of unusually high activity, as in the second half of the 20th century, and occupy about 1/10 of the time. Present day activity, since 2006, is moderate, comparable to the Sun's state during most of the past 11 000 years, suggesting that the modern Grand maximum is ending.

A thorough analysis of the reconstructed activity suggests that such periods of extreme activity occur irregularly. Although some quasi-periodicities can be found, solar activity contains a significant random component in the long-term variability. Occurrence of Grand minima and maxima cannot be predicted in a deterministic manner, only in a probabilistic sense.

CHAPTER 2.6

UNDERSTANDING SOLAR ACTIVITY

Maarit Käpylä[1], Sami Solanki[2] and Ilya Usoskin[3]

1 Introduction

While in the interior of the Sun the energy released in nuclear reactions is transported outwards mainly by radiation, in the outer 30% of our star the main energy transport mechanism is convection. This means that the fluid heated from below becomes unstable, rising and sinking bubbles are generated, and they transport heat by their motions (Figure 1). The conditions in the solar convection zone are such that the motions of the bubbles are highly turbulent, which essentially means that the matter in the convection zone is very efficiently mixed and structures at various scales (large and small) occur simultaneously. Being subject to such vigorous stirring, the strong magnetic fields concentrated during the process of the collapse of the molecular cloud leading to the formation of the Sun 4.6 billion years ago, were destroyed on a time scale much shorter than the current age of the Sun. Therefore, to explain the solar magnetic fields observed today, the magnetic field must be actively produced and regenerated by some process.

2 Solar dynamo

Such a process is provided by the inductive effect due to the moving ionised plasma in the convection zone, similar to the electromagnetic induction effect of a current flowing in an electric wire inducing magnetic field around it. The plasma motion has different sources. The first is solar rotation, which has a different angular velocity at different latitudes and also at different depths, which is referred to as differential rotation. The effect of this part of the solar flow field is to wind up any magnetic field oriented poleward (called poloidal magnetic field) into the so-called

---

[1] ReSoLVE Centre of Excellence, Computer Science Department, Aalto University, FI-00076 Aalto, Finland
[2] Max-Planck-Institut für Sonnensystemforschung, 37077 Göttingen, Germany
[3] Sodankylä Geophysical Observatory, ReSoLVE Centre of Excellence, University of Oulu, 90014 Oulu, Finland

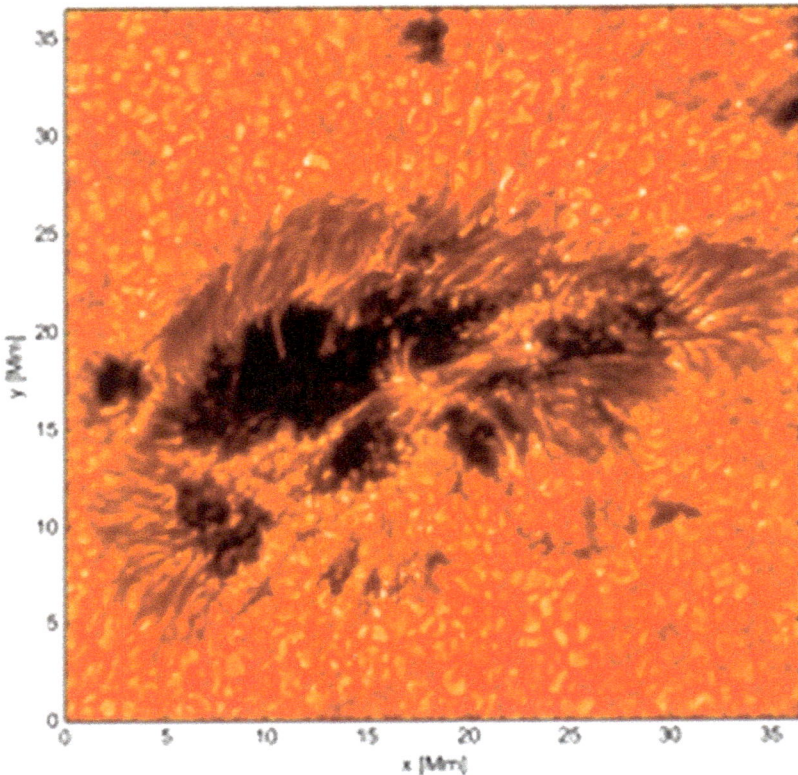

Fig. 1. TURBULENT CONVECTION. Active region observed on June 14, 2013 with the Broad Band imager at the GREGOR telescope at the Teide observatory on Tenerife (Spain). Turbulent convection is seen in the solar photosphere as a granulation pattern, which looks like a huge boiling water kettle: hot rising bubbles (brighter orange regions) are surrounded by cooler, narrow downflows (darker orange patches). In active regions, the strong magnetic field inhibits convection, suppressing the heat transport. Therefore, strongly magnetised regions appear as dark features in the photospheric image. Their sizes usually greatly exceed the typical size of a granule. Image credit: the GREGOR consortium.

toroidal field with field lines oriented along the longitude (see two leftmost panels of Figure 2) at the same time stretching out the field and thereby amplifying it. This process is called the Ω effect.

The rising convective bubbles become twisted by the Coriolis force induced by rotation, and as a sum over all the bubbles, this produces a net effect that results in small loops of magnetic field that have regained their original poleward orientation but with a reversed sign. This is called the α effect (third panel of Figure 2).

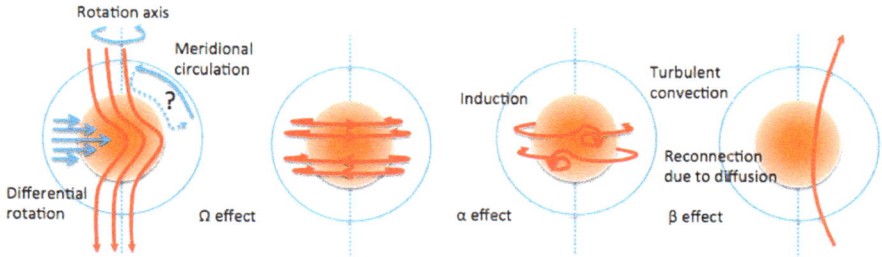

Fig. 2. DYNAMO. Schematic presentation of the dynamo cycle due to the inductive action of large-scale flows and convective turbulence according to the distributed dynamo paradigm. In flux-transport dynamo models, the field generation is localized in the bottom (Ω effect) and surface (α effect) of the convection zone. The single-cell meridional circulation pattern still lacks confirmation, indicated by the question mark in the leftmost image. In the distributed dynamo framework, field generation and destruction by turbulence occur throughout the convection zone.

With the help of the powerful turbulent mixing, called the β effect, the small loops form larger and larger ones, and finally the original, global configuration with a reversed sign is obtained (the rightmost panel of Figure 2). This process is called the hydromagnetic dynamo. One more important component of the solar flow field is meridional circulation, carrying matter and magnetic field between the polar and equatorial regions. This flow field is not a strictly required ingredient for the dynamo process to become operational, but it significantly affects it. Although at present dynamo theory is widely accepted, many details of the solar dynamo process remain uncertain.

The differential rotation profile inside the Sun is known from helioseismic studies, which indicate that most of the winding up occurs in a region localized in the bottom of the convection zone, at the interface between the fixed rotation of the solar radiative zone and the differential rotation of the solar convection zone. Helioseismology also yields information on the meridional circulation profile, although the deep-down circulation pattern is still unknown. There is a poleward directed flow near the surface, but the depth and structure of the return flow are still under debate (see the leftmost panel of Figure 2). The remaining task is to describe the inductive and destructive effects caused by the turbulent convective bubbles. These are not accessible by analytic treatment or laboratory experiments under realistic conditions. Therefore, computer simulations of turbulent convection remain the primary tool for their investigation (Figure 3). Such modelling is also very demanding, and therefore current solar dynamo models still commonly use simple parameterisations of turbulence. The problem with such an approach is that very different types of profiles and magnitudes for the turbulent quantities can satisfactorily reproduce the properties of the 22-year magnetic cycle, such as the butterfly diagram, which represents the time evolution of the surface magnetic

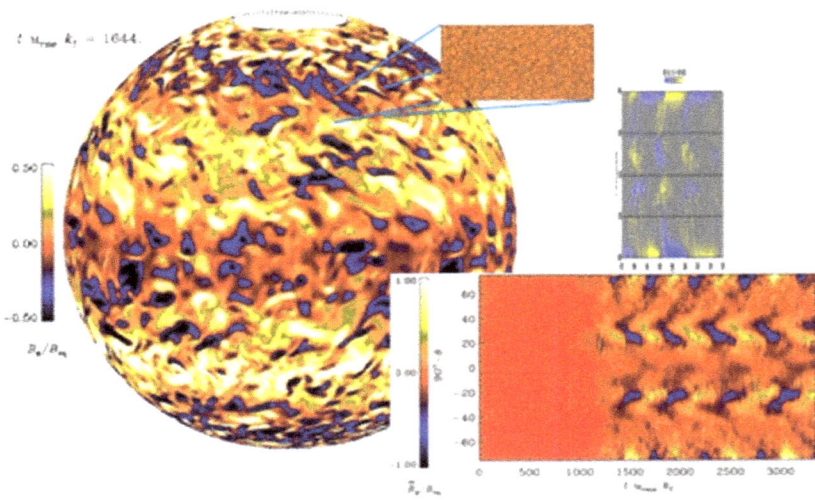

Fig. 3. DNS MODELS. DNS models of turbulent convection have recently reached into parameter regimes in which solar-like oscillatory dynamo solutions emerge. In these models, a large-scale magnetic field, one realization of the azimuthal magnetic field near the surface of the simulation domain shown on the spherical surface, is generated. The much smaller spatial scale of turbulent convection is shown (from another local DNS simulation) in the top right corner. The magnetic field evolves over time similarly to the solar magnetic field (butterfly diagram from the dynamo simulation shown on the lower right corner while the observed magnetic butterfly diagram in the middle). These simulations have been carried out with the PENCIL CODE. Image credit: Petri Käpylä, Maarit Käpylä and Axel Brandenburg (for the simulation pictures) and D. Hathaway/NASA/NSSTC (for the magnetic butterfly diagram).

field as function of time. More information of the unknowns need to be collected to pick up the model best describing the underlying physics, in which task the simulation models play a crucial role.

The presently dominant dynamo paradigm – flux transport or Babcock-Leighton type dynamo models – explains the solar cycle with two localized field generation regions, one being the layer of strong differential rotation near the bottom of the convection zone, and the other a field amplification region near the surface through the release of magnetic flux due to sunspot decay. The two layers are connected in two ways: the wound-up field is transported to the surface from the low-latitude deep layers by buoyant thin flux tubes that become twisted by

Coriolis force, while a counter clockwise conveyor belt of a single-cell meridional circulation pattern brings the generated poleward-oriented magnetic field back to the bottom, first passing through the polar regions. In the majority of these models, turbulence is regarded largely unimportant, which, among other things, implies weak mixing and a "memory" of the magnetic field over several cycles. The competing theory, the distributed dynamo scenario, postulates turbulence important throughout the convection zone, leading to shorter memory and to the fact that the meridional circulation plays a somewhat less crucial role for the results. The flux-transport dynamo models have been developed up to a point, where they have been regarded reliable enough to be used in the prediction of forthcoming solar activity (although with mixed results), whereas distributed dynamos practically lack prediction capability due to the strong turbulent mixing causing too short memory for the dynamo.

The existence of longer-term variations in the solar activity are poorly explained by either of the dynamo paradigms. The general nature of the system of equations describing the solar dynamo naturally favours the existence of chaotic solutions in the highly non-linear regime, where the Sun is inevitably operating. Nevertheless, it is still unclear which are the exact physical mechanism(s) behind the irregular behaviour; among other things, stochastic variations in the turbulent quantities and subtle nonlinear changes in the differential rotation and/or the meridional circulation have been proposed.

3 Can we predict future solar activity?

Reconstructing solar activity in the past is an important area of research and is one of the main sources of constraints on dynamo models describing the origin of solar magnetism and activity. However, for many purposes, it would be even better to predict solar activity into the future.

A few techniques have been applied. Thus, once a cycle has already started, Waldmeier's rule can be used, which says that the more steeply a cycle rises the stronger it will be. This technique can be applied typically 2–3 years into the future. For longer-term predictions, the more successful techniques are similarly empirical and use so-called precursors to determine the strength of the next cycle. A typical precursor is the strength of the magnetic field around the poles of the Sun during the minimum between two activity cycles. The more magnetic flux is present around the poles at activity minimum, the stronger the next solar activity cycle tends to be. Other techniques have typically met with less success.

One such attempt has been the usage of flux-transport dynamo models for long-term prediction of solar activity. Two main data assimilation schemes have been used, one in which the surface source term follows the sunspot number records, and another one where the modelled field is rescaled with the polar dipole field during the minimum. These methods were used to predict the current cycle 24 before it started, with conflicting results: better success was obtained with the polar dipole field method with higher turbulent mixing, while the models based on sunspot data and low turbulent mixing resulted too high activity level. At present, therefore, the prediction capability of dynamo models is still poor, and limited by the fact that the solar cycle is a result of nonlinear processes operating in the chaotic regime.

One limitation of precursor techniques is that they can roughly predict only the next solar cycle, but have little power to predict the strengths of cycles beyond that. Other techniques that claim to be able to predict on a longer term still have to demonstrate if their claims have any validity. Numerous statistical studies indicate that solar activity cannot be predicted longer than one cycle ahead because of the essentially stochastic/chaotic component. This is also supported by the direct numerical simulations of solar convection that consistently yield turbulent diffusion times of the order of the solar cycle length.

INFOBOX 2.1

ORBITAL FORCING OF GLACIAL - INTERGLACIAL CYCLES

Werner Schmutz[1] and Margit Haberreiter[1]

One of the natural climate forcings is the so-called orbital forcing, besides others, such as the variation in solar irradiance and volcanic eruptions. The orbital forcing depends on changes in the Earth's orbit around the Sun and is the result of variations of the following parameters: the orbit's eccentricity, the obliquity of the Earth's orbit with respect to the ecliptic plane, and the axial and apsidial precession, i.e., the precession of the Earth's rotation axis and the precession of the Earth's orbital ellipse around the Sun. The latter two are often combined into one process. Orbital forcing yields significantly different irradiance values for the two terrestrial hemispheres at high latitudes but has almost no effect on the mean total energy. The dominant forcing determining the ice ages is the one on the northern hemisphere.

1 Milanković cycles

An astronomical influence of the orbital forcing on climate had already been postulated in the 19th century, but a qualitative relationship was only determined at the beginning of the 20th century, mainly by Milutin Milanković. The theory that climate variations are forced by orbital variations is therefore usually referred to as *Milanković Cycles*. These cycles are schematically shown in Figure 1 and are explained in more detail below.

First, the Earth's orbit is an ellipse. The eccentricity (Figure 1a) describes the degree of oblateness (or elongation) of an ellipse. The eccentricity of the Earth's orbit ranges from 0.000055 to 0.0679, where an eccentricity of zero would indicate a perfect circle. The eccentricity varies with two main periodicities of around 100 kyr and 400 kyr.

Second, the Earth's equatorial plane is inclined with respect to the ecliptic plane. The angle of inclination is called obliquity or axial tilt, and is indicated as

---

[1] Physikalisch-Meteorologisches Observatorium Davos / World Radiation Center, Dorfstrasse 33, 7260 Davos, Switzerland

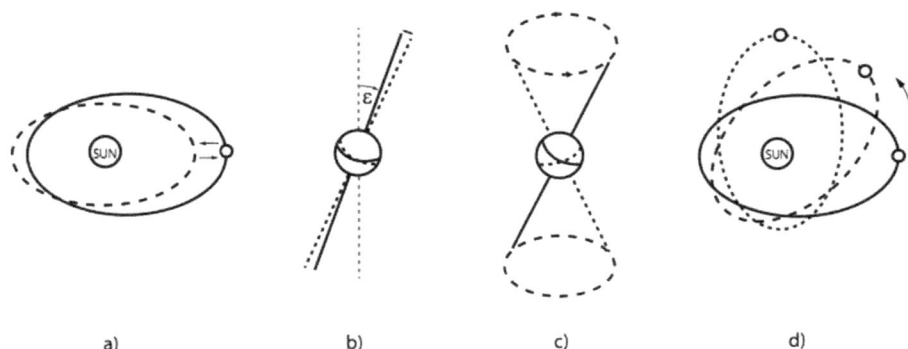

Fig. 1. Schematic illustration of the variation of the eccentricity of the Earth's orbit (a), the variation of the obliquity or inclination of the terrestrial rotation axis, ε (b), the precession of the Earth's rotation axis (c), and the apsidial precession (d); illustration courtesy of Stephan Nyeki.

ε in Figure 1b. This angle oscillates every 41 kyr from 22.1° to 24.5°. Presently, the obliquity is at 23.44°.

Third, the Earth's rotation axis is not fixed in space but describes a cone (Figure 1c) with respect to the fixed stars, i.e., it precesses. It takes 26 kyr for the Earth's rotational axis to complete one full precession cycle. Moreover, the Earth's elliptical orbit also precesses around the Sun at the rate of 112 kyr (Figure 1d), called the apsidial precession. These two precession rates lead to a combined period of ~21 kyr. As a consequence, the occurrence of perihelion, i.e., the point in the Earth's orbit where the Earth is closest to the Sun, varies from northern summer to northern winter and back to northern summer. Presently, the Earth is closest to the Sun in January, i.e., during the northern winter.

The amount of radiation reaching the Earth depends on the Sun-Earth distance. Due to the varying orbital parameters described above, the smallest Sun-Earth distance occurs at different seasons over the precession time scale. To account for this, the standard representation of the orbital forcing has been defined as the insolation at high latitude, referring to the radiation reaching Earth during northern summer, i.e., in June at 65° North.

Figure 2 shows the variation in eccentricity, climatic precession, and obliquity, along with the summer insolation and the CO_2 concentration from 200 kyr before the present to 130 kyr into the future relative to the year 1950. The projection of the CO_2 concentration based on orbital forcing for the future is based on the CO_2, i.e., without the anthropogenic CO_2 contribution, determined from the Vostok ice core (Petit et al., 1999) and the variation of the summer insolation determined by (Berger, 1978). In other words, in this analysis, the prediction of CO_2 for the future is based on the extrapolation of the past CO_2 concentration as determined from ice cores using orbital parameters, known for the past as well as for the future, as proxy. The observed CO_2 concentration for the present, including anthropogenic

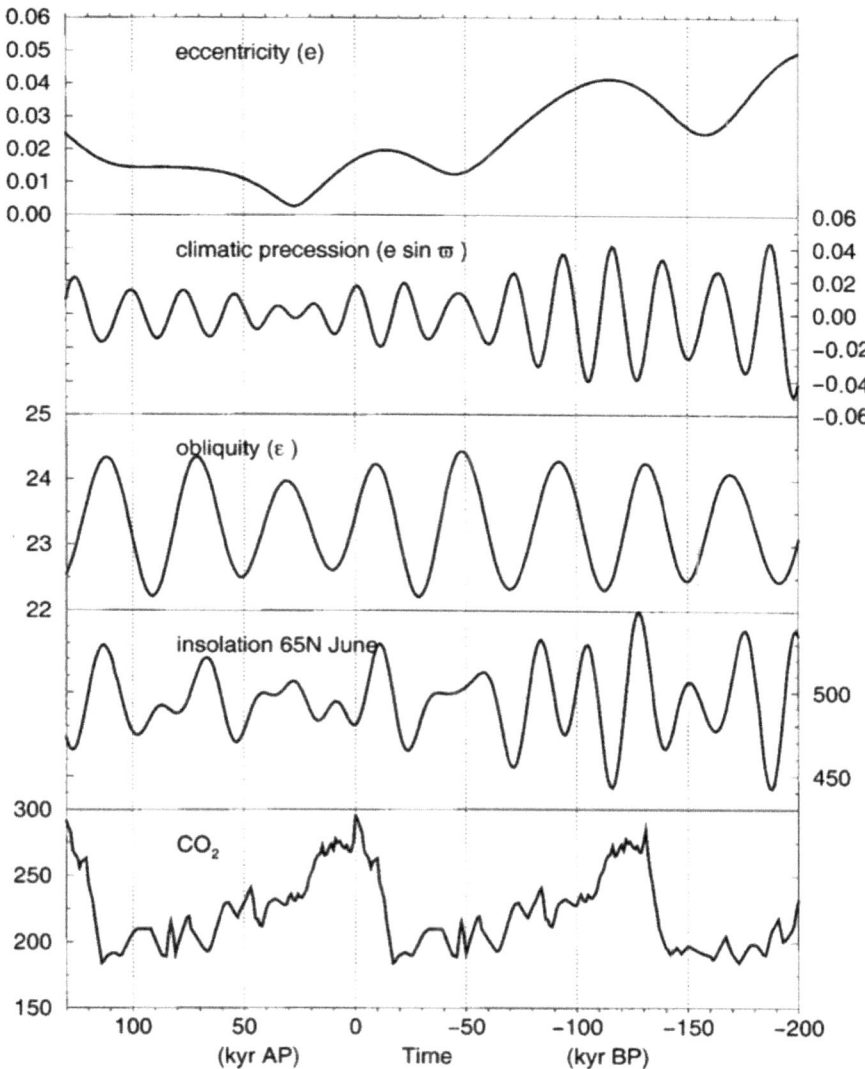

Fig. 2. Long-term variations of the Earth's eccentricity (top panel), climatic precession (second panel), obliquity (third panel), summer solstice insolation at 65° North (fourth panel), and the corresponding atmospheric CO_2 concentration (bottom panel) from 200 kyr before present (BP) to 130 kyr after present (AP). The reference year (0 kyr) is 1950. The variation of the CO_2 concentration in the lower panel is determined from the historic CO_2 concentration from Petit et al. (1999) and the orbital prediction by Berger (1978). Thus, the variation in CO_2 is purely related to the orbital effects. Note, the current CO_2 value including anthropogenic CO_2 emission is ∼400 ppmv as of February 2015. Figure courtesy Berger et al. (2003).

CO$_2$ emission, is considerably higher, i.e., ~400 ppmv as of February 2015. The panels illustrate that the summer insolation is determined by the superposition of the first three parameters. Also, it is obvious that the variation of the eccentricity (top panel), on time scales of roughly 100 kyr, dominates the peaks in the summer insolation (fourth panel), and, as a consequence, leads to an increase of the CO$_2$ concentration (bottom panel). The latter is, to a first approximation, in phase with the variation of the temperature on the Earth.

The insolation in northern summer is strongly modulated by the movement of the perihelion, precessing, as discussed above, on the combined 21-kyr time scale (Figure 1c,d). In addition, the variation in obliquity (Figure 1b) introduces an additional 41-kyr variability. Moreover, due to the multitude of planetary perturbations on the terrestrial orbit, there are many periods present in each of the variations of the three orbital parameters. The interplay of all the periods yields a seemingly chaotic picture of the combined effect of the orbital forcing on the insolation in northern summer. However, the periods are all mechanistic perturbations, and can therefore be determined precisely for the past as well as for the future; see e.g. (Berger, 1978). Based on the orbital influence alone, i.e., without the anthropogenic CO$_2$ emission, it is predicted that the Earth would be entering its next ice age within a 50-kyr time scale, when the northern summer insolation reaches its next minimum (see Figure 2, fourth panel).

The present CO$_2$ gas concentration (not illustrated in Figure 2), including the anthropogenic emission, has reached a value of ~400 ppmv as of February 2015. In fact the total CO$_2$ gas concentration has increased over the last 250 years by more than 100 ppmv, i.e., roughly the same amount it has increased over the last 10 kyr due to the orbital forcing. Thus, to obtain the total future CO$_2$ gas concentration, the value given the fourth panel of Figure 2 would need to be corrected by the increase in anthropogenic CO$_2$ emissions.

2   Correlation with climate indicators from ice cores and benthic rock

The influence of the Earth's varying orbit around the Sun on climate is established beyond reasonable doubt by the correlation between climate indicators and the orbital forcing. As already mentioned, the latter is usually represented by the insolation, i.e., solar irradiation, in the northern summer, i.e., in June at latitude 65° North.

The standard measurements to reconstruct the climate for the past are the following. First, the CO$_2$ gas concentration, which is archived in the Arctic and Antarcitc ice sheet, is a direct measurement of the greenhouse gas concentration at the time the corresponding ice layer formed.

Second, the oxygen isotope ratio (δ^{18}O) in O$_2$ reflects changes in the global ice volume and the hydrological cycle. Third, the deuterium content of the ice (δD$_{ice}$) is a proxy of the local temperature change. The age of the ice is determined according to its depth in the ice core. Even though the corresponding date of an ice layer and thus the date of the climate indicators it contains, can only be determined

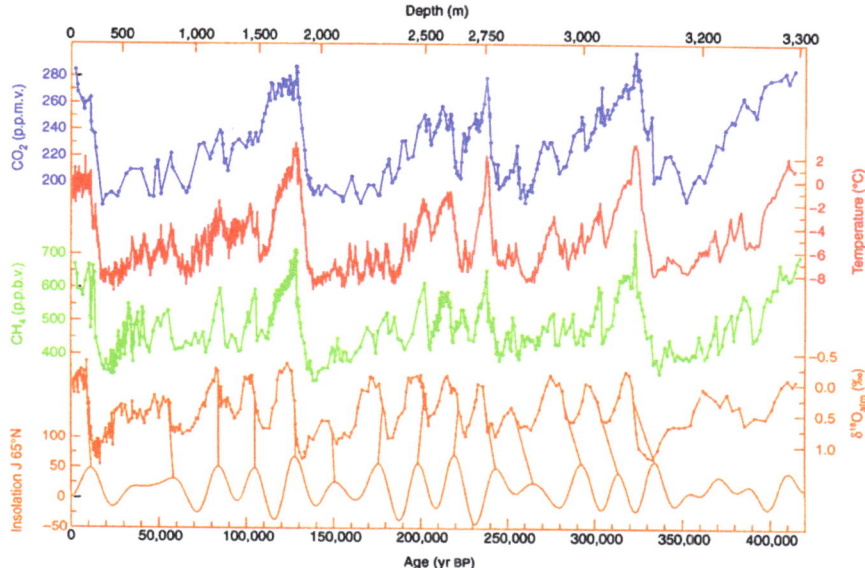

Fig. 3. Time series of climate indicators (first to forth line) determined from ice cores at Vostok station, and northern summer insolation (bottom line). The lower x-axis gives the glaciological timescale and the top x-axis the corresponding depths of the ice for: a) CO_2 (blue); b) δD_{ice} isotopic temperature of the atmosphere (red); c) CH_4 (green line); d) $\delta^{18}O_{atm}$ (brown); and e) mid-June insolation at 65° North (in Wm^2, bottom line). The mean resolution of the CO_2 (CH_4) profile is about 1500 (950) years. It goes up to about 6000 years for CO_2 in the fractured zones and in the bottom part of the record, whereas the CH_4 time resolution ranges from a few tens of years to 4500 years. The overall accuracy for CH_4 and CO_2 measurements are 20 ppbv and 2–3 ppmv, respectively; Courtesy Petit et al. (1999).

within uncertainties, they are sufficiently well known to conclude from ice cores obtained in Greenland and compared to antarctic cores, that warm and cold periods on the northern and southern hemisphere have developed in phase – at least during the last 800 kyr.

Figure 3 shows the CO_2 concentration (blue line), the temperature reconstructed from δD_{ice} (red line), CH_4 gas concentration (green line), and the oxygen isotope ratio $\delta^{18}O$ (upper brown line) from the Vostok ice core records, covering the last 400 kyr, i.e., the last four glacial–interglacial cycles. The data show that there are periods with high peak temperatures, which develop relatively quickly from a cold stage. After the temperature reaches the maximum of a warm period, it gradually decreases until the onset of the next high-temperature peak. This saw-tooth pattern repeats about every 100 kyr. At the onset of the warming there is, within the uncertainty of ice-dating, always a strong maximum of summer insolation in the northern hemisphere. This is, as discussed above, the result of a maximum in forcing resulting from a coherent positive combination of eccentricity, precession, and obliquity of the Earth's orbit, with the key drivers being a large eccentricity along with a perihelion in the northern summer. Most importantly, the repetition period of the glacial–interglacial cycles is in synchronisation with the occurrence of high eccentricity.

The ^{18}O and ^{13}C isotopes are also measured in benthic rock sediments dating back several millions of years. The synchronization between northern hemispheric summer forcing and the climatic modulations is well established down to the 41-kyr oscillation of the obliquity of the Earth's orbital axis relative to the ecliptic. Moreover, the climatic changes in the past, beyond the dating limit of ice cores, can be assessed from oxygen isotope ratios (δ^{18}O) in calcium carbonate of benthic foraminifera. These data reveal that a switch occurred from intermediate warm periods every 100 kyr – as during the last one million years – to 41 kyr, the latter being the dominant period of the Earth's obliquity variations. The transition between the two modes took place in the mid-Pleistocene, about 0.8 to 1.2 million years ago. Even though some mechanisms that are at play are only partially understood, the current understanding is that inter-glacial warm periods within the last ice ages were forced by irradiance variations resulting from the varying orbit of the Earth around the Sun.

3 Global influence of northern summer orbital forcing

Given the good correlation between orbital forcing and climatic variations, it seems obvious that the solar irradiance arriving at the terrestrial surface indeed drives the Earth's climate. However, it turns out that this relation is far from easy to understand or straightforward to model. The reasons are as follows. First, the eccentricity of the Earth's orbit varies from $\sim$0 to $\sim$0.07. Integrating the radiation energy over the year, however, it turns out that the annual energy received at the Earth does not depend on the Earth's eccentricity to first order. Only to second order does the annual energy reaching the Earth increase by a small amount for larger eccentricity. Indeed, for large eccentricity, the difference in insolation between perihelion and aphelion can be very large, i.e., up to 23 %. At the same time, the current view is that if perihelion occurs during the northern summer, the climate is forced towards a warmer global climate. Thus, the orbital forcing is not a matter of the total energy but rather its redistribution over the seasons. We note that if the correlation between climate and the insolation forcing for summer in the northern hemisphere is positive, then it has to be anti-correlated with southern summer insolation. As the warm and cold periods of the two hemispheres are varying synchronously – at least within the last 1 Mio years – it is the northern hemisphere climate which dominates over the counter-acting insolation forcing of the southern hemisphere. Most likely, this can be attributed to the much larger fractional area covered by land in the northern hemisphere with respect to the southern hemisphere.

4 The role of historic CO_2 concentration

Presently, the Earth is at perihelion in January. Thus, the positive forcing that has brought Earth into the present warm period, the Holocene, is understood to have happened half a precession period (i.e., about 10 kyr) ago, when perihelion ocurred during the northern summer. Indeed this time interval coincides with the start of the Holocene. However the question arises, why does the temperature curve (Figure 3, red curve) show this saw-tooth pattern, i.e., why does the temperature decrease relatively slowly after each sudden warming? One hypothesis is that the orbital forcing is only the trigger that releases a climate process. It appears that a thousand years of exceptionally high summer insolation in the northern hemisphere, with a negative yearly mass balance of the northern glaciers, is sufficient to completely remove the glaciation of the northern hemisphere, except for very high latitudes. Feedback mechanisms keep the land free of ice during the summer. Additionally, the warm temperatures trigger the release of CO_2, stored in the deep ocean. In principle, in this scenario of orbital forcing, the development of the Earth's climate could be completely controlled by the concentration of CO_2 and other greenhouse gases in the atmosphere. In this case, it would be the carbon cycle time scale, which is responsible for the slow temperature decline after a warm peak.

However, there are indications that this picture of the orbital forcing is not fully complete. Climate modeling, including the behavior of ice sheets reacting to orbital forcing and the understanding of the alternation between warm periods and ice ages have only recently advanced to a stage of moderate understanding of the many complicated and inter-related processes at play. Interestingly, model results suggest that limiting the orbital forcing to the trigger mechanism, as described above, is only part of the story (see e.g. Stap et al. (2014)). In fact, during the time between the inter-glacial peaks, occurring every 100 kyr, $\delta^{18}O$, the proxy for the ice volume and the hydrological cycle obtained from Antarctica, i.e., the southern hemisphere, closely follows the northern hemisphere summer forcing (see Figure 3, fourth and fifth line, respectively). It is suggested that the CO_2 gas concentration might be responsible for this global link between the northern summer orbital forcing and its corresponding $\delta^{18}O$ signal in the southern ice sheet. Thus, there are many aspects of orbital forcing, which still need to be fully understood.

INFOBOX 2.2

GRAND MINIMA AND MAXIMA OF SOLAR ACTIVITY

Ilya Usoskin[1,2] and Kalevi Mursula[2]

Solar activity depicts a great deal of variability even during the last 400 years of telescopic sunspot observations since 1610 AD (see Figure 2, Chapter 2.5). One can clearly see that there was almost no sunspots during more than a half of a century in 1645–1700 AD. It is important to understand that this lack of sunspots at this time is not due to the lack of sunspot observations. In fact, the Sun was observed by several astronomers already at this time. However, there was little to observe since the solar surface was almost spotless. Some of these early observers wrote weary comments in their notebooks: another year with no sunspots. This period is now known as the Maunder minimum, named after the British astronomer Edward Walter Maunder (1851–1928) (Eddy, 1976). Such a quiet state of the Sun (or another star), when the regular cyclic appearance of sunspots is greatly suppressed for several decades (from 40 to 200 years), is more generally called the **Grand minimum** of activity. Grand minima should not be confused with the normal sunspot minima of the 11-year solar cycle that only last for a couple of years.

Another interesting feature of the Sun in recent times is the very high level of sunspot activity during most of the last century, roughly from the 1930s to the early 2000s. This period is now called the **Modern Grand maximum** of solar activity. Grand maxima are periods when the sunspot cycles attain very high amplitudes. Presently, the Modern Grand maximum is over, and the sunspot activity level of the on-going solar cycle 24 is quite moderate, considerably lower than during the previous cycles forming the Modern Grand maximum.

Since direct sunspot observations only exist for the last 400 years, the occurrence of Grand minima and Grand maxima can be studied only by proxy data, especially the cosmogenic radionuclides ^{14}C and ^{10}Be (see Chapter 2.5). A recent study, using ^{14}C isotopes measured in tree-rings covering the whole Holocene (Figure 1), provides a list of 27 Grand minima and 19 Grand maxima during the last 12 thousand years (Usoskin et al., 2007). Grand minima can

---

[1] Sodankylä Geophysical Observatory, Oulu unit, University of Oulu, 90014 Oulu, Finland

[2] ReSoLVE Center of Excellence, Astronomy and Space Physics Research Unit, University of Oulu, Finland

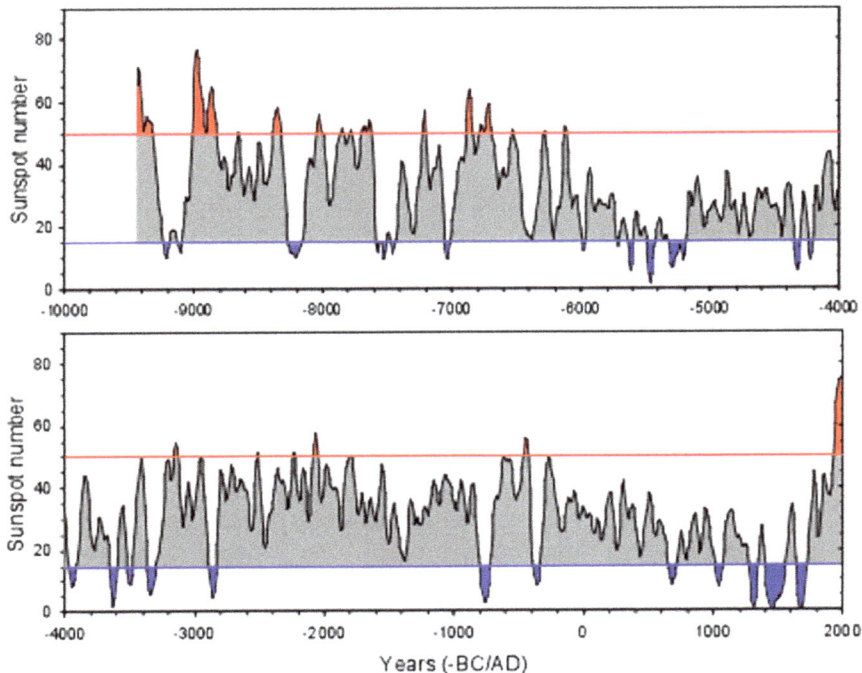

Fig. 1. Sunspot activity reconstructed from ^{14}C data throughout the Holocene. Blue and red areas denote Grand minima and maxima, respectively. The entire series is spread over two panels for better visibility (Modified after Usoskin et al. (2007)).

be robustly defined and form one distinguishable mode of the solar dynamo (Usoskin et al., 2014). They occur irregularly, which implies that these extreme states in the solar dynamo are driven by a chaotic or stochastic rather than a cyclic process. Thus, their occurrence cannot be reliably and deterministically predicted. However, Grand minima tend to cluster in groups of several minima (as, e.g., Oort, Spörer and Maunder minima in 1200–1700 AD) with roughly 200 years in-between, seem as the so-called Suess or deVries cycle. Clusters are separated by intervals of several millennia. Grand minima seem to divide into two types according to their duration: short minima, such as the Maunder minimum, which last 40–90 years and long minima, like the Spörer minimum, which last 100–170 years.

Grand maxima are more uncertain and cannot be unambiguously defined. Also, it is not known if Grand maxima form a separate dynamo mode or if they are an extreme part of the normal sunspot distribution. Grand maxima seem to occur irregularly, and without any apparent clustering or periodicity. Their duration varies between 30 and 80 years. The Modern Grand maximum in the 20th century was an example of this rare but not exceptional form of solar activity.

INFOBOX 2.3

A PRACTICAL GUIDE TO SOLAR FORCING DATA

T. Dudok de Wit[1], B. Funke[2], A. Seppälä[3], E. I. Tanskanen[3] and I. Usoksin[4]

1 Introduction

The Sun-climate connections that are discussed in this book rely on a large body of observations, which come with their uncertainties, shortcomings and pitfalls. Knowing these is crucial when it comes to assessing climate impacts.

Here, we provide a non-exhaustive list of solar forcing observables that are most frequently encountered in Sun-climate studies. For each of them, we provide some recommendations as to the data sources, the limitations to consider, etc. This practical chapter is about science in action. Therefore, the information is likely to become obsolete as new measurements are made, or old ones are reinterpreted.

2 Solar forcings

2.1 Total Solar Irradiance (TSI): direct observations

The TSI represents the total solar radiative input per unit surface, at the top of the atmosphere, for a normalised Sun-Earth distance of 1 AU. Historically, the TSI has always been the main, if not the only solar input into the atmosphere, and it remains a key ingredient of the Earth's energy budget (Trenberth et al., 2009).

Direct and continuous observations of the TSI started in November 1978, and the datasets of interest are composites that are made out of numerous observations. Most instruments agree well on relative variations in the TSI. They disagree, however, on their long-term evolution, which is still hotly debated.

---

[1] University of Orléans, France
[2] Instituto de Astrofísica de Andalucía, CSIC, Granada, Spain
[3] Finnish Meteorological Institute, Helsinki, Finland
[4] University of Oulu, Finland

The main challenge with the TSI is its extremely small relative variation with the solar cycle (less than 0.7 %), which requires high accuracy measurements. Only most recent instruments can achieve uncertainties as low as 350 ppm[5], and stabilities of the order of 10 ppm/year.

There exist today three reconstructions of the TSI, but the one that shows the best agreement with independent solar irradiance models, is the TSI composite provided by PMOD, in Davos, see `ftp://ftp.pmodwrc.ch/pub`. Its value at solar minimum is presently estimated to be 1360.8 ± 0.5 W m^{-2} (Kopp and Lean, 2011; Schmutz et al., 2013). A new composite, which should be endorsed by the scientific community, is expected to come out in 2016.

2.2 Total Solar Irradiance (TSI): indirect observations

No reliable measurements of the TSI exist prior to 1978; however, proxy-based reconstructions have become a powerful alternative for extending the TSI decades to millennia backward in time (Schmidt et al., 2010).

Early reconstructions were based on solar proxies, such as the sunspot number. However, their connection with the TSI is often too indirect to enable reliable reconstructions, especially when it comes to understanding variations on multi-decadal time scales. Today's best reconstructions are based on the concentration in natural archives of radioactive isotopes, such as ^{10}Be and ^{14}C, whose production rate is modulated by galactic cosmic ray flux, and thus by solar activity (Bard and Frank, 2006; Beer et al., 2012). The connection between the TSI and the concentration of these so-called cosmogenic isotopes involves several intermediate physical processes, and thus assumptions. Early reconstructions used an empirical relation between the two to reconstruct the TSI (Steinhilber et al., 2012). More recently, additional physical constraints have been used to refine the reconstruction (Vieira et al., 2011).

Present reconstructions of the TSI by cosmogenic isotopes allow to go back to about 9400 BC. However their time resolution cannot be much lower than about 10 years, hence excluding the observation of the Schwabe cycle. A good starting point for publicly available data sets is the National Geophysical Data Center, see `ftp://ftp.ncdc.noaa.gov/pub/data/paleo/climate_forcing/solar_variability/`

2.3 Solar spectral irradiance (SSI): direct observations

The spectrally-resolved solar irradiance (or SSI) is much more relevant for climate studies than the TSI because various wavelengths have differing impacts on the atmosphere (Ermolli et al., 2013), see Chapters 4.1 and 4.2. Unfortunately, direct SSI observations are scarce, highly fragmented in time and in wavelength. In addition, most instruments suffer from degradation.

---

[5] 1 ppm = 1 part-per-million = 0.001 %

As of today, the only existing SSI composite dataset (DeLand and Cebula, 2008) is not meant to be used for climate studies. In practice, datasets are used on a per instrument basis, with considerable differences between instruments in quality and in coverage. Several datasets are accessible at the LISIRD data center, see http://lasp.colorado.edu/lisird/. A new composite dataset will come out 2016, containing daily observations since 1980, with wavelengths spanning from 10 to 700 nm, see http://projects.pmodwrc.ch/solid/.

Several UV proxies have been developed in response to the recurrent need for characterising the solar UV forcing of the atmosphere. Most proxies reproduce the solar variability on time scales of days to months remarkably well; however, their ability to capture long-term trends is uncertain. The solar radio flux at 10.7 cm (the so-called F10.7 index) is a popular proxy for the Extreme-UV (below 120 nm). For longer wavelengths in the UV, the core-to-wing ratio of the Mg II line, which is better known as the Mg II index, and the radio flux at 30 cm, are better alternatives.

Daily values of the F10.7 index go back to 1947, and can be downloaded from ftp://ftp.ngdc.noaa.gov/STP/space-weather/solar-data/solar-features/solar-radio/noontime-flux/penticton/. Daily values of the MgII index are available since 1978 at http://www.iup.uni-bremen.de/gome/solar/.

2.4 Solar Spectral Irradiance (SSI): model reconstructions

The severe lack of reliable SSI observations has stimulated the development of various models for reconstructing the observations prior to the space age, and/or to fill in observation gaps. All these models rely on solar proxies, such as the sunspot number (going back to 1610), or cosmogenic isotopes (going back farther in time), see Chapters 2.5 and 2.6. While their agreement with direct observations give high confidence in their accuracy for reconstructions made after 1980, there are considerable uncertainties in the SSI before the space age. So, while the TSI is relatively well constrained by solar proxies (see above), the contribution of its individual bands, and in particular of the UV, remains poorly known.

This situation is improving, as new reconstructions are becoming available, based, for example, on historical of solar images (Dasi-Espuig et al., 2014).

The model reconstruction that is most widely used today for climate studies is NRLSSI, see http://lasp.colorado.edu/lisird/; in it current version, NRLSSI goes back to 1882. The SATIRE family of models is more accurate for reproducing daily/weekly variations, see http://www2.mps.mpg.de/projects/sun-climate/data.html. The most accurate of them (SATIRE-S) goes back to 1974; other versions go farther back in time, using simplifying assumptions.

2.5 Galactic cosmic rays

The Earth is continuously bombarded by highly energetic particles of extraterrestrial origin, called cosmic rays, see Chapter 2.3. Cosmic rays consist of mostly protons, accompanied by a smaller fraction of α-particles and heavier nuclei up to iron. There is also a tiny fraction of anti-protons. In addition, a small amount of electrons and positrons is also present. On average, the cosmic ray flux at the Earth's orbit is about 1 particle per cm^2 per second. The energy of cosmic rays may reach several 10^{20} eV but the flux of the most energetic particles is very small. The bulk of cosmic rays originates from our Galaxy, and mostly comes from supernova shocks. These are called galactic cosmic rays (GCR). Cosmic rays with the highest energies originate from exotic extragalactic sources.

Although the galactic cosmic ray flux is roughly constant in the interstellar space (at least at the time scale of up to a million years), the cosmic ray flux at Earth is variable because of the modulation of GCR in the Heliosphere. The latter is a region of approximately 200 astronomical units across, which is dominated by the solar magnetic field, and by the solar wind (Potgieter, 2013). The flux of GCR is greater during solar minimum, and lower during solar maximum. The level of modulation varies with the energy of the GCR, from several orders of magnitude for 100 MeV particles down to a few percent at 20 GeV energy, and vanishes at higher energies. In addition, the Earth is shielded from cosmic rays by its geomagnetic field, which provides better shielding in equatorial regions, and no shielding in polar regions. Because of their high energy, GCR can penetrate deep into the atmosphere; they constitute the main source of the atmospheric ionisation in the troposphere and stratosphere.

One of the most complete databases for accessing these data is the World Data Center for Cosmic Rays, see `http://center.stelab.nagoya-u.ac.jp/WDCCR/`. Data from European monitors are accessible through the Neutron Monitor Database, see `http://www.nmdb.eu`.

While GCR are always present at Earth, there is another sporadic component called solar energetic particles (SEP, sometimes erroneously called solar cosmic rays). These particles, which are of solar origin, mostly consist of protons, and are described below.

2.6 Solar Proton Forcing

During intense solar flares, or when large shock waves cross the interplanetary medium, solar protons can be accelerated up to high energies, giving rise to so-called *solar proton events*. Events with proton energies in excess of 10 MeV occur on average a dozen times per year, and are more frequent when the Sun is active. A small fraction of these solar particles actually consists of helium, and heavier nuclei. These particle fluxes have been routinely monitored since the 1980s by various geostationary satellites. Sometimes the energy of solar energetic particles can be as high as a several GeVs so that they can reach the ground level. Such events are called ground level enhancements (GLEs) and occur roughly ten times per solar cycle, with a peak around the maximum and declining phase of the Schwabe cycle.

For studies of solar energetic particles (see Chapters 4.5 and 4.6), the use of solar proton observations from the GOES satellites is strongly recommended.

These observations have been made readily available by the National Oceanic and Atmospheric Administration (NOAA) and all the existing data can be obtained for example from the Space Physics Interactive Data Resource (SPIDR) online data archive, see http://spidr.ngdc.noaa.gov. A database of GLE events is available at http://gle.oulu.fi.

For modelling studies wishing to include direct ionisation rates from solar protons for middle atmosphere altitudes, ionisation rates have been made available by the Solar Influences for SPARC (SOLARIS-HEPPA) activity. The ionisation rates, together with a description of how the data set was produced, and further information on how these can be converted to HO_x and NO_x production rates (Jackman et al., 2008, 2009), can be found at the SOLARIS-HEPPA homepage (http://solarisheppa.geomar.de), in the Input Data section. At the time of writing, the readily available ionisation rate dataset distributed by SOLARIS-HEPPA covers the years 1963–2013. The use of these ionisation rates for solar energetic particles is highly endorsed by the scientific community. Several of these issues are further discussed by Seppälä et al. (2014).

2.7 Energetic Electron Forcing

Although energetic electrons are usually measured by the same sets of instruments as energetic protons, their origin and their dynamics often differ substantially. Most energetic electrons originate from the Earth's radiation belts, and their energies are typically in the 0.5–10 MeV range. The measurement of energetic electron fluxes is much more challenging than that of energetic protons, and considerable efforts are presently put into the making of long records, starting with the early observations of 1978.

Precipitating energetic electrons from auroral and radiation belt sources (as described in Chapter 4.5) frequently cause enhanced ionisation in the atmosphere. Due to lack of continuous observations (see for example Rodger et al. (2010)), datasets, such as those used for solar protons are currently not available, although alternative approaches are being developed. As these become available they will be included in the above mentioned SOLARIS-HEPPA activity recommended inputs (http://solarisheppa.geomar.de).

As a best estimate for the impact of electron precipitation some models use parameterisations that are based on indices of geomagnetic activity, such as the Kp or Ap index to either describe ionisation (in the case of some models) or NO_x production (in other models) from electron precipitation. These indices can be obtained, for example, from the SPIDR online data archive (http://spidr.ngdc.noaa.gov/spidr/). In high top models (models with upper limit in the thermosphere, such as the CESM/WACCM), it is possible to use a parameterisation for thermospheric ionisation from auroral electrons that is based on the Kp index (Marsh et al., 2007). For lower top models (models with upper limit in the mesosphere), auroral production of NO_x and the subsequent descent from higher altitudes to the model domain can be described as an upper boundary condition. These boundary conditions are usually semi-empirical, so that they concur with observations for last decade or so. It should be noted however, that the assumption for the upper boundary condition for NO_x production is not ideal for situations where exceptional meteorological conditions are present. This typically occurs during sudden stratospheric warming and elevated stratopause events (Funke et al., 2014). An example of a semi-empirical Ap-driven NO_x upper boundary for models is given by Baumgaertner et al. (2009). An updated Ap-driven NO_x upper boundary based on MIPAS observations (Funke et al., 2014) is currently under development and will be made available on the SOLARIS-HEPPA webpage (see above) by the end of 2015.

2.8 Solar wind data

The first samples of solar wind were taken in 1959 by Luna spacecraft (Luna 1, 2 and 3) which measured the solar wind ion flux by Faraday cups. Since then, the solar wind has been sampled and monitored by a large amount of scientific spacecraft, including Mariner (1962), HEOS-1 (1968), ISEE-2 (1977), Ulysses (1990) and SOHO (1995). Continuous solar wind measurements have been made at the L1 Lagrange point (located between the Sun and Earth), mainly by the Wind (1994) and ACE (1997) spacecraft, both of which are still functional.

Most solar wind spacecraft carry a standard set of instruments that measure proton and electron flux and energy, magnetic field, solar wind composition, temperature, and plasma waves. Each of these reveals a specific aspect of the way the Earth's environment responds to the varying solar wind, and so the impact of the solar wind cannot be reduced to one single observable. These datasets are publicly available at the individual mission web sites, such as http://www.srl.caltech.edu/ACE/, and http://wind.nasa.gov/. The most important observables have been merged into composite datasets spanning from the 1970s till today; they are available from the Coordinated Data Analysis Web, CDAWeb (http://cdaweb.gsfc.nasa.gov/), and from the Space Physics Interactive Data Resource SPIDR (http://spidr.ngdc.noaa.gov/spidr/). Several tools have been developed to locate the spacecraft in the solar wind. One such tool was developed within the International Solar-Terrestrial Physics programme http://pwg.gsfc.nasa.gov/orbits/.

2.9 Geomagnetic data

The geomagnetic field has been known to exist since the compass was invented in China around 100 AD. The first multipoint measurements are from the end of 1700s when Alexander von Humboldt first time measured changes of the geomagnetic field simultaneously at the several locations and described the observed changes as geomagnetic storms. Nowadays, geomagnetic field measurements are produced by hundreds of observatories, research institutes and universities around the globe. Their data are delivered mainly by individual data producers.

For practical purposes, *geomagnetic indices* are frequently used instead of direct geomagnetic observations. These indices quantify specific properties of the geomagnetic field, such as its fluctuation level, rather than its absolute value. Most indices also combine observations that are made at specific geomagnetic latitudes, thereby enhancing the signature of specific sources of the varying geomagnetic field at a planetary scale (Menvielle and Berthelier, 1991). The Ap index, for example, is a measure of the general level of geomagnetic activity over the globe, and is provided every 3 hours. The Kp index is quasi-logarithmic index of the 3-hourly range in magnetic activity relative to an assumed quiet-day, and is computed from 13 geomagnetic observatories between 44 and 60° northern or southern geomagnetic latitude. Other quantities can be found, for example, at the National Geophysical Data Center (http://www.ngdc.noaa.gov/stp/geomag/indices.html).

There are two reasons for which such geomagnetic indices are of direct interest to Sun-climate connection studies. First, since geomagnetic fluctuations directly affect the rate of precipitation of energetic particles into the atmosphere, they may be considered as proxies for the latter. Secondly, geomagnetic fluctuations are ultimately driven by the varying solar wind, and may thus also serve as indirect tracers of solar activity. Some indices have been continuously recorded for more than a century, and thus offer precious historical records of the response of the Earth's magnetic field to solar activity, see for example (Svalgaard and Cliver, 2007; Lockwood and Owens, 2011).

Geomagnetic data are collected by large international initiatives, such as INTERMAGNET and EPOS. The international real-time magnetic observatory network (INTERMAGNET) started developing in 1986 when the first discussion began on ways to better transfer the data between the participating institutes. INTERMAGNET data are available as magnetograms or as digital data files (http://www.intermagnet.org/). The main indices can also be obtained from SPIDR (http://spidr.ngdc.noaa.gov/spidr/).

The European plate observing system (EPOS) collects magnetic, seismic and other solid Earth measurements to solve how the Earth response to the disturbances from terrestrial (earthquakes, volcanoes and tectonics) and extraterrestrial sources, such as solar storms and solar wind disturbances. EPOS is in a process of better integrating existing geoscientific research infrastructures and improved services will be available in coming years (http://www.epos-eu.org/). In Europe, the largest international magnetometer network IMAGE (Tanskanen, 2009) provides data and data products via http://space.fmi.fi/image/.

Existing geomagnetic data sets are suitable for studies covering months to years, while we lack good quality datasets for studies covering centuries and beyond. In particular, issues such as station relocation, or instrument replacement must be considered very carefully when interpreting long-term changes, e.g. (Clilverd et al., 2005). Several on-going projects are addressing this need for homogeneous and corrected data sets suitable for the long-term Sun-climate coupling studies. The Substorm Zoo is a different initiative that provides long-term datasets from the entire heliosphere together with browser-based visualisation and social data analysis tools (http://www.substormzoo.org).

Longer-term reconstructions of the geomagnetic field on millennial time scales are important for properly modelling the connection between cosmogenic isotopes and solar activity, see Chapter 2.2. For such long time scales, no information is available about the fluctuation level. However, reconstructions of the total field strength are possible, see Korte et al. (2011).

Further reading

Kamide, Y. and Chian, A. C.-L.: 2007, Handbook of the Solar-Terrestrial Environment, Springer Verlag, Berlin.

Mironova, I. et al. 2015, Energetic Particle Influence on the Earth's Atmosphere, Space Science Reviews, in press, doi: 10.1007/s11214-015-0185-4

References of Part II

Bard, E., and M. Frank. Climate change and solar variability: What's new under the sun? *Earth and Planetary Science Letters*, **248**, 1–14, 2006. DOI:10.1016/j.epsl.2006.06.016.

Baumgaertner, A. J. G., P. Jöckel, and C. Brühl. Energetic particle precipitation in ECHAM5/MESSy1 - Part 1: Downward transport of upper atmospheric NO_x produced by low energy electrons. *Atmospheric Chemistry & Physics*, **9**, 2729–2740, 2009.

Bazilevskaya, G. A., I. G. Usoskin, E. O. Flückiger, R. G. Harrison, L. Desorgher, et al. Cosmic Ray Induced Ion Production in the Atmosphere. *Space Science Reviews*, **137**, 149–173, 2008. DOI:10.1007/s11214-008-9339-y.

Beer, J., K. McCracken, and R. von Steiger. Cosmogenic Radionuclides. Springer Verlag, Berlin, 2012.

Berger, A., M. F. Loutre, and M. Crucifix. The Earth's Climate in the Next Hundred Thousand years (100 kyr). *Surveys in Geophysics*, **24**, 117–138, 2003.

Berger, A. L. Long-Term Variations of Daily Insolation and Quaternary Climatic Changes. *Journal of Atmospheric Sciences*, **35**, 2362–2367, 1978. DOI:10.1175/1520-0469(1978)035¡2362:LTVODI¿2.0.CO;2.

Clette, F., L. Svalgaard, J. M. Vaquero, and E. W. Cliver. Revisiting the Sunspot Number. A 400-Year Perspective on the Solar Cycle. *Space Science Reviews*, **186**, 35–103, 2014. DOI:10.1007/s11214-014-0074-2.

Clilverd, M. A., E. Clarke, T. Ulich, J. Linthe, and H. Rishbeth. Reconstructing the long-term aa index. *Journal of Geophysical Research*, **110**, 7205, 2005. DOI:10.1029/2004JA010762.

Dasi-Espuig, M., J. Jiang, N. A. Krivova, and S. K. Solanki. Modelling total solar irradiance since 1878 from simulated magnetograms. *Astronomy and Astrophysics*, **570**, A23, 2014. DOI:10.1051/0004-6361/201424290.

DeLand, M. T., and R. P. Cebula. Creation of a composite solar ultraviolet irradiance data set. *Journal of Geophysical Research (Space Physics)*, **113**, 11103, 2008. DOI:10.1029/2008JA013401.

Desorgher, L., E. O. Flückiger, M. Gurtner, M. R. Moser, and R. Bütikofer. Atmocosmics: a Geant 4 Code for Computing the Interaction of Cosmic Rays with the Earth's Atmosphere. *International Journal of Modern Physics A*, **20**, 6802–6804, 2005. DOI:10.1142/S0217751X05030132.

Dorman, L. I., ed. Cosmic Rays in the Earth's Atmosphere and Underground, vol. 303 of *Astrophysics and Space Science Library*, 2004.

Eddy, J. A. The Maunder Minimum. *Science*, **192**, 1189–1202, 1976. DOI:10.1126/science.192.4245.1189.

Ermolli, I., K. Matthes, T. Dudok de Wit, N. A. Krivova, K. Tourpali, et al. Recent variability of the solar spectral irradiance and its impact on climate modelling. *Atmospheric Chemistry and Physics*, **13**, 3945–3977, 2013. DOI:10.5194/acp-13-3945-2013, 1303.5577.

Forbush, S. E. On the Effects in Cosmic-Ray Intensity Observed During the Recent Magnetic Storm. *Physical Review*, **51**, 1108–1109, 1937. DOI:10.1103/PhysRev.51.1108.3.

Funke, B., M. López-Puertas, L. Holt, C. E. Randall, G. P. Stiller, and T. von Clarmann. Hemispheric distributions and interannual variability of NOy produced by energetic particle precipitation in 2002–2012. *Journal of Geophysical Research: Atmospheres*, **119**(23), 2014. DOI:10.1002/2014JD022423, http://dx.doi.org/10.1002/2014JD022423.

Haberreiter, M., W. Schmutz, and Kosovichev, A. G. Solving the discrepancy between the seismic and photospheric solar radius. *Astrophysical Journal*, **675**, L53-L56, 2008.

Harder, J. W., J. M. Fontenla, P. Pilewskie, E. C. Richard, and T. N. Woods. Trends in solar spectral irradiance variability in the visible and infrared. *Geophysical Research Letters*, **36**, 2009. DOI:10.1029/2008GL036797.

Jackman, C. H., D. R. Marsh, F. M. Vitt, R. R. Garcia, E. L. Fleming, et al. Short- and medium-term atmospheric constituent effects of very large solar proton events. *Atmospheric Chemistry & Physics*, **8**, 765–785, 2008.

Jackman, C. H., D. R. Marsh, F. M. Vitt, R. R. Garcia, C. E. Randall, E. L. Fleming, and S. M. Frith. Long-term middle atmospheric influence of very large solar proton events. *Journal of Geophysical Research (Atmospheres)*, **114**, D11304, 2009. DOI:10.1029/2008JD011415.

Judge, P. G., G. W. Lockwood, R. R. Radick, G. W. Henry, A. I. Shapiro, W. Schmutz, and C. Lindsey. Confronting a solar irradiance reconstruction with solar and stellar data. *Astronomy and Astrophysics*, **544**, A88, 2012. DOI:10.1051/0004-6361/201218903.

Kopp, G., and J. L. Lean. A new, lower value of total solar irradiance: Evidence and climate significance. *Geophysical Research Letters*, **38**, 1706, 2011. DOI:10.1029/2010GL045777.

Korte, M., C. Constable, F. Donadini, and R. Holme. Reconstructing the Holocene geomagnetic field. *Earth and Planetary Science Letters*, **312**, 497–505, 2011. DOI:10.1016/j.epsl.2011.10.031.

Krivova, N. A., L. E. A. Vieira, and S. K. Solanki. Reconstruction of solar spectral irradiance since the Maunder minimum. *Journal of Geophysical Research (Space Physics)*, **115**(A12), 2010. http://dx.doi.org/10.1029/2010JA015431.

Lockwood, M., and M. J. Owens. Centennial changes in the heliospheric magnetic field and open solar flux: The consensus view from geomagnetic data and cosmogenic isotopes and its implications. *Journal of Geophysical Research (Space Physics)*, **116**, 4109, 2011. DOI:10.1029/2010JA016220.

Marsh, D. R., R. R. Garcia, D. E. Kinnison, B. A. Boville, F. Sassi, S. C. Solomon, and K. Matthes. Modeling the whole atmosphere response to solar cycle changes in radiative and geomagnetic forcing. *Journal of Geophysical Research (Atmospheres)*, **112**, 23306, 2007. DOI:10.1029/2006JD008306.

Menvielle, M., and A. Berthelier. The K-derived planetary indices - Description and availability. *Reviews of Geophysics*, **29**, 415–432, 1991.

Petit, J. R., J. Jouzel, D. Raynaud, N. I. Barkov, J.-M. Barnola, et al. Climate and atmospheric history of the past 420,000 years from the Vostok ice core, Antarctica. *Nature*, **399**, 429–436, 1999. DOI:10.1038/20859.

Porter, H. S., C. H. Jackman, and A. E. S. Green. Efficiencies for production of atomic nitrogen and oxygen by relativistic proton impact in air. *Journal of Chemical Physics*, **65**, 154–167, 1976. DOI:10.1063/1.432812.

Potgieter, M. Solar Modulation of Cosmic Rays. *Living Reviews in Solar Physics*, **10**, 3, 2013. DOI:10.12942/lrsp-2013-3, 1306.4421.

Quack, M., M.-B. Kallenrode, M. von Koenig, K. Kuenzi, J. Burrows, B. Heber, and E. Wolff. Ground level events and consequences for stratospheric chemistry. *International Cosmic Ray Conference*, **10**, 4023, 2001.

Rodger, C. J., M. A. Clilverd, J. C. Green, and M. M. Lam. Use of POES SEM-2 observations to examine radiation belt dynamics and energetic electron precipitation into the atmosphere. *Journal of Geophysical Research (Space Physics)*, **115**, 4202, 2010. DOI:10.1029/2008JA014023.

Schmidt, G. A., J. H. Jungclaus, C. M. Ammann, E. Bard, P. Braconnot, et al. Climate forcing reconstructions for use in PMIP simulations of the last millennium (v1.1). *Geoscientific Model Development Discussions*, **3**(3), 1549–1586, 2010. DOI:10.5194/gmdd-3-1549-2010, http://www.geosci-model-dev-discuss.net/3/1549/2010/.

Schmutz, W., A. Fehlmann, W. Finsterle, G. Kopp, and G. Thuillier. Total solar irradiance measurements with PREMOS/PICARD. In *American Institute of Physics Conference Series*, vol. 1531 of *American Institute of Physics Conference Series*, 624–627, 2013. DOI:10.1063/1.4804847.

Schrijver, C. J., W. C. Livingston, T. N. Woods, and R. A. Mewaldt. The minimal solar activity in 2008-2009 and its implications for long-term climate modeling. *Geophysical Research Letters*, **38**, L06701, 2011. DOI:10.1029/2011GL046658.

Seppälä, A., K. Matthes, C. E. Randall, and I. A. Mironova. What is the solar influence on climate? Overview of activities during CAWSES-II. *Progress in Earth and Planetary Science*, **1**, 24, 2014. DOI:10.1186/s40645-014-0024-3.

Shapiro, A. I., W. Schmutz, E. Rozanov, M. Schoell, M. Haberreiter, A. V. Shapiro, and S. Nyeki. A new approach to the long-term reconstruction of the solar irradiance leads to large historical solar forcing. *Astronomy & Astrophysics*, **529**, A67, 2011. DOI:10.1051/0004-6361/201016173.

Solanki, S. K., N. A. Krivova, and J. D. Haigh. Solar Irradiance Variability and Climate. *Annual Review of Astronomy and Astrophysics*, **51**, 311–351, 2013. DOI:10.1146/annurev-astro-082812-141007, 1306.2770.

Stap, L. B., R. S. W. van de Wal, B. de Boer, R. Bintanja, and L. J. Lourens. Interaction of ice sheets and climate during the past 800 000 years. *Climate of the Past*, **10**, 2135–2152, 2014. DOI:10.5194/cp-10-2135-2014.

Steinhilber, F., J. A. Abreu, J. Beer, I. Brunner, M. Christl, et al. 9,400 years of cosmic radiation and solar activity from ice cores and tree rings. *Proceedings of the National Academy of Sciences*, **109**(16), 5967–5971, 2012. DOI:10.1073/pnas.1118965109, http://www.pnas.org/content/109/16/5967.full.pdf+html, http://www.pnas.org/content/109/16/5967.abstract.

Steinhilber, F., J. Beer, and C. Fröhlich. Total solar irradiance during the Holocene. *Geophysical Research Letters*, **36**, 2009. DOI:10.1029/2009GL040142.

Svalgaard, L., and E. W. Cliver. Long-term geomagnetic indices and their use in inferring solar wind parameters in the past. *Adv. Space Research*, **40**, 1112–1120, 2007. DOI:10.1016/j.asr.2007.06.066.

Tanskanen, E. I. A comprehensive high-throughput analysis of substorms observed by IMAGE magnetometer network: Years 1993-2003 examined. *Journal of Geophysical Research (Space Physics)*, **114**, 5204, 2009. DOI:10.1029/2008JA013682.

Trenberth, K. E., J. T. Fasullo, and J. Kiehl. Earth's Global Energy Budget. *Bulletin of the American Meteorological Society*, **90**, 311–324, 2009. DOI:10.1175/2008BAMS2634.1.

Usoskin, I. G., G. Hulot, Y. Gallet, R. Roth, A. Licht, F. Joos, G. A. Kovaltsov, E. Thébault, and A. Khokhlov. Evidence for distinct modes of solar activity. *Astron. Astrophys.*, **562**, L10, 2014. DOI:10.1051/0004-6361/201423391, 1402.4720.

Usoskin, I. G., and G. A. Kovaltsov. Cosmic ray induced ionization in the atmosphere: Full modeling and practical applications. *Journal of Geophysical Research (Atmospheres)*, **111**, 21206, 2006. DOI:10.1029/2006JD007150.

Usoskin, I. G., G. A. Kovaltsov, and I. A. Mironova. Cosmic ray induced ionization model CRAC:CRII: An extension to the upper atmosphere. *Journal of Geophysical Research (Atmospheres)*, **115**, 10302, 2010. DOI:10.1029/2009JD013142.

Usoskin, I. G., S. K. Solanki, and G. A. Kovaltsov. Grand minima and maxima of solar activity: new observational constraints. *Astron. Astrophys.*, **471**, 301–309, 2007. DOI:10.1051/0004-6361:20077704, arXiv:0706.0385.

Vieira, L. E. A., S. K. Solanki, N. A. Krivova, and I. Usoskin. Evolution of the solar irradiance during the Holocene. *Astronomy and Astrophysics*, **531**, A6, 2011. DOI:10.1051/0004-6361/201015843, 1103.4958.

Yeo, K. L., N. A. Krivova, and S. K. Solanki. Solar Cycle Variation in Solar Irradiance. *Space Science Reviews*, **186**, 137–167, 2014. DOI:10.1007/s11214-014-0061-7.

Part III

DETECTING SOLAR INFLUENCE ON CLIMATE

CHAPTER 3.1

OBSERVATIONS ON PALEOCLIMATIC TIME SCALES

Kristoffer Rypdal[1] and Tine Nilsen[1]

1 Introduction

For the last 11 700 years, human civilisation has developed under stable and warm climatic conditions in the interglacial period called the Holocene. During the Pleistocene (2 588 000 years before present up to the Holocene), there were many glacial periods, also known as ice ages, followed by shorter interglacials. There is little evidence that variation of solar irradiation output has played an important role in trigging the shifts between glaciations and interglacials, but changes in irradiation impinging on the Earth due to changes in the Earth's orbit around the Sun definitely have (see Box 2.1). We do not know much about solar variability beyond the Holocene, with one important exception; the Sun has increased its irradiance by about 25% since the Earth was formed 4.5 billion years ago. This is called *the faint young Sun paradox*, since the young Sun with today's atmospheric composition would leave the Earth frozen and unable to sustain life. The solution to the paradox obviously involves an evolution of the atmosphere, from one dominated by greenhouse gases to one rich in oxygen and only traces of carbon dioxide and methane. This transition was driven by the sudden evolution of microbes that obtain their energy from photosynthesis. These bacteria transform CO_2 and water into sugars and oxygen, and hence created an atmosphere rich in oxygen with low concentrations of greenhouse gases.

It is believed that on time scales of billions of years, the combination of an increasingly intensive Sun, drifts and collisions of continents, and evolution of the biosphere have contributed to the variations of the climate. It has been shown by analysis of a variety of so-called *proxy data* (to be defined below) that there is climate variability on time scales from years up to hundreds of million years. There are different drivers of climate change operating on different time scales, but the climate state is not just a simple function of the drivers, but rather the result of a

---

[1] Department of Mathematics and Statistics, UiT - The Arctic University of Norway, N-9037 Tromsø, Norway

complex interplay between external forcing and internal processes in the climate system. There is, for instance, no simple linear relation between slowly increasing solar irradiance and the global temperature over the lifetime of the Earth.

To learn more about this complex interplay, we need reliable quantitative proxy-based estimates of climates of the past. This chapter deals with the most used proxies involved in such studies, and the methods employed to extract climate-related information from them. Since this book is about the impact of solar activity, we will focus on reconstruction of global-scale surface temperature of the Holocene, and in particular the last two millennia. Beyond this time span, we do not at present have sufficiently accurate reconstructions of temperature and solar forcing to make meaningful inferences about their connection.

2 An overview of proxies

In the study of past climates, proxies are preserved physical characteristics of the past that stand in for direct measurements to the climatic conditions. Examples of proxies include ice cores, tree rings, sub-fossil pollen, boreholes, corals, lake and ocean sediments, and cave stalagmites. The character of deposition of the proxy material has been influenced by the climatic conditions of the time in which they were laid down or grew. Chemical traces produced by climatic changes, such as quantities of particular isotopes, can be recovered from proxies. Some proxies, such as gas bubbles trapped in ice, enable traces of the ancient atmosphere to be recovered and measured directly to provide a history of fluctuations in the composition of the Earth's atmosphere. Systematic cross-verification between proxy indicators is necessary for accuracy in readings and record-keeping. In general, proxies must be calibrated against modern instrumental records to yield a quantitative reconstruction of past climate.

2.1 Ice cores

Ice cores are recovered by drilling through the Greenland and Antarctic ice sheets, glaciers in North American regions, islands of the North Atlantic and Arctic Oceans, and alpine, tropical and sub-tropical locations. Measuring oxygen isotope composition in water molecules allows estimation of past temperatures and snow accumulations. The heavier isotope (^{18}O) condenses more readily as temperature decreases and falls as precipitation, while the lighter isotope (^{16}O) can fall in even colder conditions. In addition to oxygen isotopes, water contains hydrogen isotopes, ^{1}H and ^{2}H, which are also used as temperature proxies. The best dated series are based on sub-annual sampling of ice cores and the counting of seasonal ice layers. Such series may have absolute dating errors as small as a few years in a millennium. Dating may be performed using for instance volcanic ash layers with assumed dates.

2.2 Sediment cores

Marine sediment cores are widely used for reconstructing past climate. One of the common approaches is to extract and study the marine microfossils that are preserved in the sediments. Carbonate deposits from foraminifera and coccolithopores are examples of abundant microfossils that are good indicators of past environmetal conditions found in deep-sea sediments. Diatoms are also of great importance for reconstructing past climate, they are unicellular, photosynthetic algae with a siliceous shell. The general assumption is that the down-core composition of diatomic microfossil assemblages is related to past environmental conditions at the core site. A number of statistical techniques are elaborated to convert assemblages to past estimates of hydrographic conditions, including sea-surface temperature at the study site. In lake sediment cores, remains of microorganisms, such as diatoms, foraminifera, microbiota, and pollen within sediment can indicate changes in past climate, since each species has a limited range of habitable conditions.

2.3 Tree rings

Dendroclimatology is the science of determining past climates from properties of the annual tree rings. Rings are wider when conditions favor growth, narrower when times are difficult. Other properties of the annual rings, such as maximum latewood density have been shown to be better proxies than simple ring width. Using tree rings, local climates can be reconstructed for hundreds to thousands of years. By combining multiple tree-ring studies, sometimes with other climate proxy records, one can make inferences about past regional and global climates.

2.4 Corals

Palaoclimate reconstructions from corals provide insights into the past variability of the tropical and sub-tropical oceans and atmosphere, prior to the instrumental period, at annual or seasonal resolutions, making them a key addition to terrestrial information. The corals used for paleoclimate reconstruction grow throughout the tropics in relatively shallow waters, often living for several centuries. Accurate annual age estimates are possible for most sites using a combination of annual variations in skeletal density and geochemical parameters. Paleoclimate reconstructions from corals generally rely on geochemical characteristics of the coral skeleton, such as temporal variations in trace elements or stable isotopes.

2.5 Speleothems

Speleothems are mineral deposits formed from groundwater within underground caverns. Stalagmites, stalactites, and other forms may be annually banded or contain compounds that can be radiometrically dated. Thickness of depositional layers or isotopic records can be used to determine past climate conditions.

2.6 Borehole measurements

Borehole data are direct measurements of temperature from boreholes drilled into the Earth's crust. Departures from the expected increase in temperature with depth can be interpreted in terms of changes in temperature at the surface in the past, which have slowly diffused downward, warming or cooling layers below the surface. Reconstructions show substantial sensitivity to assumptions that are needed to convert the temperature profiles to ground surface temperature changes, hence borehole data are most useful for climate reconstructions over the last five centuries.

2.7 Uncertainties of proxy-based paleoclimate reconstructions

There are multiple sources to uncertainties when proxy data are being used to reconstruct past climate conditions. During field work, the sampling procedure itself may disturb the record of interest. Marine sediment cores may for instance be compressed by the coring equipment during sampling, and the stratigraphy could be slightly disturbed. Furthermore, when the desirable proxy data material has been extracted and processed, the assumptions that convert the raw proxy data to the final climatic variable or condition are of great importance. *Transfer functions* are used to convert the climate-related variable to the desired quantity, and these functions describe the assumed relationship between the two quantities. The relationship can be tested against present-day conditions, and often it is found that additional factors must be taken into consideration. For example, let us look at a proxy for paleoglaciation; the oxygen isotope ratios in marine carbonate deposits. The carbonate deposits are extracted from marine or lake sediment cores, and are microfossils of foraminifera which generally produce their carbonate shell from the surrounding ocean water. It is assumed that the oxygen isotope ratios in the carbonate reflect the oxygen isotope ratio in the ocean at the time the shell was formed. The oxygen isotope ratio in the ocean is dependent on temperature, salinity and effects related to upwelling. The effect of the different components is difficult to isolate, and hence today such records are considered to reflect global paleoglaciation and not ocean temperature isolated. Meanwhile, studies have shown that certain types of foraminifera produce a shell which is not equivalent to the isotopic composition in the ocean, and that different species living in the same environment record different oxygen isotope values. Such differences between species are called vital effects, and these effects are corrected before reconstructions are created. Yet another source of uncertainty is imperfections of the age model. Proxy data which are annually layered is not a problem, but e.g., marine sediments are virtually impossibly to date with high accuracy. A linear sedimentation rate is generally assumed, but this is in lack of better knowledge about the sedimentation rate in the area. Particular features in the sediments may help in the dating process, such as volcanic ash layers and dust. Each volcanic eruption produces ash and larger particles with a specific chemical "fingerprint", so if the timing of the eruption is known, then the sediments where the ash is found can be dated accurately.

2.8 Simulating proxies in climate models

State-of-the-art climate models are not only used to predict future climate changes, but also to simulate climate in the past and present. Modelling the present climate is important for validation of the climate models, by comparison of model output with observations. The available proxy data material in paleoclimate studies indicate that climate has varied considerably also earlier in Earth's history, but the nature and mechanisms of such changes are in many cases poorly understood. Paleoclimate model simulations are therefore used to study past climate dynamics, but the validation of such simulations is difficult because direct measurements of the true climatic variables do not exist further back than a few centuries. We only have proxy-based reconstructions to rely on when studying longer time periods, where climatic variables, such as temperature are estimated from proxy data using transfer functions. As mentioned above, the transfer functions are based on assumptions and involve uncertainties. An alternative method for validating the paleoclimate simulations is called the "forward proxy modelling technique". This method involves simulating the actual proxy data in the model. The modelled proxy can be compared with true proxy data, and hence the problem with transfer functions is eliminated. A number of climate models currently have stable water isotope diagnostics implemented, which are useful both for validating experiments but also for testing the spatial relationship between changes in the heavy oxygen istotope ($\delta^{18}O$) and temperature. A model study by Sturm et al. (2010) shows that the altitude effect makes $\delta^{18}O$ decrease more with height than the temperature, and the same is true when moving from coastal to continental sites during winter (the continental effect).

3 Solar activity reconstructions from cosmogenic isotopes

Chapters 2.1 and 2.2 in this book describe reconstructions of solar activity based on sunspot observations combined with models that relate solar irradiance to such observations. Systematic sunspot observations, however, have been recorded only after Galileo made his first telescopic observations in AD 1610. Reconstructions for the more distant past has been made feasible by the observation that the sunspot number (SSN) is strongly correlated with the concentration of certain cosmogenic isotopes in tree rings and ice cores. We consider two solar activity reconstructions, Solanki et al. (2004), which is based on ^{14}C in mid-latitude tree rings, and Steinhilber et al. (2009), which is based on ^{10}Be, measured in polar ice cores. Both proxies measure intergalactic cosmic ray flux, which is correlated with solar activity. The reason for this correlation is that the solar wind is stronger during high solar activity, and this increases the strength of the interplanetary magnetic field. This field deflects galactic cosmic rays (GCR) and reduces the GCR-flux impinging on the Earth's atmosphere and thus reduces the production of the radio-isotopes ^{14}C and ^{10}Be. Solanki et al. (2004) reconstruct the sunspot number, while Steinhilber et al. (2009) reconstruct total solar irradiance (TSI), and are based on physical modelling of the whole chain of processes from activity on the Sun to the accumulation of the isotopes in the tree rings and ice. This modelling is quite different for the two proxies. For instance, ^{14}C production on mid-latitudes is strongly influenced by the almost horizontal Earth magnetic field, which has been increasing in strength throughout the Holocene. This influence can be corrected for, but more difficult is the complex biological interactions of Carbon, which make it problematic to relate the concentration in tree rings directly to the concentration in the atmosphere at a given time. ^{10}Be deposited in polar ice cores, on the other hand, is less influenced by the almost vertical magnetic field at high latitudes and is directly washed out of the atmosphere and deposited as snow at the actual site.

The two reconstructions span most of the Holocene and are shown in Figure 1a. They are quite similar on the time scales of the millennial oscillations, and many (but not all) of the century scale fluctuations are present in both curves, although sometimes with a phase shift of up to a century in one direction or the other. Power spectral analysis of time series like these has been used to detect cycles in the solar activity beyond the 11-yr solar cycle (the Schwabe cycle). Examples are a 70–90 year periodicity called the Gleissberg cycle and the 210-year de Vries cycle. The cycles are very weak, however, and very arcane spectral methods are needed to detect the spectral peaks. The reality of many of these cycles are still subject to some controversy.

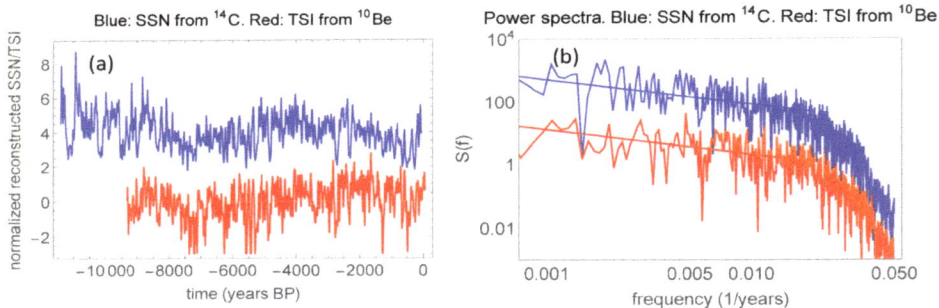

Fig. 1. (a) Blue curve: The reconstructed sunspot number from the isotope ^{14}C as presented in Solanki et al. (2004). Red curve: The reconstructed solar irradiance from the isotope ^{10}Be as presented in Steinhilber et al. (2009). (b) The power spectral densities of the time series in (a). The slopes of the fitted straight lines are $\beta = 0.8$.

The power spectra contain interesting information beyond the cycles, as shown in Figure 1a. Here, we have plotted on logarithmic axes a very simple estimate of the power spectral density called the periodogram for the two reconstructions. The rapid loss of power for frequencies $f > 1/50 \text{ yr}^{-1}$ (periods shorter than 50 yr) just reflects that the reconstructions (which are given with 10-year resolution) really are smooth on time scales shorter than 50 yr. For $f < 1/50 \text{ yr}^{-1}$, both spectral plots can be fitted by a straight line with negative slope of $\beta \approx 0.8$. Since both axes are logarithmic, this means that the power spectra on time scales slower than 50-years has the power-law form $S(f) \sim f^{-\beta}$. Time series with such spectra and $\beta > 0$ are said to be long-range dependent since they exhibit stronger variability on the long time scales as compared to random noise.

Interestingly, the Northern hemisphere temperature reconstructions discussed in the next section also exhibit power spectra of this type with $\beta \approx 0.8$. This should perhaps indicate that the long-range dependence seen in the temperature reflects the dependence of the solar driver. However, paleoclimatic model simulations exhibit the same long-range dependence even in model runs without a solar driver, as shown by Østvand et al. (2014), suggesting that the long-range dependence in solar activity and global climate are properties of the internal dynamics in the Sun and the Earth's climate system, respectively.

4 Hemispheric and global reconstructions

Raw proxy data must be transferred into climatological quantities like temperature, precipitation, ice cover, sea level and so on. Some of the methods are specific for each proxy, but also rely heavily on general statistical methods like regression analysis (see Chapter 3.9). If we want to understand the climate impact of external forcings like solar irradiance and volcanic aerosols, we cannot be satisfied with local reconstructions of the climate at the proxy sites. It is necessary to reconstruct a large-scale climate field by employing multivariate regression methods, so-called climate field reconstructions. Such methods have been applied both to filling spatial gaps in early instrumental climate datasets and to the problem of reconstructing past climate patterns. Independently, multiple temperature reconstructions for the past millenium have been produced, and they show the same evolution of the global or hemispheric mean surface temperature, namely a weakly cooling trend up to AD 1850, and an abrupt increase in temperature over the last century. The techniques for such reconstructions are diverse and rather arcane and disputes about their validity gave rise to the highly politicised "hockey-stick" controversy.

4.1 Temperature reconstructions

In Figure 2a, we show a graph of the $\delta^{18}O$ proxy for the Greenland temperature as recovered from the GRIP ice core. Prior to 11.7 kyr before present (BP), this graph shows a cold, but very unstable, climate ridden by sudden warming events where temperature in the region could rise more than 10° C in less than a decade followed by a slow recovery of the cold state over a few centuries. Altogether 25 such events took place over the last glacial period. The glaciation ended with a sudden transition into the warm Holocene seen to the right in the figure. The graph is smoothed by a 30-year filter to make it appear less noisy. The onsets of the warming events are even sharper than apparent in the figure, and demonstrate the remarkable time resolution that can be obtained from ice cores. One drawback with ice cores is that they are limited to those regions that are covered by ice sheets or glaciers today, and hence cannot alone be used for climate field reconstruction. However, the ice cores from Antarctica show that the warming events detected from GRIP were anti-correlated with Antarctic temperatures, and hence they did not represent massive changes in global temperature.

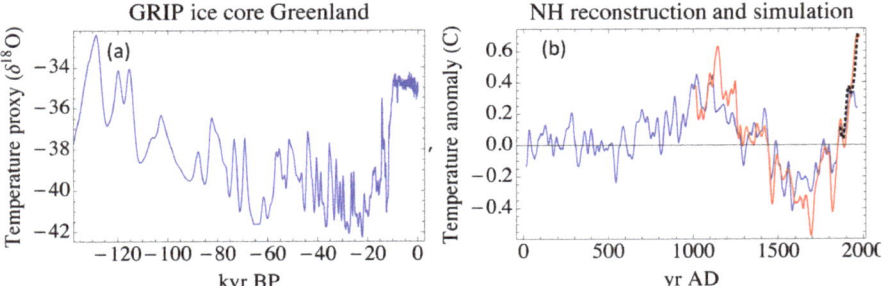

Fig. 2. (a) A temperature proxy from the GRIP ice core showing variations in Greenland temperatures during the last ice age and the Holocene. (b) Blue curve: the Moberg et al. (2005) NH temperature reconstruction. Red curve: the NH temperature derived from the ECHO-G Erik1 simulation driven by solar and volcanic forcing. Black-dotted curve: the instrumental global temperature record. The temperatures are represented as anomalies relative to the mean of the Moberg curve over the period AD 0–500. All records have been smoothed by a 30-year low-pass filter.

Up till very recently, reconstructions of climate fields and mean temperatures on global scale have been limited to the Northern hemisphere (NH). This is because the main landmasses and the majority of proxy archives are located in this hemisphere. The first NH reconstructions published in the late 1990s were obtained by techniques involving Principal Component Analysis (PCA; see Chapter 3.9), and were criticised for suppressing variance on long time scales. This was an issue because the methods could potentially suppress the temperature difference between the Medieval Warm Period (AD 800–1100) and Little Ice Age (AD 1550–1850). A two-millennium long reconstruction based on other techniques was published by Moberg et al. (2005), which was supposed to give a better representation of this difference. This reconstruction is shown as the blue curves in Figure 2b. A large number of millennium-long NH reconstructions based on different sets of proxy archives and statistical methods have been published, and despite some differences the large-scale features agree, rather well. Many millenium-long climate model simulations have also been performed under the Paleoclimate Modelling Intercomparison Project (PMIP3), and the results agree well with the reconstructions. An example is shown by the red curve in Figure 2b.

4.2 PAGES 2k and climate field reconstructions

The past 2000 years of climate change have been reconstructed in more detail than ever before by the PAGES 2k project. The 2k Network of the Past Global Changes (PAGES) project aims to generate a globally encompassing, high-resolution regional synthesis of climate variability for the last 2000 years. The results reveal interesting regional differences between the different continents, but also important common trends. The global average of the new reconstruction looks like the original "hockey stick." It is based on 511 climate archives from around the world. The two main results are a confirmation that current global surface temperatures are higher than at any time in the past 1400 years, and that the Medieval Warm Period and Little Ice Age were not globally synchronised events. The period from around AD 830 to 1100 generally encompassed a sustained warm interval the Northern hemisphere, but in South America and Australasia, it occurred from around AD 1160 to 1370. Thus the maximum observed around AD 1000 is absent in the 2k global reconstruction. Instead the planet has been subject to a long-term cooling trend over the two millennia, terminating with present-day global warming over the last 150 years.

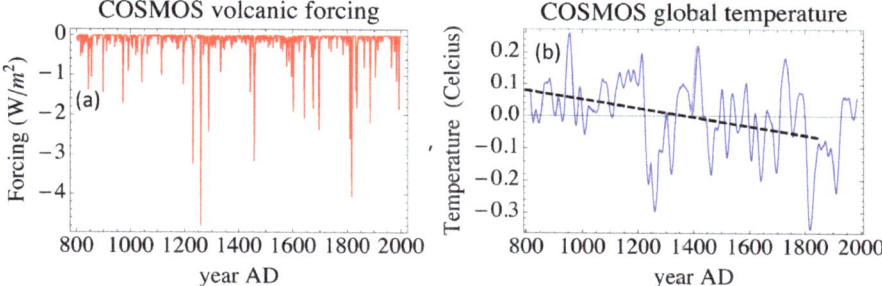

Fig. 3. (a) The reconstructed volcanic forcing used in a COSMCS model run. (b) The global temperature in a COSMOS simulation with the only varying forcing being volcanic aerosols. The result is 30-year low-pass filtered. The black dotted line is the linear trend for the period from the start of the MWP to the end of the LIA (AD 800–1850).

4.3 What caused the cooling trend?

The reconstructions themselves do not attribute the cooling trend to solar irradiance changes or any other forcing. There are at least three candidates, the Sun, volcanoes, and orbital forcing. Long climate simulations with reconstructed forcing are required to answer this question. The orbital forcing has experienced a weak steady decline over the Holocene and is known with high accuracy. The forcing due to volcanic aerosols is estimated with some accuracy from sulphur trapped in ice cores, while the solar forcing reconstructions are still very uncertain. In Figure 3, we show a reconstruction of the volcanic forcing over the last two millennia and the global temperature derived from a climate model. The response to this forcing and the trend over the period AD 800–1850 is shown in Figure 3b. This trend is compatible with the trend in the 2k global reconstruction, and shows that clustering in volcanic activity over last 1200 years may be sufficient to explain the difference between the medieval warmth and the cold seventeenth century observed in the most recent reconstructions. More accurate reconstructions of both solar and volcanic forcing are required before a reliable quantitative assessment of the Sun's role in this cooling trend can be given.

Further reading

Bender, M. L.: 2013, *Paleoclimate, A Princeton Primer in Climate*, Princeton, Princeton University Press.

Lowe, J. J., and Walker, M. J. C.: 1997, *Reconstructing Quaternary Environments*, New York Pearson Prentice Hall.

Mann, M. E.: 2012, *The Hockey Stick and the Climate Wars - Dispatches from the Front Line*, New York, Columbia University Press.

Sturm, C., Zhang, Q., and Noone, D.: 2010, *An introduction to stable water isotopes in climate models: benefits of forward proxy modelling for paleoclimatology*, Climate of the Past, 6 (2010), pp. 115–129.

CHAPTER 3.2

GROUND-BASED OBSERVATIONS

Franz-Josef Lübken[1] and Keri A. Nicoll[2]

1 Introduction

Surface and airborne observations of meteorological parameters have been made for many centuries, producing large datasets which span significant periods of time. Although satellites provide more comprehensive global coverage than ground-based observations, their short period of operation means that detecting the effect of long term changes in solar variability on climate is difficult, not least due to the additional complication of increased anthropogenic activity during the period of satellite measurements, which is likely to mask solar signals. Ground-based observations therefore provide a vital existing data source of the investigation of the effects of long term solar changes on climate, however, as with any long-term data source, caveats exist. For example, ground-based instruments are normally upgraded from time to time to take advantage of contemporary technological developments. This might influence the statistics, e.g., when more sensitive instruments are applied. Generally speaking, trends are not monotonous in time because of non-uniform forcing and/or non-linear interactions. Physical and chemical parameters in the upper atmosphere are typically impacted by the 11-year variation of solar radiation. This variation is small for the total solar radiation ($\sim$0.1%) but is large (up to a factor of 100) when UV and EUV radiation is considered (see Chapter 4.1). Therefore, ground-based techniques should be in place for several decades in order to distinguish long term trends from variations due to the solar cycle.

Ground-based measurements provide information at one location only. More information may be needed from several stations or from satellites to get a complete picture. Furthermore, trends of parameters measured from the ground may be influenced by several processes, sometimes with opposite effects. In this chapter, we concentrate on two altitude regions which can be measured using ground-based techniques: the troposphere–stratosphere region (0–50 km), and the mesosphere–lower thermosphere region (MLT, 50–150 km). We include sensors

---

[1] Leibniz-Institute of Atmospheric Physics, 18225 Külungsborn, Germany
[2] Department of Meteorology, University of Reading, UK

measuring atmospheric parameters directly (in-situ) at the ground or from balloons, and techniques which are based on remote sensing from the ground. In-situ measurements are also available in the MLT from sounding rockets, but they only occur infrequently and are not considered further here. A summary of techniques is presented in Tables 1 and 2.

Table 1. Summary of meterorological parameters and the most common measurement techniques used to observe them routinely in the troposphere and stratosphere. Only direct methods (not proxy methods) of observation are included.

Parameter	Sensor/Technique	Remarks
Temperature	- liquid-in-glass thermometer - thermograph - electrical resistance thermometer - thermocouple	Mounted at a height between 1.2-2 m above ground level, and sheltered from direct exposure to sun. Temperature above the surface measured by electrical thermometers on meteorological radiosondes.
Sea Surface Temperature	- sample of sea surface water with bucket - reading temperature of the condenser intake water - exposing electrical thermometer to sea water temperature - infrared radiometer mounted on ship to look down on sea surface	No standard device for measuring sea surface temperature due to huge diversity in ship size/speed, cost, ease of operation and maintenance.
Pressure (sea level)	- electronic barometer - mercury barometer - aneroid barometer	Should be placed in an environment where external effects will not lead to measurement errors. Pressure above the surface measured on meteorological radiosondes.
Humidity	- pyschrometer (Assman, whirling or screen) - hair hygrometer - chilled mirror dew point cell - electrical resistive and capacitive hygrometer	Screen pyschrometer and hair and electrical hygrometers should be mounted inside thermometer screen at height of 1.2-2 m above the surface. Relative humidity above the surface measured on meteorological radiosondes.
Wind - speed - direction	- cup/propeller anemometer - wind vane	Standard exposure of wind instruments is over level terrain, 10 m above the surface. Pitot-tubes, hot wire and sonic anemometers, as well as radar and lidar can all be used to measure wind properties, but are not yet that common in routine meteorological networks. Wind above the surface is also measured by GPS from meteorological radiosondes

Solar radiation - direct - global - diffuse - net global - longwave - total	- pyrheliometer - pyranometer - pyranometer - net pyranometer - pyrgeometer - pyradiometer	Radiation sensors must be sited in locations with freedom from obstructions to the solar beam at all times and seasons of the year, and correct alignment of radiation sensors, as well as regular calibration is essential.
Cloud	- synoptic (cloud amount, type) - cloud base recorer/laser ceilometer (cloud amount, height)	Synoptic observations are visual observations made by an observer at the surface. Cloud radars can also be used to measure cloud properties above the surface but these are primarily research instruments and not yet used in widespread global operational meteorology.
Rainfall	- manual rain gauge - electronic rain gauge (tipping bucket, weighing or float) - X, C or S band weather radar (GHz)	Gauges are typically mounted between ground and 1.5 m above the surface.
Ozone - total ozone - vertical profile	- Dobson ozone spectrophotomer - Brewer spectrophotometer - M-124 filter ozonemeter - ozonesonde (from balloon) - lidar - microwave radiometer	Dobson and M-124 instruments stored indoors and transported outside to take a measurement. Brewer instrument permanently mounted outdoors. Lidar technique only operates at night and when no cloud cover is present.
Electric field	- radioactive probe - electric field mill	Typically mounted between 1–3 m above the ground. Measurements must be corrected for distortion of the electric field by the mounting pole and other surrounding metal objects.
Lightning	- thunder days (local lightning presence) - VLF direction finder (lightning location and strike rate) - VLF time of arrival receivers (lightning location and strike rate)	Ground-based lightning receivers are typically deployed in networks consisting of many sensors at a variety of locations to ensure the best coverage.

Table 2. Summary of ground-based remote sensing techniques for the mesosphere/lower thermosphere.

	Technique	Parameters	Approx. height coverage
Lidars	Rayleigh	Neutral density, temperatures aerosols, winds	Ground to upper mesosphere
	Resonance	Na, K, Fe density, temperatures aerosols, winds	80–120 km
Radars	MF (0.3–3 MHz)	Winds, electron densities	60–100 km
	Meteor	Winds, meteor heights	80–110 km
	VHF (30–300 MHz)	Aerosols, winds	80–90 km
	Incoherent (UHF, >300 MHz)	Ionospheric (electrons, ions)	80 to several hundred km
Ionosondes HF (1–40 MHz)		Electron densities	100–400 km

2 Ground-based observations in the troposphere and stratosphere

Here, we describe observations of the troposphere and stratosphere. We focus on meteorological parameters in which solar influences have already been detected: temperature, sea surface temperature, rainfall, ozone, cloud and atmospheric electricity. Table 1 details commonly measured meteorological parameters in the troposphere as well as their measurement techniques. More information can be found, e.g., from WMO (1988).

2.1 Temperature

Air temperature is perhaps the most commonly measured meteorological parameter, typically measured at a height of 1.5 m using a thermometer, housed inside a ventilated screen to protect the thermometer from overheating from the effects of direct solar radiation. Traditionally thermometers used mercury or alcohol in glass, but nowadays electronic thermometers using resistive methods are common. The longest instrumental time series of daily temperature measurements is the Central England Temperature series, dating back to 1772 (see Figure 1). Although such time series are invaluable for detecting long term trends in data Bremer and Berger (2002), it is important to consider that changes in thermometer instrumentation, meteorological site characteristics, and the state of the protective screen can all contribute to unwanted biases in temperature data. This is particularly true for upper air measurements of temperature made from meteorological radiosonde balloons, which have seen significant changes in the development and accuracy of the temperature sensors since their inception in the 1940s. This has resulted in inhomogeneities within the long-term time series, which must be

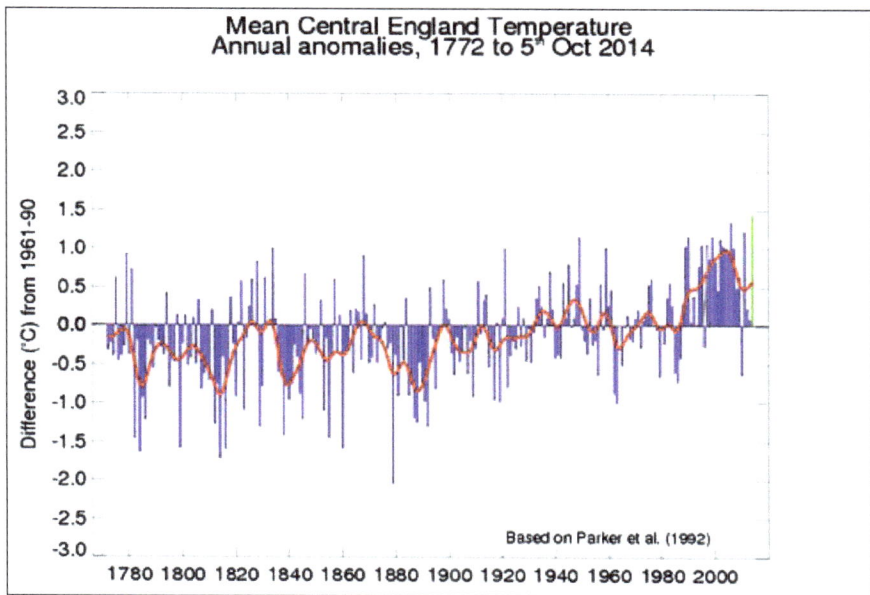

Fig. 1. Annual temperature anomalies from the longest time series of temperature measurements in the world – the Central England Temperature series, which is representative of a roughly triangular area of the United Kingdom enclosed by Lancashire, London and Bristol. Daily values date back to 1772, and monthly values to 1659. Anomalies shown are relative to the 1961–1990 average and the red line is a 21-point binomial filter, which is roughly equivalent to a 10-year running mean (Source: The Hadley Centre, United Kingdom Meteorological Office).

removed (often using re-analysis data, see Chapter 3.4) before any long-term changes can be analysed. Accurate documenting of changes in sensor (whether at the surface or on balloons) and site characteristics (e.g., exposure changes due to growing trees or erection of buildings) is therefore absolutely essential to maintain an accurate long-term temperature record.

2.2 Sea surface temperature

In-situ measurements of sea surface temperature (SST) take place within the near surface layer typically between a few centimeters and 20 meters below the sea surface. A variety of techniques have been used to measure SSTs, including from ships, and from moored and drifting buoys. Although in-situ measurements of SST date back to 1662, from the International Comprehensive Ocean-Atmosphere Data Set (ICOADS), inconsistencies in the data exist back to the 1900s due to the number of different techniques used and the depth at which measurements were taken, requiring the application of detailed correction algorithms to SST data.

2.3 Rainfall

Many studies, dating back to the late 1800s, have addressed the potential relationship between solar activity and rainfall. Despite the long history, there is still significant debate in the literature about whether such a relationship exists. This is in part related to sampling issues in that rainfall has been historically measured using rain gauges (simple devices which collect precipitation and store it until it is emptied manually or automatically), which are point measurements. Rainfall varies by a large amount spatially, therefore rain gauges do not always provide reliable regional means, as accurate measurements require a dense network of stations. The worldwide rain-gauge network is also not distributed evenly over the globe (see Figure 2), with no measurements over oceans, and very few in developing countries, such as Africa. Furthermore, rain gauges systematically underestimate precipitation, by up to approximately 30%. These limitations, coupled with frequent re-design of instrumentation over the years, mean that rain gauge measurements are extremely difficult to interpret in terms of long term climate change. Radar (RAdio Detection And Ranging) techniques can provide a much more spatially representative measurement of rainfall than rain gauges, but they have significant uncertainties associated with the conversion from reflectivity to rainfall rate. Although the number of weather radars used in operational systems is increasing all the time, there is still far from global coverage. Many of the early studies into links between solar activity and rainfall used lake level and river flow changes as a proxy for rainfall, which provide much larger coverage area than a single rain gauge, but are not without their complications. Lakes lose water in many ways, including evaporation and stream run-off, they can become dammed or create new channels, meaning that every lake must be treated individually and immense care must be given to the mechanisms which control its water levels.

2.4 Ozone

Ozone measurements are typically made using the Dobson spectrophotometers, which have provided data back to the late 1950s, and can be used to measure both total column ozone and vertical profiles of ozone. The Dobson spectrophotometer measures the ratios of UV light at two pairs of wavelengths. Within each pair, one wavelength absorbs ozone strongly, the other weakly, so that the ozone absorption can be determined (see WMO, 1988, for more detail). The geographic coverage of stations has improved recently, but the historical records are overwhelmingly biased to the northern mid-latitudes. Understanding how the long term calibrations of the instruments change with time is essential when dealing with long term changes of a few percent in a decade. Therefore, the calibration of Dobson instruments has been maintained through a program of intercomparisons traceable to an international standard instrument. Improvements in the quality of the Dobson data record have been made over the past few decades, partly through reanalysis of data and using satellite data, as well as improving the quality of the measurements as they are made.

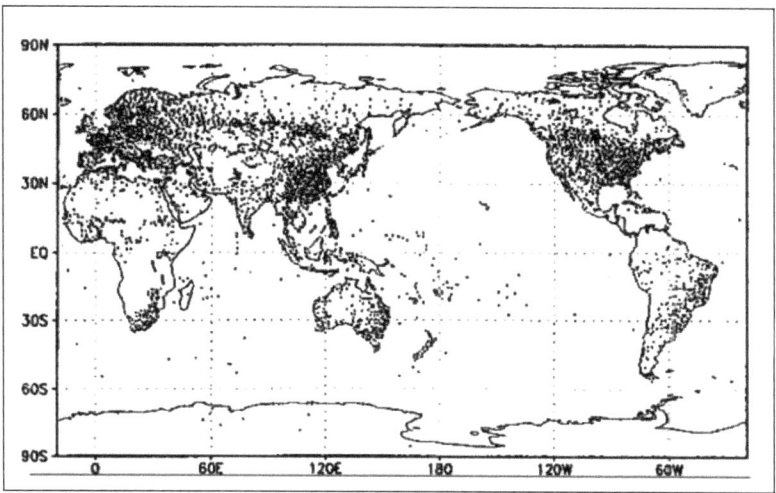

Fig. 2. Typical global distribution of rain gauge reports (for June 1998). Reproduced from Janowiak and Xie (1999). © American Meteorological Society. Used with permission.

Additional vertical profiles of ozone up to altitudes of ~30 km are made using 'ozonesondes', which employ an electrochemical sensor alongside a meteorological radiosonde (see Figure 3). Launches are typically made once per week at a number of locations around the world. The main instrumental problems arise from the quality control in manufacture and preflight preparation of the electrochemical sensors, as a different sensor must be prepared for each flight. Inter-comparisons demonstrate that the quality of the ozonesonde data is good in the stratosphere, but not so reliable in the troposphere.

2.5 Clouds

Studies looking for connections between solar activity and clouds (see Chapters 4.7–4.8) have used a number of direct and indirect proxies for cloud measurement, both from satellites (see Chapter 3.3) and ground-based observations. Table 3 describes some of the advantages and disadvantages of various cloud detection methods. Direct measurements of clouds include synoptic observations from human observers, as well as active detection methods, including cloud base recorders, lidars (Laser Induced Detection And Ranging) and radars. Synoptic observations involve the subjective classification of cloud cover (in oktas, or eighths of the sky) and cloud type by a human observer. Although observers are often consistent in their cloud estimates, they can show a significant personal bias which makes it difficult to extract meaningful long term information from a single measurement location. Using data from multiple sites, or a single site where many observers contribute to the observations may negate such effects. An alternative method of cloud detection employs an automated cloud base recorder (ceilometer), which

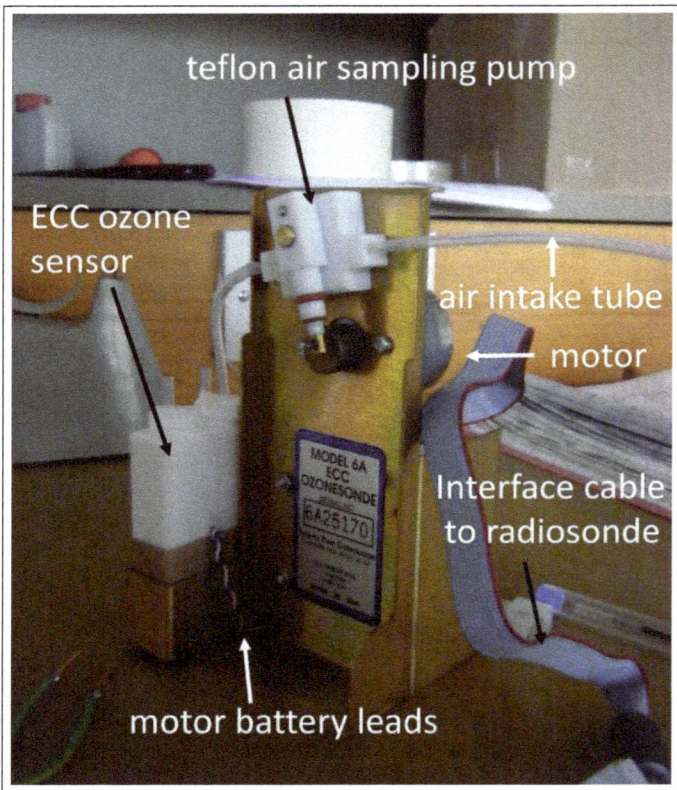

Fig. 3. Photograph of an ozone sonde, whereby ambient air is drawn (by a pump) into an Electrochemical Concentration Cell (ECC) containing potassium iodide solution. A reaction between ozone and the potassium iodide generates a current in the cell, from which the concentration of ozone can be deduced. These devices are flown on free balloons alongside meteorological radiosondes and have been in operation since the 1960s.

uses an upward pointing laser beam to transmit pulses of near-infrared radiation towards clouds, and receives the reflected signal. Cloud base recorder observations are typically available from the 1950s onwards, however, developments in instrumentation mean that the original recorders (which used searchlight beams and rotating shutters), have now been replaced with more modern laser ceilometers. Therefore, as with all long term data series, care must be taken when investigating long term changes. Unlike satellites, which provide information about cloud tops, and are subject to uncertainties regarding retrieval algorithms, cloud base recorders provide high vertical resolution ($\sim$10 m) detailed information on cloud base properties.

Due to the relatively short period of satellite and direct ground based observations, a number of studies have used proxies for cloud cover instead of direct

Table 3. Summary of advantages and disadvantages for ground-based cloud detection methods previously used for identifying solar influences on climate.

	Cloud observation type	Advantages	Disadvantages
Direct	Synoptic observations	Long time series of measurements (back to 1880s), available worldwide	Subject to observer biases
Direct	Cloud base recorder	High resolution measurements of cloud base height	Only sensitive to lowest cloud layer (higher layers obscured)
Proxy	Diffuse fraction	Measurements available worldwide	Determination of cloud type is difficult
Proxy	Sunshine recorder	Long time series of measurements (back to 1890s), available worldwide	Quantitative cloud fraction and cloud type not available

measurements. These include sunshine records dating back to the 1800s, and solar radiation measurements. Perhaps the longest time series of indirect cloud measurements is from sunshine records, using the Campbell-Stokes sunshine recorder. This provides a measurement of the duration of bright sunlight by making a burn mark on a treated burn card, by focusing solar radiation through a spherical glass lens. A study comparing synoptic cloud observations with sunshine recorder data in Ireland concluded that sunshine recorders provide a more reliable measurement of long term trends in cloud cover than synoptic observations due to observer biases, however information about cloud type and fraction are unlikely to be obtained through this method. An alternative proxy for cloud cover uses surface measurements of solar radiation by calculating the Diffuse Fraction (DF, Duchon and O'Malley, 1999), the ratio of diffuse solar radiation (that scattered by clouds and aerosol) to total global solar radiation. DF measurements are sensitive to cloud amount and cloud thickness, providing a proxy for cloud cover, as well as discriminating between convective and stratiform cloud by considering the variability in DF. Since surface measurements of solar radiation are available at a wide variety of meteorological stations worldwide, many of which have been operating for long periods of time, the DF method increases the number of surface observations of cloud cover available to investigate solar variability effects on clouds.

2.6 Atmospheric electricity

Measurements of atmospheric electrical parameters (including the atmospheric electric field, or potential gradient, PG; and vertical conduction current, J_c) exist at a handful of locations around the world, but the small number of observations make it difficult to progress knowledge in this area. In addition, atmospheric electrical parameters are particularly sensitive to local meteorological conditions, as well as aerosol concentrations, therefore any solar signals (see Chapters 4.7–4.8) can only be detected on 'fair weather days' exempt from these influences.

A long time series of hourly PG and J_c measurements exists from the UK Met office (dating from early 20th century to 1984), but higher temporal resolution measurements (of order seconds and minutes) and more measurement sites are required to better understand the mechanisms discussed in Chapters 4.7–4.8.

Assessing the role that solar variability plays in respect to global thunderstorms and lightning is extremely difficult, and mostly due to our inability to accurately measure global lightning activity. Most operational methods of lightning detection use the Very Low Frequency (VLF) radio pulses generated by lightning, using a network of radio receivers at a variety of locations around the world to determine the location of the lightning strike (e.g., World Wide Lightning Location Network (WWLLN); UK Met Office ATDNET; Earth Networks Total Lightning Network (ENTLN); North American Lightning Detection Network (NALDN)). Such systems have been in operation since the 1980s but constant upgrading of the networks (i.e., by adding more sensors to increase coverage and improving processing algorithms) has meant that more recent measurements are much more sensitive to lightning strikes, therefore global lightning activity over the past few decades has appeared to increase, but this is mostly due to improvements in instrumentation and processing techniques. An example of this for the WWLLN is shown in Figure 4. For the purposes of studying long term change in lightning data, it is therefore necessary to only take into account data which were obtained during periods when networks are stable, which is typically only for periods of a few years. At the moment, it is therefore only really possible to utilise data from lightning detection networks to study short term solar effects on lightning.

3 Ground-based observations in the mesosphere – lower thermosphere (MLT)

For the purpose of measuring the mesosphere and lower thermosphere region, some relatively simple remote sensing observations have been available for many years. More sophisticated measurements of temperatures, winds, and ice particles in the MLT by lidars and radars have now been available for a few decades (see e.g., Lübken et al., 2013).

Ground-based remote sensing provides important information about the thermal, dynamical, and compositorial structure of the MLT. In-situ techniques normally allow the most accurate measurements but suffer from e.g., sporadic sampling. On the other hand, remote sensing techniques can be applied nearly continuously, although with some limitations. For example, optical methods require clear sky conditions. Remote sensing relies on the analysis of electromagnetic radiation which has interacted with the atmosphere. A technique is called 'passive' if the source of the radiation is external (Sun, stars, galactic radiation) or if the radiation is created within the atmosphere for example by chemo-luminescent reactions (e.g., hydroxyl, OH*). 'Active' remote sensing relies on sources of electromagnetic radiation which are somehow controlled by the experimenter, e.g., lidars, radars, and ionosondes.

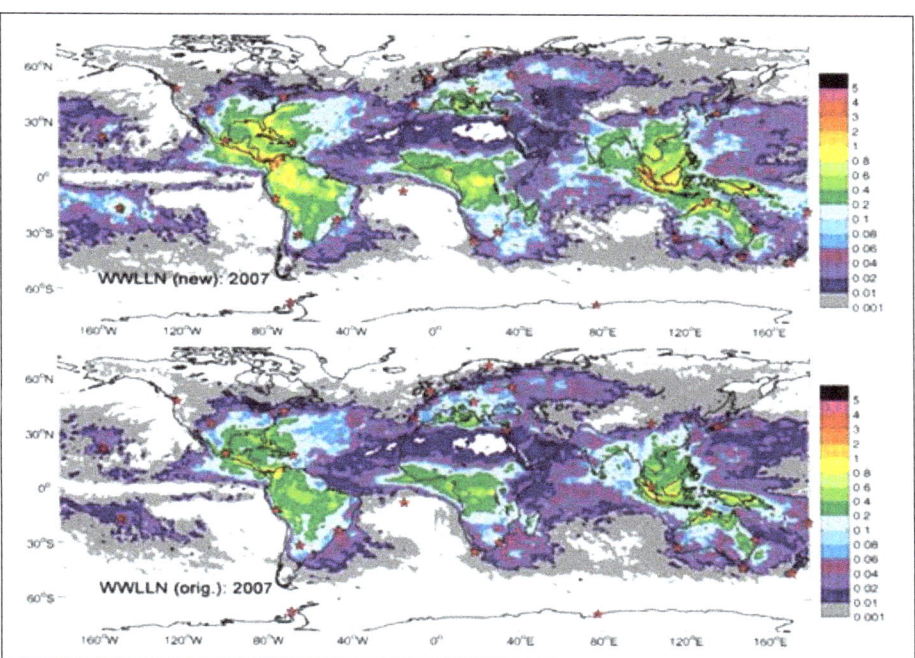

Fig. 4. Example of how the sensitivity of lightning location networks have changed during recent years. The lower plot shows global lightning strikes per km and per year for 2007 from the World Wide Lightning Location Network (WWLLN) using the original processing algorithm, and the upper plot the same for a new algorithm which increased the number of lightning strikes detected by 63%. The location of the WWLLN stations are shown as stars. Reprinted with permission from Rodger et al. (2009). Copyright 2009, AIP Publishing LLC.

3.1 Neutral air temperature

Here we describe the main ground-based remote techniques to measure temperatures in the MLT, namely lidars and hydroxyl spectrometry (OH*). Furthermore, indirect information about temperatures is deduced from the reflection height of radio waves. It should be kept in mind that long term temperature trends in the mesosphere are due to a combination of effects with different time signatures. For example, (i) the non-linear increase in the concentration of CO_2 and (ii) stratospheric ozone variation on time scales of years and decades due to anthropogenic destruction and recovery, both of which affect temperatures in the mesosphere (Lübken et al., 2013).

Lidars emit laser pulses into the atmosphere and detect the backscattered signal as a function of time and altitude (see Figure 5). Typically 30 pulses/second with a pulse duration of 11 ns and length of 3 m are emitted with wavelength from the UV to the infrared. Standard lidars apply a Nd:YAG laser and operate

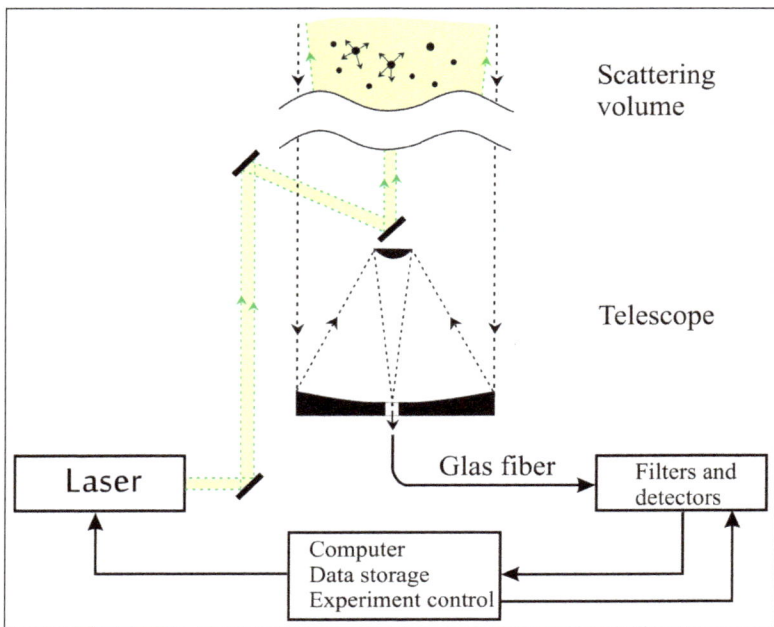

Fig. 5. Principle of lidar technique: a laser pulse is emitted into the atmosphere where it is scattered by molecules and aerosols. The backscattered photons are collected by a telescope and measured by detectors. The scattering altitude is derived from the time between emission and reception and the speed of light. A small field of view of the telescope and various filters in the detection system are required for making measurements during daylight when the solar background signal is large.

at 532 nm (green), but various other types are also in operation, e.g., alexendrite lasers. Typically one trillion photons are emitted per pulse but only a few of them are received after scattering in the uppermost altitudes at 90–100 km. The power within a pulse is typically 40 MW. Atmospheric parameters, such as temperatures and aerosol content are derived from the backscattered signal. Lidars deduce a density profile up to approximately 90–100 km from scattering of laser light on molecules and atoms which is called Rayleigh scattering. Relative densities are converted to temperatures by applying the equation of hydrostatic equilibrium. Temperatures are also obtained from the Doppler broadening of resonance scattering of laser light from metal atoms (K, Na, Fe) being present at roughly 80–130 km due to the ablation of meteors. The longest record of Rayleigh lidar temperatures is available at the Observatory of Haute-Provence (since 1978), which is frequently used for trend analysis (Keckhut et al., 2001). Lidars with similar capabilities but less history are available at several stations, some of which are listed in Table 4. Transportable lidars have been developed for applications at remote locations and on aircrafts (see Table 4). Lidars measure temperatures

Table 4. Locations of selected Lidar observation sites. Transportable lidars are shown with *.

Name	Latitude
Ny-Ålesund	79°N
Spitsbergen*	78°N
ALOMAR	69°N
Sondrestrom	67°N
Poker Flat	65°N
Kühlungsborn	54°N
Haute-Provence	44°N
London (Ontario)	43°N
Fort Collins	41°N
Starfire Optical Range*	35°N
Tenerife*	28°N
Réunion	22°S
Davis*	68°S
Davis and Syowa	69°S
South Pole*	90°S

with high temporal and altitudinal coverage of typically 15–30 minutes and 200–300 m, respectively, which also allows analysis of short term variations caused by, e.g., gravity waves.

Atmospheric temperatures can also be derived from ground-based detection and spectrometric analysis of infrared radiation received from approximately 85 km. These emissions are due to the chemoluminscent reaction: $O_3 + H \rightarrow O_2 + OH^*$, where the de-excitation of OH^* leads to an entire spectrum of vibrational/rotational emission lines. This spectrum is analysed by ground-based spectrometers to deduce temperatures. This 'passive' technique does not provide altitude information, but it is known that OH^* is formed in a limited height range around 85 km only. Several decades of temperature records are available at several stations (see e.g., Offermann et al., 2010).

A third technique providing information about temperature changes in the mesosphere comes from the analysis of radio wave reflection in the lower ionosphere at a pressure of ∼0.006 hPa (∼82 km) (for more information see e.g., Bremer and Berger, 2002, and references therein). This 'phase height technique' has been applied since the late 1950s. It provides the longest, quasi-continuous data record of (indirect) temperature information from the mesosphere. In addition to neutral temperature, radars also provide electron and ion temperatures in the thermosphere analysing the spectral characteristics of incoherent scattering.

3.2 Winds

The main ground-based remote sensing technique for obtaining winds in the MLT uses radars. They deduce winds from the Doppler shift of the scattered signal, basically permanently and independent of weather conditions. Radars operate similarly to lidars but at different wavelengths (kHz to many MHz) and with antennas for transmitting and receiving radio waves being scattered on free electrons in the ionosphere. While here we focus on the MLT, it should be noted that radars are frequently applied to observe winds and other parameters in the troposphere, but these have only relatively recently been deployed for operational meteorological purposes. Scattering occurs on individual electrons ('incoherent scattering') or on spatial structures in the electron gas with scales on the order of $\lambda/2$ (λ = radar wavelength). For radars in the MF, HF, and VHF frequency ranges (see Table 2), this corresponds to spatial scales from 500 m to 0.5 m. Several processes form spatial structures of this size in the plasma, including layering due to absorption of sunlight, meteor trails, and turbulence. The altitude range covered by radars in the MLT region is typically 60–100 km for MF, 80–110 km for HF, 80–90 km for VHF, and 80 km to several hundred kilometers for incoherent scatter radars. Sophisticated techniques have been developed to further improve the accuracy and spatial/temporal coverage of radars, and to derive further atmospheric parameters, for example turbulence or electron densities. Radars for atmospheric applications are available at several stations worldwide, from Arctic to Antarctic latitudes.

Recently, the lidar technique has been further developed to obtain winds in the stratosphere and mesosphere from the Doppler shift of the scattered signal (Baumgarten, 2010). It is very challenging to measure this frequency shift, particularly during daylight conditions, but this is the only method to measure winds permanently between the lower stratosphere and the ionosphere.

3.3 Ice clouds in the summer mesopause region

The summer mesopause at middle and high latitudes is the coldest place on Earth with temperatures as low as 120 K. Although the abundance of water molecules is small at these altitudes, they may form ice particles under these special circumstances. These ice particles can be seen by the naked eye from the ground and are known as 'noctilucent clouds' (NLC). The first sightings in the late 19th century were able to determine their altitude (typically 83 km) using triangulation methods. Since NLC are sensitive to background conditions, such as temperatures and water vapor changes, and they constitute the longest data record of ground-based observations in the MLT, they are attractive for long term studies. Modern lidars detect NLC particles at any location, even at polar latitudes where visual sightings are not possible because of distortion from sunlight. The first lidar detection of NLC was performed at ALOMAR in 1989 and several thousands of hours of data with detailed information about NLC are now available (Fiedler et al., 2009). NLC have also been observed from satellites since 1979. Surprisingly, they still appear at the same altitude as 120 years ago which imposes a major constraint on model studies of long term changes (Figure 6). It is not clear at the moment, how these observational facts can be brought into agreement with expectations deduced from increase of CO_2, CH_4 and H_2O.

Radars detect strong backscattered signals from the summer mesopause region at middle and polar latitudes, which are called 'polar mesosphere summer echoes'

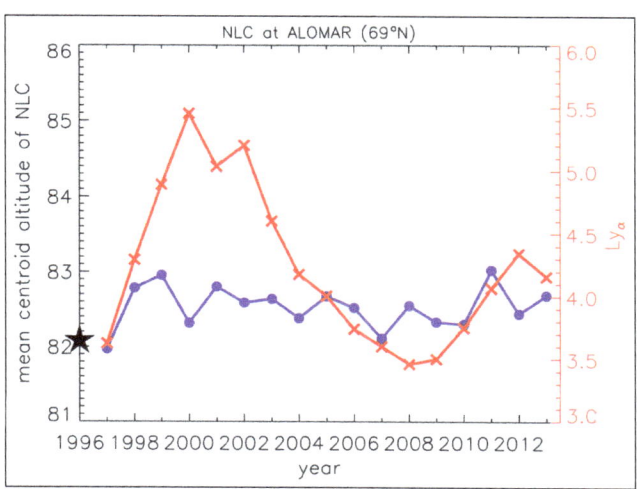

Fig. 6. Mean centroid altitude of noctilucent clouds (NLC) from the Lidar at ALOMAR (69°N). This Figure is an update of Fiedler et al. (2009). Only strong NLC are shown, more precisely those with volume backscatter coefficients $\beta_{max} > 13 \cdot 10^{-10}$ m^{-1} sr^{-1}. Red line: Lyman-alpha flux (in 10^{11} photons s^{-1} cm^{-2}) which is a measure of solar activity. The star on the left axis marks historical measurements from optical triangulations at the end of the 19th century.

(PMSE). These echoes rely on the impact of charged ice particles on the diffusion of free electrons in a turbulent background field. PMSE were first observed in 1979 and some long-term effects are detected. The interpretation is complicated, however, since various parameters control the existence of PMSE, including the production of free electrons by solar radiation, temperatures, and turbulence.

3.4 Ionospheric parameters

Long-term measurements of ionospheric parameters are available from ionosondes, which transmit a HF radio signal scanning in frequency from typically 1–40 MHz. The reflection height of these waves depends on frequency and occurs in the E- and F-region at typically 100–400 km. For the ordinary wave, the reflection height is determined by the plasma frequency which is proportional to electron density. Ionosondes measure the reflection height as a function of frequency and derive characteristic quantities, such as the frequencies where the reflection in the E, F1, and F2 layers occurs (foE, foF1, foF2), and the altitude of the F-region peak in electron concentration (hmF2). Electron densities in the lower thermosphere depend on various parameters, e.g., solar and geomagnetic activity, and composition. Some ionosondes have been in operation for decades and allow long-term studies.

CHAPTER 3.3

SATELLITE OBSERVATIONS

Donal Murtagh[1], Annika Seppälä[2] and Erkki Kyrölä[2]

1 Introduction

Satellite data have become an essential part of all atmospheric research since the first weather satellites were sent into space in the 1960s. Data contribute continuously to numerical weather predictions and are consistently processed in re-analysis datasets (see Chapter 3.4). In addition, the horizontal and vertical distributions of many individual atmospheric gases as well as observations of the oceans, land and cryosphere are available from a number of different instruments on board an array of satellites. For the latest information on Earth Observation space missions, see for example http://database.eohandbook.com and https://eoportal.org for databases of missions and instruments, http://www.esa.int for the European Space Agency (ESA) missions and http://www.nasa.gov for the National Aeronautics and Space Administration (NASA) missions.

For the purpose of detecting solar influence on the atmosphere, one of the key elements is ozone (see e.g., Chapters 4.1 and 4.6). Due to ozone's importance, we will focus in this chapter on the different ways atmospheric ozone has been observed from space. Note that many of the satellites observing ozone also observe other atmospheric parameters using the basic observation techniques we present in the following. At the end, we will also give examples of some of the many other variables observed by satellite instruments.

For some of the satellite data products, more than 4 solar cycles in time are covered with varying quality, while for others, only limited time series are available. In this chapter, we have chosen to highlight the instruments and missions that have yielded time series of at least 5 years, but note that more information is available, particularly from the space agencies (see weblinks given above). Some of the shorter missions are also listed towards the end of the chapter (see in Figure 3).

---

[1] Chalmers University of Technology, Göteborg, Sweden
[2] Earth Observation, Finnish Meteorological Institute, Helsinki, Finland

2 Ozone, O_3

2.1 Ozone column

Ozone column refers to the total amount of ozone in a column from the surface to the top of the atmosphere. Sometimes the column can also be defined as a more limited vertical extent, such as the "stratospheric ozone column". Ozone column data are usually derived using the so-called backscatter UV method (see Figure 1) where sunlight scattered from atmospheric molecules is measured from the nadir direction. Ozone is retrieved from pairs of wavelengths where at one wavelength ozone absorbs strongly and at other wavelength absorption is weak. This methodology has been used since the 1970s with the early BUV (Backscatter UV Spectrometer) instrument on Nimbus-4, SBUV/TOMS (Solar Backscatter Ultraviolet Radiometer/Total Ozone Mapping Spectrometer) on NIMBUS-7 and the SBUV/2 series of instruments on NOAA satellites. The same measurement method has been used by the GOME instrument (Global Ozone Monitoring Experiment) on ERS-2, GOME-2 on EUMETSAT's MetOp satellites and the OMI instrument (Ozone Monitoring Instrument) instrument flying on the NASA EOS-Aura satellite. The SCIAMACHY instrument (SCanning Imaging Absorption spectroMeter for Atmospheric CHartographY) on ENVISAT also made measurements in the nadir direction for deriving the total ozone column.

The first GOME was launched on ESA's ERS-2 satellite in 1995. It measures the spectrum of backscattered sunlight between 240 nm and 790 nm in four overlapping bands. GOME is nadir (directly downwards) scanning instrument with a pixel size of 40 km × 40 km to 40 km × 320 km, and it can scan the surface of the whole globe in 3 days. To continue GOME's observations, its big brother SCIAMACHY was launched in 2002 on board ESA's Envisat and continued to observe the atmosphere until the end of the Envisat mission in 2012. A follow-up GOME instrument, the GOME-2, is now part of the operational European meteorological satellites known as MetOp-satellites. Figure 2 shows an example of the spectral wavelength regions covered by GOME and how these are used for the detection of various species.

2.2 Ozone profiles

Ozone profiles are measured from space by a variety of techniques ranging from solar occultation to thermal emission measurement (see Figure 1), each with its own sampling and coverage features and issues.

2.2.1 Solar occultation

The strength of the solar emission provides an excellent light source for absorption spectroscopy. The principle of the measurement is simple. The solar spectrum, taken above the atmosphere, serves as a reference for the application of the Lambert-Beer's law to extract column densities at varying tangent altitudes as the Sun sets or rises through the atmosphere on the Earth's limb. Finally vertical

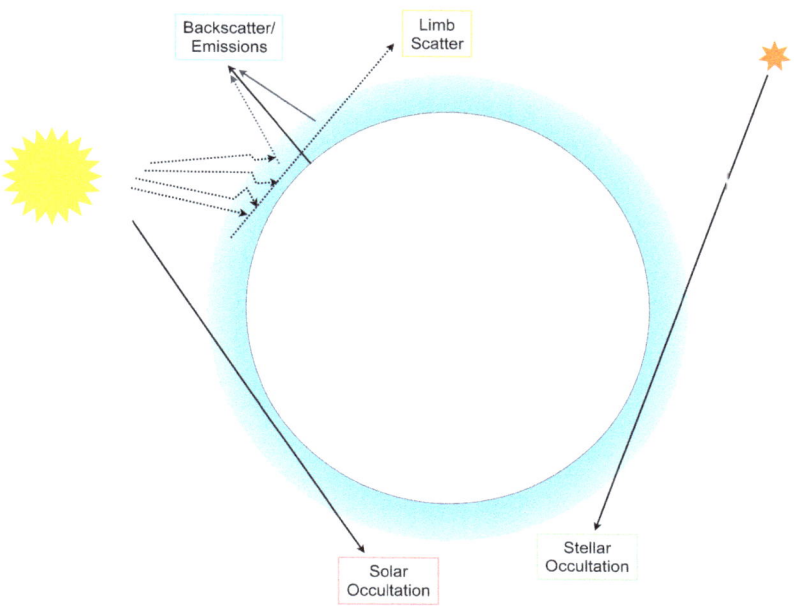

Fig. 1. Different methods for observing atmospheric composition from satellites.

profile of chemical species and aerosols are derived. The main disadvantage of this technique is the limited coverage with only 15 sunrises and 15 sunsets per day for a polar orbiting satellite in low Earth orbit. The latitudinal coverage is also limited, being a function of time of year and whether or not the satellite orbit is sun-synchronous. The latter implies fairly limited latitude coverage to the high latitude polar regions.

The Stratospheric Aerosol and Gas Experiment (SAGE) mission began with the first SAGE instrument, SAGE-I, which performed near global measurements of ozone, aerosol extinction (at 0.45 and 1.0 µm), and NO_2 from 1979–1981. Its sister instrument, SAGE-II, was launched aboard the ERBS (Earth Radiation Budget Satellite) in October 1984 and operated for over 21 years until it was powered-off in August 2005. The instrument had seven channels cantered at wavelengths ranging from 0.385 to 1.02 µm and provided profiles of ozone, aerosol extinction (at 0.385, 0.453, 0.525, and 1.02 µm), H_2O, and NO_2 from the upper troposphere to the middle stratosphere. The data sets provided by these two instruments are among the longest with respect to stratospheric ozone, aerosol, and water vapour (Damadeo et al., 2013). The SAGE mission was supplemented by a third instrument, the SAGE-III, which provided profile measurements of O_3, aerosol extinction, H_2O, NO_2, NO_3, OClO, clouds, temperature and pressure in the mesosphere, stratosphere, and upper troposphere with a high vertical

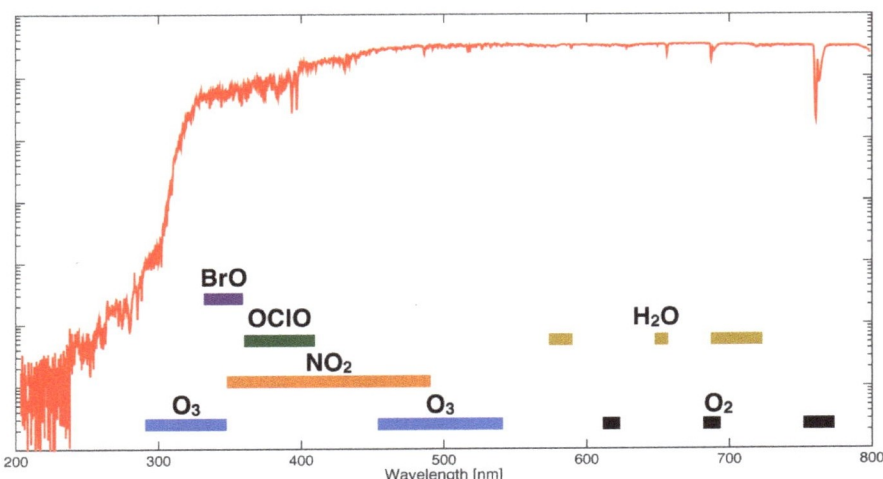

Fig. 2. Different methods for observing atmospheric composition from satellites.

resolution. This instrument was flown on board the Russian Meteor-3M platform, which was launched in December 2001 and ceased operation in March 2006.

In the infrared (IR) region, the longest time series of ozone measurements are from the HALOE (Halogen Occultation Experiment) instrument covering the period 1991–2005 and the later ACE/FTS instrument from 2003 to present (at time of writing in 2014). The HALOE instrument aboard NASA's Upper Atmosphere Research Satellite (UARS) measured vertical profiles of middle atmospheric constituents, including ozone, and temperature from September 1991 until November 2005. In addition to ozone, HALOE measured the vertical distributions of HCl, HF, CH_4, NO, H_2O, NO_2, as well as temperature, which was derived from CO_2 absorption. Due to the occultation viewing geometry and the UARS orbit, the latitudinal coverage of HALOE observations was from 80°S to 80°N.

The Atmospheric Chemistry Experiment Fourier Transform Spectrometer (ACE/FTS) is a high-resolution Fourier transform spectrometer on the Canadian ACE (also known as SCISAT) satellite, launched in August 2003 (Bernath et al., 2005). The instrument concept was based on an earlier instrument, which flew on the Space Shuttle in the 1990s, although the ACE-FTS instrument has been considerably improved and miniaturised. Due to the ACE orbit, solar occultation observations are performed mainly at high and middle latitudes, making the observations ideal for studying the polar regions (see e.g., Chapter 4.6), and only during even-numbered months in the tropics. A consistent set of 37 microwindows around 10 µm is used for the ozone retrievals. A large number of other atmospheric constituents are also measured at the same time.

2.2.2 Stellar occultation

Some of the coverage issues related to solar occultation can be resolved by using stars as the light source while at the same time retaining the inherent self-calibration of the technique. The instrument that has exploited this method over a substantial period is GOMOS (Global Ozone Monitoring by Occultation of Stars) on Envisat.

GOMOS is a medium resolution spectrometer measuring in the ultraviolet, visible and near-infrared spectral regions using the stellar occultation technique (see Figure 1). The instrument consists of four spectrometers covering the 248–371 nm, 387–693 nm, 750–776 nm and 915–956 nm spectral ranges, allowing for observations of O_3, NO_2, NO_3, atmospheric density and aerosols, as well as O_2 and H_2O, from the upper troposphere up to 120 km, but the actual altitude coverage depends on the observed species, the characteristics of the star and illumination conditions in the atmosphere. Additionally, two fast photometers sampling at a frequency of 1 kHz in two wavelength regions are used to correct for perturbations from scintillation effects and to determine vertical profiles of temperature with extremely high vertical resolution (200 m).

2.2.3 Limb scattered solar

This is a relatively new technique although it was successfully used in the 1980s on the Solar Mesosphere Explorer to derive NO_2 profiles. The first long time series of limb scatter ozone measurements come from the OSIRIS (Optical Spectrograph and InfraRed Imager System) instrument on the Odin satellite and the earlier mentioned SCIAMACHY instrument. The methodology is similar to the backscatter UV and solar occultation, except that the reference spectrum is often taken from the upper atmosphere and the altitude scanning is obtained by pointing the instrument at different tangent altitudes on the Earth's limb and observing the scattered solar radiation (see Figure 1). Measurements are taken globally but are limited to the daytime part of the orbit.

2.2.4 Thermal emission

Thermal emission in the infrared or microwave wavelength region is another powerful technique that is applied to derive profiles of multiple species in the atmosphere. The technique relies on emission from thermally excited vibration-rotation states (IR) or rotational states (microwave) being received by the instrument as it is scanned over the Earth's limb (see Figure 2). Fourier transform spectrometers are often used in the IR while heterodyne spectrometers are used in the microwave region.

In the IR wavelengths, ozone data are available from several instrument with MIPAS (Michelson Interferometer for Passive Atmospheric Sounding) and SABER (Sounding of the Atmosphere using Broadband Emission Radiometry) providing the longest datasets. MIPAS was launched on Envisat in February 2002. A large number of species are recovered from the data that typically cover the altitudes of 6–68 km. There are however special high vertical resolution modes and an extended scan mode providing data with improved resolution or extending to higher altitudes. SABER was launched in December 2001 on board the TIMED spacecraft. SABER measures ozone from 20 to 80 km in the infrared region and, additionally, mesospheric ozone is also derived from the $O_2(^1\Delta)$ emission that in daytime is dominated by photo-dissociation of ozone itself and subsequent reaction of photo-dissociation products (Mlynczak et al., 2007).

Millimetre wave limb sounding instruments dedicated to middle atmospheric research were successfully employed in the early 1990s. The first instrument of this type, the Microwave Limb Sounder (MLS), was launched on board the earlier mentioned UARS satellite in September 1991. MLS observed the atmospheric limb perpendicular to the UARS orbit track. Regular UARS yaw maneuvers allowed either the high southern latitudes 34° N–81° S or northern latitudes 81° N–34° S to be observed.

The first limb observations in the sub-millimetre wavelength range from space came from the Sub-Millimetre Radiometer (SMR) flying on board the Odin satellite, which was launched in February 2001 (Murtagh et al., 2002). The SMR instrument consists of four mechanically cooled single sideband Schottky-diode receivers in the 486–581 GHz range and one millimetre receiver at 119 GHz (for O_2). Measurements of several spectral lines belonging to different species are observed. The atmospheric observation modes were time-shared with astronomical observations until 2008 after which the satellite focused on atmospheric observations. Ozone is measured from about 15–70 km with a resolution of 2–3 km.

The second MLS instrument (see above) was launched on NASA's EOS-Aura satellite in July 2004, equipped with radiometers operating in the millimetre, sub-millimetre, and far infrared spectral ranges (Waters et al., 2006). The Aura-MLS measures ozone as well a several other species from the tropopause to 70 km.

3 Other minor species

Ozone is the central element in the atmospheric chemistry and dynamics and its role in screening UV-radiation is essential. The main ozone loss reactions are driven by constituents of the O_x, NO_x, HO_x, and Cl_x families. These processes depend also on the solar flux. Additionally, in the polar areas, energetic particle precipitation (see Chapters 4.5-4.6) from solar storms influence the HO_x and NO_x families and therefore ozone balance. Other essential constituents include CO_2, CH_4 and H_2O, which are key elements in the radiative balance of the Earth (the greenhouse effect). Most of the instruments mentioned above also measure these constituents. Climatologies and information about these measurements can be found from http://www.sparc-climate.org/data-center/data-access/sparc-data-initiative/.

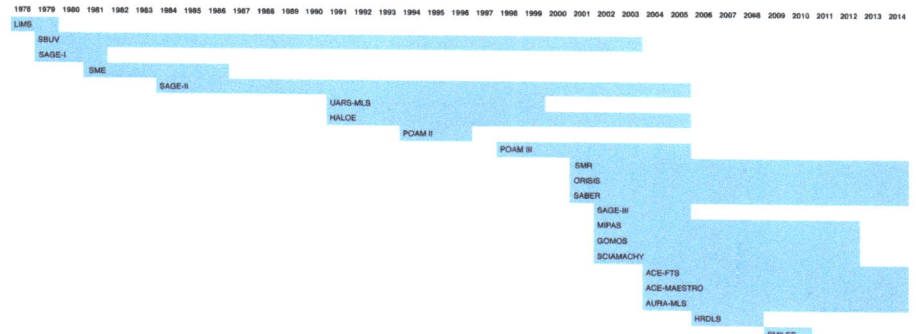

Fig. 3. Temporal coverage of satellite observations (instrument name given in the graph) of ozone and other minor species from 1978 onwards.

4 Examples of other satellite observed parameters

In addition to measurements connected to the chemical composition of the atmosphere, satellites provide data on the solar and terrestrial radiation, thermal structure, clouds and dynamics of the atmosphere.

4.1 Radiation and thermal structure

At the top of the atmosphere, the irradiance from the Sun and the outgoing terrestrial radiation (reflected solar light and emitted infrared radiation) determine the equilibrium temperature of the Earth. The total solar irradiance (i.e., the solar constant) has been monitored by several instruments since 1978. The UARS satellite included the ACRIM II instrument (Active Cavity Radiometer Irradiance Monitor II), which provided data from 1991 to 2001 with one second temporal resolution. Some instruments have also measured the spectral distribution of the solar irradiance. These measurements have concentrated on the UV-part of the spectrum that does not reach the Earth's surface because of the screening of the ozone layer. SOLSTICE (Solar/Stellar Irradiance Comparison Experiment) onboard UARS was a three-channel grating spectrometer for measuring solar and stellar irradiation. The three overlapping channels are 119–190 nm, 170–320 nm, and 280–420 nm. Measurements of reflected (scattered) solar radiation is used to retrieve ozone columns by instruments discussed above (GOME, OMI etc.). More information on the solar spectrum and observations of the solar irradiance is given in Chapter 2.2.

The outgoing longwave terrestrial radiation is measured continuously by numerous operational weather satellites. From these measurements, one can determine the surface temperatures and the temperature profiles in the atmosphere.

Several limb-viewing satellite instruments discussed earlier, including MIPAS and ACE-FTS, are able to retrieve high-resolution temperature profiles. Instruments recording GPS-occultations can also determine temperature profiles from the bending angle of the GPS-signal. EUMETSAT's MetOp-satellites use the GRAS-instruments for these measurements.

4.2 Clouds and aerosols

The connection between solar variability and clouds is discussed in Chapter 4.7. One of the main cloud datasets comes from the International Satellite Cloud Climatology Project (ISCCP), which was established in 1982 as part of the World Climate Research Program (WCRP). Weather satellite radiance measurements are used to infer the global distribution of clouds, their properties, and their diurnal, seasonal and interannual variations. Data are currently available for the period 1983–2009.

CloudSat together with CALIPSO (Cloud-Aerosol Lidar and Infrared Pathfinder Satellite Observations) were launched in April 2006. CloudSat carries a cloud profiling radar operating at 94 GHz with ground resolution of about 1.5 × 1.5 km and 500 m in the vertical. CALIPSO carries a lidar (see Chapter 3.2) operating at 532 and 1064 nm, pulsing at about 20 Hz. Cross polarisation measurements are also made for the visible channel. Products for both aerosol and cloud are available with about 1 km along track resolution and 60 m in the vertical.

4.3 Dynamics

Operational weather satellites determine lower atmosphere winds tracking the movements of clouds, ozone or water vapour.

Observing upper atmosphere atmospheric dynamic variables like winds from satellites has turned out to be difficult. The upper atmosphere winds have been observed by two instruments on the UARS satellite (1991–2005). The WINDII instruments employed several natural airglow emission lines and the Doppler Michelson interferometry to measure winds in the altitude range 80–300 km. The HRDI (High-resolution Doppler imager) used absorption and emission of O_2 to recover horizontal winds. The TIDI instrument (TIMED Doppler interferometer) on the TIMED satellite measures upper atmosphere wind using the same techniques as WINDII and HRDI.

5 Conclusion

Although here we chose to highlight ozone observations, a multitude of variables relevant for solar influences on the atmosphere and climate can be observed from space. Contrasting to ground-based observations (see Chapter 3.2), satellite observations provide a larger horizontal coverage around the Earth, which has provided vital insights to global responses to variability in solar activity. However, satellite missions are costly and have long lead times, as it typically takes a decade to plan and build a satellite instrument before it is launched (itself including risks) and can start making observations. There is no doubt that satellite observations have provided invaluable information about solar influence on the Earth system (see Part 4 of this book). Currently many of these observations, vital for establishing the Sun's role as a natural source for climate variability are under threat. Scientist are now relying on ageing satellite missions, many of which are at the end of their lifetimes having spent ten years or over orbiting the Earth (Figure 3), facing serious deterioration from ageing as well as running out of fuel. With no replacement mission planned or approved for funding at present, there is no continuation of vital space based observational datasets in sight.

Further reading

European Space Agency (ESA) http://www.esa.int

The ESA Committee on Earth Observation Satellites database http://database.eohandbook.com, Earth Observation Portal https://eoportal.org

National Aeronautics and Space Administration (NASA) http://www.nasa.gov

CHAPTER 3.4

REANALYSIS DATA

Peter Thejll[1] and Hans Gleisner[2]

1 What are reanalysis data?

Compare weather- or climate-model output to observations: Not only are the model outputs regularly gridded in time and space but are also complete. The availability of observations depends on the coverage, in space and time, of global observing systems, which includes weather stations, radiosondes, airplanes, and instruments onboard satellites. Despite this wealth of data, the coverage is not complete and data quality may vary with time. Some form of data selection and interpolation, or other processing, can bring observations into a uniform shape, which is appealing from a data analysis point of view. Purely mathematical interpolation can do this, but does not provide any extra information, as it is unphysical. In reanalysis, physical models are used to provide the extra information, resulting in a physically consistent description.

Models used for reanalysis are meteorological numerical weather prediction (NWP) models that assimilate available observations and ensure that the numerical solution differs as little as possible from the observations where these are present, and offers physically consistent values for data sparse regions. The models start at a state given by previous solutions, calculate one time step forward, compare the solution to observations that are relevant for the time, calculate the error, go back and modify the starting condition, and so on: by iteration, a solution (the 're-analysis') is achieved in which the model and the observations are as consistent as possible to within their respective errors. The process is then repeated for the next time step. Reanalysis grew out of the NWP field since NWP models base their forecasts on current observations and a physical model. As part of the data assimilation, a physics-based quality control is performed. Reanalysis models use similar model software but assimilate historical data – weather station

---

[1] Danish Climate Centre, DMI, Lyngbyvej 100, DK-2100 Copenhagen, Denmark
[2] Centre for Atmospheric Physics and Observations, DMI, Denmark

data going back more than a century is now being assimilated. As more data describing the past are revealed in digitisation and archival work, new reanalyses can be calculated. At the same time, the descriptions of physical processes in the models are improved by researchers and new re-analyses of the updated body of data become available. This is computationally intensive and expensive and therefore only a limited number of reanalysis projects exist – the most important are described below.

In principle, paleo-climate data in the form of climate sensitive proxies could also be used for reanalysis if the inherent smoothing in proxy data, and the peculiar forms of climate sensitivity (e.g., non-linearities, and presence for some seasons only) in the proxies were represented by the model.

2 Global reanalysis data sets

There is a limited range of rather widely used reanalysis datasets available, but the number is growing. While most reanalyses cover the global atmosphere, there are also reanalyses for the oceans and for the land surfaces. More narrowly focused reanalyses, for polar regions or for details of the hydrological cycle, are also available. The data differ in temporal and spatial resolution, in the time span covered, and in the set of observations that were assimilated by the model. This constitutes a fundamental limitation to the reanalyses – land- and sea-surface data records go back to the mid-19th century, while regular weather balloon data are available only from the 1950s. The 1970's saw a rapid increase of the number and quality of satellite-based data. Hence, many reanalyses begin in 1979, which is used as the starting point of the "satellite era".

There are three major reanalysis projects: the European ERA project, the US NCEP/NCAR project, and the Japanese JRA project. In addition, several specialised reanalysis projects exist.

The European atmospheric reanalysis *ERA-40* (Uppala et al., 2005), produced at the European Centre for Medium-range Weather Forecasts (ECMWF), covers the 45-year time period from September 1957 to August 2002. It has a 6-hour time resolution and covers 60 vertical levels up to an altitude of 64 km (0.1 hPa). The follow-on reanalysis *ERA-Interim* starts in 1979, is being continued in real time, and has a similar vertical resolution and coverage as *ERA-40* (Dee et al., 2011). In addition, a set of completely new reanalyses are currently in preparation through the *ERA-CLIM*[3] project led by ECMWF. The final global reanalysis, covering the whole 20th century, 1900–2010, is not expected to be finalised before 2017. However, the results of a few specialised reanalyses, e.g., a high-resolution land-surface reanalysis and a 20th century atmospheric reanalysis based only on surface observations, are already available. The *ERA-CLIM* reanalyses constitute significant advancements, both in terms of resolution, time coverage (the whole

---

[3] http://www.era-clim.eu

20th century, 1900–2010), and in the systematic handling and error characterisation of the observational data. For more in-depth information on ECMWF reanalyses, see lecture notes[4] prepared by the ECMWF.

The US NCEP/NCAR reanalyses come in two different versions: *NCEP/NCAR I*, which covers the time span from 1948 to the present, and *NCEP/NCAR II*, which only covers the major satellite era after 1979 (Kalnay et al., 1996). Data from these reanalyses are available on 17 pressure levels up to 10 hPa. The latter reanalysis incorporates more observational data and addresses some deficiencies in the older data assimilation system. For the satellite era, 1979 and onward, there is also the *NCEP CFSR* reanalysis (Saha et al., 2010), which is unusual in the sense that the reanalysis is done with a fully coupled atmosphere-ocean-land surface-sea ice model to provide an optimal estimate of the state of these coupled domains.

The Japanese Meteorological Agency has also produced a set of global atmospheric reanalyses: the 55-year reanalysis JRA-55 with 60 vertical levels, and the 25-year reanalysis JRA-25 with 40 vertical levels (Onogi et al., 2007). The former starts in 1958 and the latter in 1979. Both are currently undergoing an update to the present.

While these three reanalysis projects are the most widely used, there are also others available, some of them more specialised. MERRA[5] is a NASA reanalysis covering the satellite era. It has partly a focus on the hydrological cycle and water vapour, and has been used extensively for water vapour budget studies. NOAA has produced a 20th century reanalysis[6], covering the time 1871–2012. It has a 6-hourly time resolution, covers both the troposphere and part of the stratosphere, and data are provided on a 2° grid. A more specialised dataset is the Arctic System Reanalysis (ASR)[7], covering the Arctic region at a high spatial resolution. The arctic contents of the major reanalysis products are revieweded in Lindsay et al. (2014).

3 Limitations

Reanalysis data can be no better than the underlying observations, or than the models that process the data. On the modelling side, limitations exist because of the resolution chosen in the model and the parameterisations used for sub-grid processes. Since different reanalyses use somewhat overlapping, but not identical input data and methods, the quality of the reanalyses differ. Efforts are made to systematically assess and inter-compare reanalyses, e.g., as in the Ocean

---

[4] http://www.ecmwf.int/en/training-course-ecmwf/eumetsat-nwp-saf-satellite-data-assimilation
[5] http://gmao.gsfc.nasa.gov/merra/
[6] http://www.esrl.noaa.gov/psd/data/20thC_Rean/
[7] http://polarmet.osu.edu/ASR/

Synthesis/Reanalysis Intercomparison Project[8], and in the SPARC intercomparison project (S-RIP)[9].

For atmospheric reanalyses, a typical model-caused limitation is the number of vertical levels included in the software – both number of levels and the height reached are important, as is the treatment of the upper boundary. At the upper limit, there is typically a 'fade to nothing' condition imposed: the ionosphere and thermosphere, etc. are simply not part of the model, so potential influences originating in those parts of the near-Earth environment cannot be propagated into the stratosphere and troposphere. Solar UV irradiance does interact with the stratospheric ozone and might couple to the troposphere via wave coupling (Haigh, 1994), but if the model is not given observations for the UV flux, this will not be part of the reanalysis dataset.

The inputs describing tropospheric and stratospheric conditions are limited to a subset of what is possible and includes temperature, pressure, and humidity. Satellite and radio occultation (RO) data can be assimilated with methods based on forward modelling from the meteorological model to quantities the satellites measure, such as microwave fluxes and phase-changes in radio waves.

The observations come from surface observations, radiosondes (i.e., weather balloons), aircraft, satellites and also radio-occultations between GPS satellites. Well-known inhomogeneities are present in reanalyses due to the gradual and time-dependent introduction of satellite, aircraft and RO data, and the disappearance of surface and radiosonde observations. Reanalysis output contains many quantities not all of which are primary (such as pressure and temperature). Secondary variables are those (for instance, snow depth, and other surface fluxes) that depend on the state of the system but are not provided as observational input or are only indirectly connected to input observations – they may thus be 'consistent' but are not used to define the solution (Kalnay et al., 1996).

4 Reanalysis data in Sun-Climate research

Reanalysis data have been used extensively to detect and characterise the relations between solar activity and atmospheric structure. Most of these studies have focused on variability related to the solar cycle, but also day-to-day variability and periodicities related to the solar rotation rate have been studied. The standard approach has been to detect statistically significant correlations between an indicator of solar activity or solar irradiance, and various atmospheric quantities, through multiple regression, also allowing for an impact from greenhouse gasses, volcanic emissions, or internal variability in the climate system.

Examples include studies of solar cycle-related variability in zonally averaged temperatures and Hadley cell structure (Gleisner and Thejll, 2003; van Loon et al.,

---

[8] http://www.clivar.org/organization/gsop/resources/ocean-synthesisreanalysis-intercomparison-project
[9] http://s-rip.ees.hokudai.ac.jp/

2007), in the strength and location of mid-latitude jet winds (Haigh et al., 2005), in troposphere-stratosphere wind changes apparently coupled to ozone dynamics (Crooks and Gray, 2005), and solar cycle variability in the height of the pressure surface at 300 hPa (Brönnimann et al., 2007). These studies used either *NCEP/NCAR* reanalyses or *ERA-40*, and applied various types of multivariate regression. There are also examples of spectral methods being used to study solar effects in reanalysis data, (e.g., Mayr et al. (2007) and Mayr et al. (2009)). Not only the free atmosphere – troposphere and stratosphere – but also surface temperature variations have been investigated, e.g., by Soon et al. (2011).

The last decade has seen a wealth of studies on Sun-climate relations using reanalysis data sets (see, e.g., Gray et al. (2012) and Mitchell et al. (2014) for overviews of studies and datasets). As the reanalysis methods improve, and the satellite data records become longer, there is considerable potential, not only for an improved characterisation of Sun-climate relations, but also for a better understanding of the underlying physical mechanisms.

5 Accessing reanalysis data

A good starting point for selecting and accessing reanalysis data is {reanalysis.org}[10], which provides overview summaries of the most widely used datasets. Getting reanalysis data is relatively simple – they are provided over the Internet and can be downloaded simply. The NCEP is particularly simple to pull down, while the ERA and JRA require a password (easily enough obtained in a day or so). ERA-interim[11], and NCEP[12] are easily reached over the Internet. A summary of most access options is given at reanalysis.org[13]. Once downloaded you have to have the right sort of software for reading the files. Formats include NetCDF and GRIB, and there exists software to read these formats, or to transform one into the other. CDO[14] is a tool for converting GRIB format to NetCDF (amongst many other things). There are libraries in many programming languages for reading NetCDF and GRIB: IDL provides simple ways to read NetCDF files; FORTRAN can[15] also do it, with patience; Python has a NUMPY interface[16]; and Java also has a library[17]. 'NCL' and 'GrADS' are programmes that allow visualisation of data, mathematical manipulation and the ability to read NetCDF files. The ESRL at NOAA[18] provides access to several such tools. Online services

---

[10] http://reanalysis.org
[11] http://data-portal.ecmwf.int/data/d/interim_moda/
[12] http://www.esrl.noaa.gov/psd/data/gridded/data.ncep.reanalysis.pressure.html
[13] http://www.reanalysis.org/atmosphere/how-obtainplotanalyze-data
[14] http://www.studytrails.com/blog/install-climate-data-operator-cdo-with-netcdf-grib2-and-hdf5-support/
[15] http://www.unidata.ucar.edu/software/netcdf/examples/programs/
[16] http://code.google.com/p/netcdf4-python/
[17] http://www.unidata.ucar.edu/software/thredds/current/netcdf-java/
[18] http://www.esrl.noaa.gov/psd/data/gridded/tools.html

allow access and inspection of reanalysis datasets, such as the KNMI climate explorer[19].

6 Take-Home message

Reanalyses are a convenient source of realistic, high-resolution data about the climate system. Global coverage extends to one century, although the best data are available since about the start of the satellite era – for just the Northern Hemisphere somewhat earlier.

---

[19] http://climexp.knmi.nl/start.cgi?id=someone@somewhere

CHAPTER 3.5

UNCERTAINTIES AND UNKNOWNS IN ATMOSPHERIC OBSERVATIONS: HOW DO THEY AFFECT THE SOLAR SIGNAL IDENTIFICATION?

Bernd Funke[1], Annika Seppälä[2] and Thomas von Clarmann[3]

1 Introduction

A large variety of observations, used to identify the solar fingerprint on Earth's atmosphere, have been discussed in the previous chapters. All of them have limitations and uncertainties which have to be considered in their analysis. The main difficulties in the identification of solar signals from atmospheric measurements are related to three principal observational limitations:

- Instrument precision and resolution. Solar-induced variations of atmospheric observables are often too small to be resolved within the instrumental precision. Also, drifts may arise from instrumental degradation over longer time periods.

- Temporal coverage of observational records. Particularly satellite data, being of highest utility for solar signal detection on a global scale, are available only for the last three decades, a period too short for an unequivocal attribution of observed signals to mid- and long-term solar variability.

- Interference with other sources of variability (see Chapter 4.3). Atmospheric variability is typically dominated by other sources than solar variability. The extraction of the solar signal is thus often challenging, particularly if other sources of variability show an apparent correlation with solar activity.

The relevance of the latter two limitations depends on the timescale of solar activity variations. Transient solar events such as solar proton events produced

---

[1] Instituto de Astrofísica de Andalucía, CSIC, Granada, Spain
[2] Earth Observation Unit, Finnish Meteorological Institute, Helsinki, Finland
[3] Karlsruhe Institute of Technology, Institute for Meteorology and Climate Research, Karlsruhe, Germany

by solar flares leave a clear fingerprint in the chemical composition of the atmosphere. In such cases, the attribution of the atmospheric changes to a solar event is relatively easy. Similarly, the response of several short-lived trace gases (such as atomic oxygen and nitric oxide in the upper mesosphere and lower thermosphere) to the 27-day solar rotation can be unequivocally identified by time series analysis from observational records, provided the signal is large enough and the temporal coverage is sufficiently long to provide a statistically robust dataset.

The situation is much more complicated for the Schwabe cycle, which affects the atmosphere by solar irradiation variations and by the modulation of energetic particle fluxes. Atmospheric responses are small and generally masked by the background variability. Furthermore, long-term datasets covering at least one full Schwabe cycle (ideally several cycles) are required. However, such datasets are often not available. Direct chemical responses to solar variations (e.g., the ozone response in the tropical upper stratosphere to solar irradiance variations in the ultraviolet or the reactive nitrogen response in polar winter to particle flux variations) are the most promising candidates to search for Schwabe cycle signals in observations, as they follow directly the solar forcing variations. Indirect responses occurring by transport of processed air masses and circulation changes in response to solar activity are more difficult to detect because of the time delays introduced by these coupling mechanisms. Particularly in the latter case, interferences with internal atmospheric variability drivers (a problem known as aliasing) are a major issue in time series analysis of both observational and model data. As an example, a recent study (Chiodo et al., 2014) suggests that a large fraction of the apparent solar-induced lower stratospheric temperature and ozone response, inferred by means of multilinear regression analysis of a 44-year model simulation, could be due to two major volcanic eruptions (i.e., El Chichón in 1982 and Mt. Pinatubo in 1991), which are concurrent with periods of high solar activity (see Figure 1).

Each type of observation is confronted with different problems and limitations, which will be discussed separately in the following sections. Hence, it is clear that the accurate solar signal identification in the atmosphere benefits form the synergistic use of multiple observation techniques.

2 Proxy-based observations

In the context of this chapter, we refer to proxies as preserved physical characteristics of the past that stand in for direct measurements, to enable scientists to reconstruct the atmospheric conditions that prevailed during much of the Earth's history (see also Chapter 2.1). Proxy-based observations, such as temperature reconstructions from tree rings or nitrate depositions in ice cores, the latter potentially providing information on atmospheric ionisation by cosmic rays, are used to identify solar signals on very long time scales.

However, the principal problem of these reconstructions is that the relationship of the proxies with the respective atmospheric parameters can only be established at present when direct observations of the latter are available. A drift or

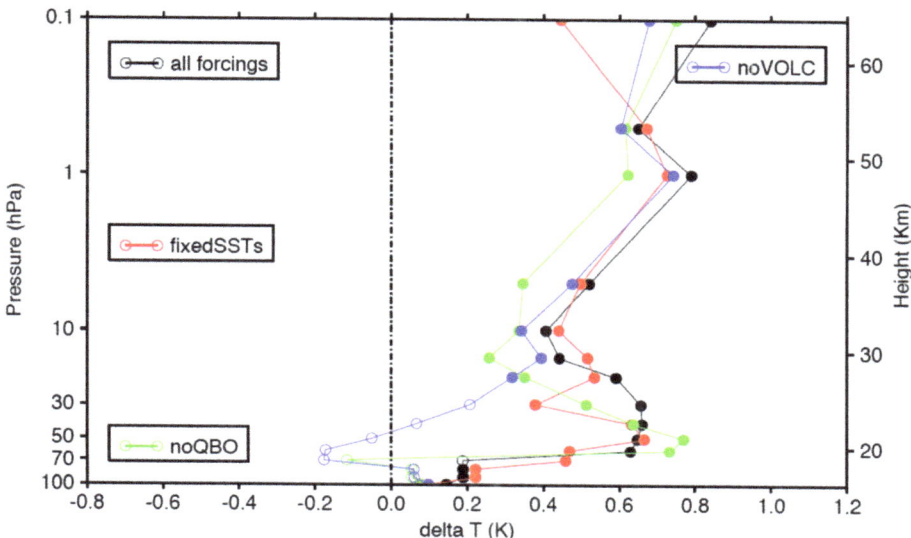

Fig. 1. Solar signal (response to a 2-σ variation of the 240–270 nm UV flux, corresponding to about 70% of the average Schwabe cycle variation) in the tropical zonal mean temperature as estimated from multiple linear regression analysis of a 44-year climate model simulation including all non-solar forcings (all forcings), assuming fixed sea surface temperatures (fixed SSTs), as well as excluding the quasi-biennal oscillation (no QBO) and volcanic eruptions (no VOLC). Note that the modelled lower stratospheric solar signal vanishes when excluding volcanic eruptions. Reprinted from Chiodo et al. (2014).

modulation of this relationship by other climate forcings (e.g., volcanoes or variations in green house gas concentrations) and by variations of the Earth's magnetic field over longer historical periods cannot be excluded. This is particularly important if such modulations show a correlation with the solar signal to be extracted.

3 In situ observations

In situ instruments such as radiosondes provide very accurate observations of tropospheric and lower stratospheric meteorological parameters potentially affected by solar variability. Some in situ observations (e.g, surface temperature and sea level pressure measurements) have been made routinely for more than a century.

However, these local measurements are often not representative for the atmospheric mean state even on a regional scale. Therefore, data assimilation techniques (see Chapter 3.4) are required to reconstruct global synoptic fields from a large amount of individual measurements. Furthermore, both the accuracy and the geographic coverage of historic data are generally not comparable to modern standards, making the analysis of long-term time series with respect to solar signals highly challenging. Evidently, it is much more difficult to obtain in situ observations in the middle atmosphere. High-altitude balloons and rockets have been used for stratospheric and mesospheric air sampling only on a sporadic basis during the last decades. Due to their sparseness, these measurements are of limited use for solar signal detection in the middle stratosphere and mesosphere.

4 Remote sensing observations

Remote sensing techniques allow to infer many atmospheric parameters (i.e., temperature, composition, clouds, etc.) by observing the radiance emitted, transmitted or scattered by the Earth's atmosphere from a remote location. This radiance modulation is taking place over the entire light path observed by the instrument. Remote sensing observations are thus integrated over large air volumes, but the information on small-scale structures is generally lost. In contrast to in situ observations, they only provide an indirect measure of the observables of interest, because the latter have to be inferred from the observed radiances by inverting the radiative transfer equation.

4.1 Atmospheric soundings from the ground

Ground-based remote sensing instruments (see Chapter 3.2) provide long-term observations of temperature, trace gases, and cloud parameters, potentially affected by solar variability. For example, ground-based total ozone column observations are available since the beginning of the last century. Calibration and substitution of individual instruments is much easier on the ground than in space, guaranteeing stable long-term records that are required for the detection of decadal variability related to solar variations. Nowadays, global networks of inter-calibrated ground-based instruments provide measurements in all continents and climate zones.

However, the limited vertical resolution of some ground-based instruments, resulting from the upward looking geometry, complicates the identification of altitude-dependent solar responses, as in the case of tropical stratospheric ozone.

4.2 Satellite observations

Space instruments (see Chapter 3.3) provide observations of atmospheric key parameters with a large spatial coverage. Hence, they are most suited for solar signal detection on a global scale. However, atmospheric observations from space are available only since the late 1970s, covering less than three Schwabe cycles. Moreover, the typical lifetime of space missions does rarely exceed 5–10 years. Long-term observational records, required for the detection of Schwabe cycle signals, can therefore only be obtained by combining data from multiple instruments.

Since inter-calibration of different satellite sounders is much more complicated than that of ground-based instruments, the construction of multi-instrument climate records is not straightforward and requires complex mathematical tools.

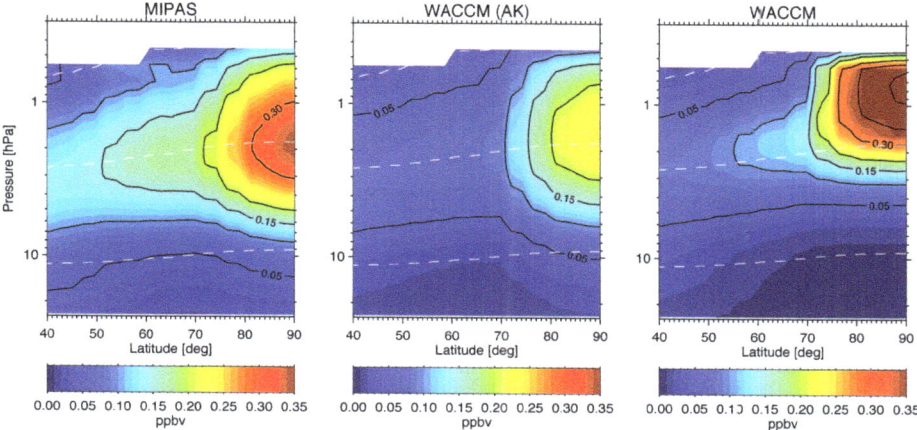

Fig. 2. Observed (left) and modelled (middle) hypochlorous acid (HOCl) response to the October/November 2003 solar proton event in the Arctic polar stratosphere. Only after taking into account the instrument response function by applying MIPAS averaging kernels to the model data (middle), the vertical structures of observed and modelled responses become comparable and a slight underestimation of the model result shows up. Reprinted from Funke et al. (2011).

The temporal overlap of successive equivalent instruments is crucial and has to be taken into account in the planning of space missions. The continuity of satellite data records in the future is vital to the understanding of the atmospheric response to solar variations on a longer time scale. Current space missions targeting stratospheric and mesospheric observations have ended recently, e.g., ENVISAT, or will be phased out over the next few years. As a consequence, an observational gap is expected in the second half of this decade, which will seriously harm the continuity of observational long-term records.

For some space instruments, the sampling pattern is not uniform, which causes additional complication (i.e., sampling errors, see Toohey et al. (2013)). This is particularly true for solar occultation instruments, which can measure only during local sunrise or sunset, or for instruments in a slowly drifting orbit. The latter is of particular importance if the orbit characteristics change on timescales comparable to the Schwabe cycle.

In contrast to nadir-sounders, which look from the satellite through the atmosphere to Earth's surface, limb emission, limb scattering and occultation instruments look horizontally through the atmosphere. By varying the viewing angle, a good altitude resolution of the signal can be obtained and vertically-structured solar signals in temperature and composition can be extracted from the measurements.

However, remote sensing instruments, in contrast to in situ instruments, always provide a smeared or even distorted picture of reality. This deviation can be

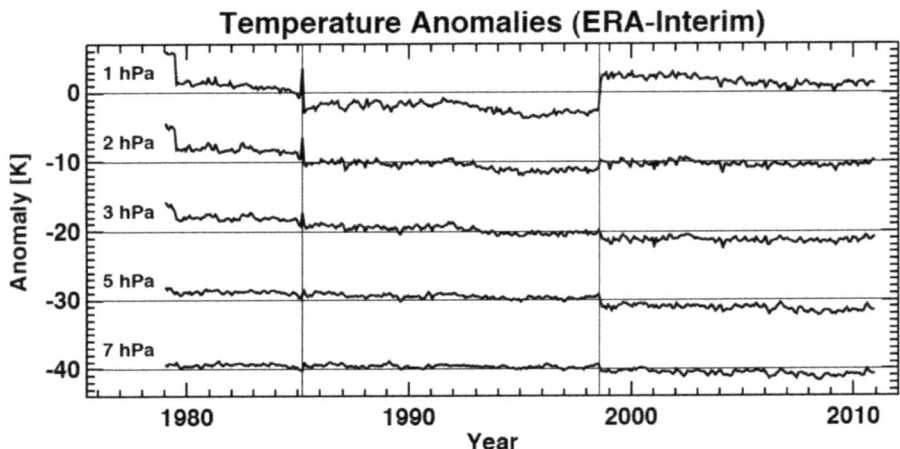

Fig. 3. Global and monthly mean ERA-Interim temperature anomalies due to changes in satellite radiance data used in the assimilation at the top five pressure levels in the stratosphere. The curves at 2, 3, 5 and 7 hPa are shifted by 10 K with respect to the curve above. The jump times in March 1985 and August 1998 are denoted by the thin vertical lines. Reprinted from McLandress et al. (2014).

taken into account when measured data are compared to data simulated by climate models by application of the instrument response function, expressed by the so-called averaging kernel (Rodgers, 2000), to the model data. This is particularly important for nadir measurements with their limited altitude resolution, but can also be important for the better-resolving limb sounders. Funke et al. (2011) compared HOCl measurements from the MIPAS limb emission spectrometer to model results and found large discrepancies which could be removed by the application of the grossly asymmetric MIPAS averaging kernel which shifted the maximum of the concentration profile (see Figure 2).

5 Re-analysis data

Consistent distributions of pressure, temperature, humidity, trace gas concentrations and others are compiled in the so-called 're-analysis data' (see Chapter 3.4), which are obtained by assimilation of observed data in numerical weather prediction (NWP) models. Observations used in data assimilation cover a wide range of in situ, ground-based, and satellite data. In this sense, the re-analysis of assimilated meteorological and chemical fields is a powerful tool to make use of complementary information derived from different observational techniques in a synergistic manner.

On the other hand, it is evident that all limitations and uncertainties encountered in the observations will propagate into the re-analysis data products. Also, discontinuities may arise when new instruments are included into the data assimilation system (see Figure 3), which may lead to artefacts in time series analysis with respect to solar signals. An additional problem arises from the influence of the NWP model itself, which might suffer from an incomplete or erroneous representation of atmospheric processes involved in solar signal propagation. Particularly at high altitudes, where observational data is sparse, this can lead to important deviations of solar responses in the re-analysis data.

Further reading

Rodgers, C. D. 2000, *Inverse Methods for Atmospheric Sounding: Theory and Practice*, World Scientific Publishing Co. Pte. Ltd, Singapore.

Toohey, M. et al. 2013, Characterizing sampling biases in the trace gas climatologies of the SPARC Data Initiative, J. Geophys. Res. (Atmospheres), 118, 11847-11862.

Wilks, D. S. 2011, Statistical Methods in the Atmospheric Sciences: an Introduction, International Geophysics. Academic Press, San Diego, 3rd edition.

CHAPTER 3.6

NUMERICAL MODELS OF ATMOSPHERE AND OCEAN

Pekka T. Verronen[1] and Hauke Schmidt[2]

1 Introduction

Numerical models of atmosphere and oceans are frequently used for forecasting, e.g., of weather on a time horizon of up to about 10 days, or for projecting climate under the assumption of certain scenarios of future technical and economic developments. In addition, numerical modelling is an important tool for atmosphere and ocean research (see also Infobox 3.1 and 3.2). Measurements often provide only an indirect or partial view on the properties and processes with a limited geographical and/or altitude coverage. Interpretation of such observations can be greatly facilitated by using suitable models, leading to a better understanding of the atmosphere/ocean system. The other way round, model results may provide information, which can be used to guide the planning of future observations and laboratory work. In comparison to measurements, modelling can provide a relatively comprehensive representation of atmosphere and ocean. It is important to understand, however, that there is no perfect model. Not all processes are well understood, there are likely some that are not yet known to us, and simplifying assumptions have to be made because the processes are many, they happen on a large range of temporal and spatial scales, their interactions are complex, and computing power is limited. Therefore, validation of model results using measurements is vital.

Numerical modelling became practical with the emergence of electronic computers. When the first computer ENIAC was built in the 1940s, one of the first applications was weather forecasting (Jacobson, 1999). From the 1970s on, model development was inspired by several environmental problems, such as ozone depletion (widely referred to as the ozone hole) and global climate change. Over time, models have become more and more complex. Nowadays, climate models can

---

[1] Earth Observation Unit, Finnish Meteorological Institute, PO Box 503, 00101 Helsinki, Finland
[2] Max Planck Institute for Meteorology, Bundesstr. 53, 20146 Hamburg, Germany

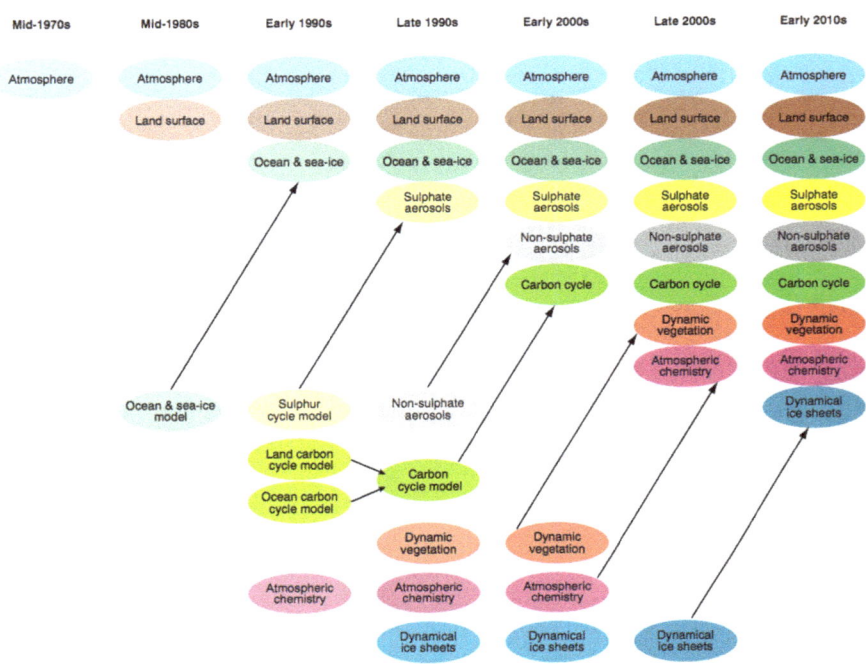

Fig. 1. Chronology of climate model development, Fig 1.11 from Stocker (2011).

include interactions between atmosphere, ocean, land, ice, aerosols, and biosphere (Figure 1).

An atmospheric model may simulate a combination of radiative, gas, aerosol, dynamical, transport, and cloud processes and the interactions between them (Jacobson, 1999). In the model, these are described by physical, chemical, and dynamical equations (see Infobox 3.2). While many of the equations are based on basic physical principles, some processes need to be parameterised (i.e., simplified and often based on empirical relationships). The atmospheric processes take place on various scales, ranging from millimetres and less (molecular) to 10 000 km and more (planetary). Thus, one of the challenges of atmospheric modelling is to combine these processes of different scales mathematically in an efficient way. Parameterisations are in particular needed for processes taking place on smaller spatial scales than can be resolved by the model. Depending on the purpose of the model, some of the processes may be omitted in order to save computing time (Stocker, 2011). One example of such omission would be the chemical reactions between ionic and neutral species in the middle atmosphere.

Another way to reduce the computing time is to restrict the spatial dimensions or resolution of the model (see Table 1 for an explanation of typical model dimensions). Today, a range of models exists with different combinations of dimensions

Table 1. Model dimension combinations explained. Note that the dimension(s) and examples given are typical but other combinations could be used too.

Type	Dimension(s)	Examples
0-dimensional (0-D)	N/A	Chemistry in a confined atmospheric box (thus sometimes called a "box model"), without connections to outside
1-dimensional (1-D)	Altitude	Chemistry in a vertical stack of 0-D boxes with vertical transport connection between boxes; radiation and convection in an atmospheric column
2-dimensional (2-D)	Altitude, Latitude	Chemistry in horizontal row of 1-D stacks with transport connections in the vertical and meridional dimensions
3-dimensional (3-D)	Altitude, Latitude, Longitude	Chemistry in longitudinal sets of 2-D rows with transport in the vertical, meridional and zonal dimension; atmospheric general circulation modeling

ranging from 0-D (global or local average) to 3-D (resolving latitude-longitude-altitude structure) for the atmosphere and oceans (Stocker, 2011). Although a 3-D model can treat the dynamical and transport processes more realistically, there are still some specific purposes for which fewer dimensions are more practical. For example, preliminary testing of potentially important processes or feedback is much faster with less complex models. The required complexity also depends on the science question of interest. For example, if the primary interest is in a certain atmospheric process, it could be possible to represent the ocean with a less detailed model without affecting the results. Moreover, to identify which processes are relevant for a certain simulated or observed feature, it may be useful to identify the least complex model configuration that can reproduce it. However, there are no set rules for useful simplifications and the decision requires deep understanding of the models, their processes, and the physical basis.

2 Atmospheric chemistry models

Atmospheric composition and chemistry plays an important role in our climate system e.g., through the greenhouse effect: without it the global average temperature would be about $-18°C$ instead of $+15°C$ (Schneider, 1992). Thus, it is important to understand the variability and long-term change of the main greenhouse gases H_2O, CO_2, CH_4, and O_3 and other species that affect these through chemical reactions.

Pure atmospheric chemistry models (not including dynamics) may have just one altitude level (0-D, box model, see Table 1 for explanations) or use several, vertically connected altitude levels (1-D). They are thus models that do not resolve the latitude-longitude structure. Their advantage is that they can be used to represent chemistry in such detail and complexity that the computing time would become too long for any practical application if the the same amount of detail was included in models representing more dimensions. This kind of models is still used, e.g., for the ionospheric D-region (at 60–90 km altitude), modelling of which requires consideration of tens of ions and hundreds of ionic reactions. Obviously, species that are strongly affected by atmospheric dynamics and transport cannot be well represented.

Chemical Transport Models (CTM) are used to study specific chemical species by simulating the temporal variation caused by chemical production/loss and atmospheric dynamics. The main input to CTMs is solar radiation and energetic particles. Since CTMs are more focused on the chemistry than the dynamics, these models do not compute the dynamics (temperature, winds) from solar input but prescribe winds and temperatures from another model, typically driven by measured data. This makes the results more comparable to episodic observations, e.g., ozone, which may be representative only for specific conditions. An example of a CTM application is stratospheric ozone, which requires a representation of a large number of chemical species and their reactions, including heterogeneous aerosol chemistry.

3 Climate models

Earth's climate is influenced by a variety of factors, including e.g., solar radiation, atmospheric composition, and land use. Depending on the problem to address, a modern climate model may need to include all these and interactions between land, sea, air, and biosphere. As one example, it is now understood that the stratosphere plays an important part in understanding tropospheric climate.

Although there is no generally accepted definition of the term "climate model", it is in general applied only when the model consists of at least coupled general circulation models (GCMs) of atmosphere and ocean (AOGCM). Atmosphere GCMs are used to simulate atmospheric dynamics on a planetary scale. The main input is solar radiation, which is used to calculate the energy input, thermal balance, and the resulting fluid dynamics (e.g., winds). The difference to CTMs is that standard atmospheric GCMs compute prognostic equations for dynamical parameters, such as wind, temperature and water in different phase states, but typically do not incorporate detailed chemistry of atmospheric species. Thus they require radiatively active composition fields, such as carbon dioxide, ozone, and methane but these are imposed in a very simple way, usually with zonally averaged or monthly averaged climatological values that do not specifically

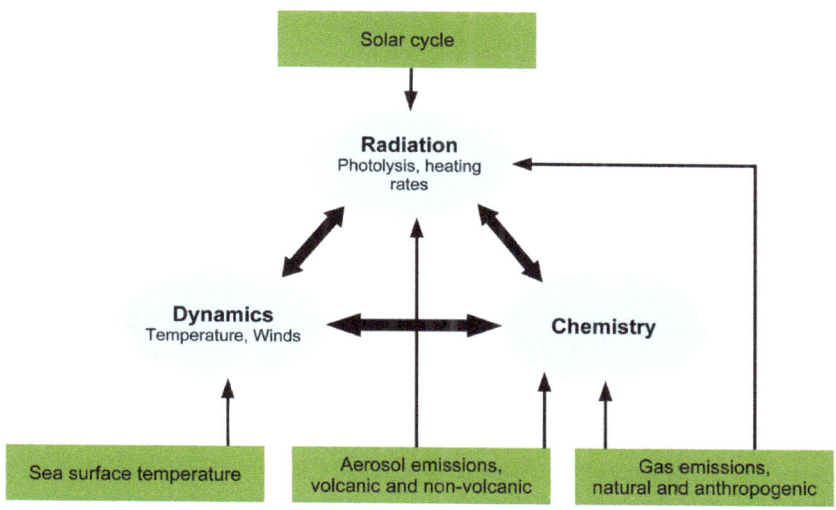

Fig. 2. Basic structure of a CCM and external forcings.

reflect the meteorological conditions. Atmospheric GCMs were first developed for the purpose of weather forecasting. GCMs used for studies of climate processes on decadal to centennial scales are traditionally often offsprings of weather forecast models. Climate models examining longer timescales need to include oceans which contribute significantly to the long-term variability of the climate system.

So-called Chemistry Climate Models (CCMs) typically combine an atmospheric GCM and a CTM. Hence, they include major modules constituting a GCM (the dynamical core, diabatic physics, e.g., radiation, and the transport scheme) plus the chemistry and microphysics modules to describe the chemical composition change[3]. These are linked through the effect of radiation on dynamics, interactions of radiation and chemistry through photolysis of species and absorption and scattering of radiation, and dynamics affecting chemistry through the transport of chemical constituents and the dependence of reaction rates on temperature (Figure 2) together with land and ice models. Thus, CCMs allow for a representation of inter-atmospheric feedbacks, e.g., between stratospheric ozone and tropospheric climate.

The term Chemistry Climate Model is sometimes considered as misleading because traditionally these models are not coupled to ocean GCMs and hence,

---

[3] SPARC Report on the Evaluation of Chemistry Climate Models, Edited by V. Eyring, T. Shepherd and D. Waugh, SPARC, June 2010.

may be more appropriately named atmospheric general circulation and chemistry models. The inclusion of components like atmospheric chemistry or the carbon cycle into coupled AOGCMs has led to the use of the term Earth System model (ESM; see Chapter 3.7). Increasing available computing power has over time on the one hand allowed to make atmospheric and climate models more complex by including new processes. On the other hand, it has enabled the use of models with higher spatial resolution.

4 Solar input needed for atmospheric modeling

Models require solar input as the main driver of photochemistry and, via radiative heating, also for dynamics. The altitude range of a model should determine which set of solar input is included. For instance, models covering only the lowermost atmosphere (troposphere, i.e., the "weather region") only require visible and infrared solar radiation as input. These wavelengths are typically represented as the so-called total solar irradiance (TSI), which is solar radiation integrated over the full spectrum. Ultraviolet and X-ray radiation are heavily absorbed already in the middle and upper atmosphere, so their influence can only be represented in its full extent in a model extending up to the upper atmosphere and the solar spectrum has to be divided into wavelength bins (SSI, solar spectral irradiance), which are then absorbed at different altitudes according to their energy and the height-dependent atmospheric composition. The situation is similar for solar energetic particles: the altitude of atmospheric forcing is dependent upon the particle energy with most of the forcing taking place in the upper stratosphere, mesosphere, and lower thermosphere. Thus, the consideration of "full" solar input requires a so-called whole atmosphere model. In practice, however, to save computing time, many models cover the atmosphere only partially in altitude. For example, many climate models only consider troposphere and stratosphere, and are typically excluding effects of energetic particle precipitation, X-ray radiation, and a large part of the ultraviolet radiation.

In the satellite era, solar irradiance and energetic particle fluxes have been measured from satellites, although temporal and wavelength gaps over the years are sometimes significant. Going beyond into the past requires usage of the so-called solar proxies, i.e., measurements that can be used to estimate the solar forcing. Such proxies include sunspot numbers, magnetic indices, and cosmogenic isotopes from e.g., ice core drilling (Gray et al., 2010). The challenge there is, however, to describe the connection between proxies and solar irradiance. This is done using reconstruction models which typically combine a number of proxies as input (see e.g., Chapter 2.5).

5 Future developments in climate modeling

Future developments in climate modeling will likely be substantially influenced by the expected increase in computer power as it has been the case in the past. With this increase, modelers can improve on the one hand the resolution of the models and on the other hand increase their complexity (see Figure 1 for the past evolution). While typical atmospheric components of climate models today still use horizontal resolutions of the order of 100 km, eventually, climate simulations with grid spacings of the order of 1 km will become feasible. This will allow to gradually avoid the necessity of parametrisations for e.g., clouds, convection, or small-scale dynamical disturbances. The current technical development is rather dominated by an increase of the number of computing processors than by an increasing speed of single processors. This requires a better parallelisation of computer codes, which favours high spatial resolutions. Simulations on longer time scales (e.g., for full glacial cycles) will benefit less from this development. Similar to atmospheric modeling, also the ocean components of climate models will be able to resolve finer scale structures, like ocean eddies. In terms of increasing model complexity, one may for example imagine the inclusion of a larger number of species in chemical modules, a better distinction of different aerosol or vegetation types or the inclusion of biogeochemical cycles for so far often ignored components like nitrogen or phosphorous. However, history has shown that higher complexity does not in general reduce the spread between models performing the same numerical experiments, as e.g., scenarios for the future climate evolution. The understanding of modeling results also gets more difficult for more complex models. As a consequence, Bony et al. (2013) recommend to "recognise the necessity of better understanding how the Earth system works in terms of basic physical principles as elucidated through the use of a spectrum of models, theories and concepts of different complexities". Very simple numerical models should hence keep or even regain their role in climate research. With the focus of a better understanding of sun-climate connections, this means that the modeling groups should not only concentrate on comparing responses to solar forcing in complex models where it is in general difficult to trace back the causes for differing responses but should also try to isolate key processes in very idealised studies.

CHAPTER 3.7

FROM CLIMATE TO EARTH SYSTEM MODELS

Ingo Kirchner and Kerstin Prömmel[1]

1 Climate sensitivity and feedback

The key drivers of the climate system are the Sun, orbital variations, volcanoes and human activities. The Sun heats the Earth, the orbital variations modify the available energy budget, volcanoes and humans change the composition of the air and the surface conditions. The dissipation and advection ensure an adjustment of arising imbalances. However, each component of the Earth system (Peixoto, 1992; Mc Guffie, 2005) has its own adjustment time. The coupling between all components generates internal variations in a more or less chaotic way (Peixoto, 1992). The quasi-random interplay of all forcing factors causes the non-steady state of the climate. The variations have a wide spectrum in space and time and they influence each other. In spite of the quasi-chaotic behaviour of the climate system, the energy flow can be more or less organised. As a result, a fingerprint of a forcing scenario (e.g., differential surface heating as a function of albedo) is visible in the field variables (e.g., near surface temperature of air). The response of forcing changes (e.g., reduction of a net radiative forcing) can be measured as temperature change and is called climate sensitivity.

Changes in Earth system properties can be caused by many different processes (from micro- to macro scale). For example, any single process can lead to an increase or decrease of local temperature. But these single (independent) effects tend to cancel each other out, and their net effect is a slow change. The phase shift between the forcings and their periodicity or aperiodicity creates quasi-stationary changes of the climate. A typical example of counteracting feedback is the temperature-water-vapour feedback (Figure 1, left side). The increase of CO_2 warms the atmosphere due to the greenhouse effect. A higher temperature leads to more water vapour (positive feedback). The water vapour will increase the greenhouse effect on one side, but on the other side, it causes more clouds. The clouds reduce the incoming radiation and cool down the atmosphere.

---

[1] Freie Universität Berlin, Institut für Meteorologie, Germany

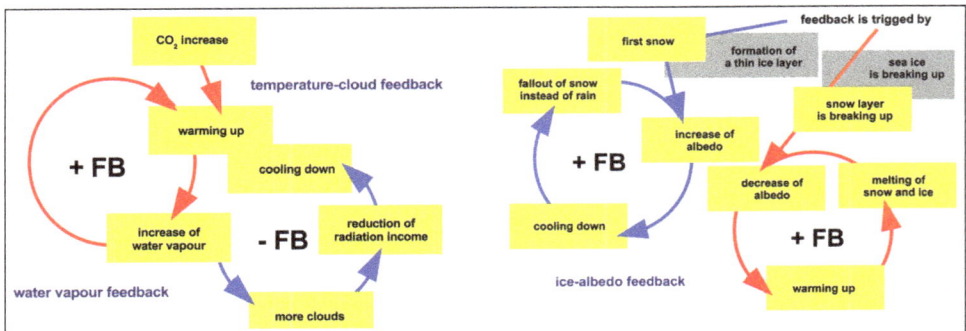

Fig. 1. Feedbacks, left hand side: interaction between a positive and negative feedback, example temperature-water-vapour feedback; right hand side: triggering a positive or negative feedback, example ice-albedo feedback, blue color means cooling and red means heating

This negative feedback will balance the temperature changes (blue circle). Another example of positive feedback is given by the snow-albedo interaction (Figure 1, right hand side). E.g., a small impact in autum might create a snow layer. The increased albedo cools and stabilises the atmosphere, and the next precipitation will fall down as snow. And the same positive feedback causes a speed-up of snow melt in spring. The temperature changes are biased not only due to physical processes. For example in the stratosphere, the temperature is controlled by the ozone concentration. The height of maximum ozone follows the balance between the available solar radiation and the intensity of the chemical reactions. It is the prime task of climate models to adequately simulate the balance between all such feedbacks. Experiments with controlled forcing scenarios allow the investigation of all processes and serve to verify the simulations.

2 Climate model versus "Earth System Model"

The simulation of the climate of the Earth can be realised from different perspectives. While the global energy balance can be used to describe the global (temperature) mean, it is insufficient to describe the local climate of a forest. In addition to the range of spatial scales, small systematic perturbations in the temporal scales become dominant and interact with each other. In order to represent climate sensitivity and feedback, all major components of the Earth system are relevant. The different classes of climate models are well discussed in (Mc Guffie, 2005). The strength of the interaction between radiation, dynamics and surface processes are an indication of the complexity of the model. Without interaction only a single process will be modelled. A model at the top of the climate-model pyramid has all interactions implemented, that's the Earth System Model (ESM). There is no common definition of the term "Earth System Model". In the climate community, "ESM" means the representation of the Earth system with all its physical, chemical and biogeochemical aspects ((IPCC, 2014), Box TS.4 on page 75). These aspects are considered in the different components, such as the atmosphere, the ocean, the sea ice, and the land surface. All subsystems and their interactions (Figure 2) are implemented as computer programs. In practice, an ESM is constructed from a number of multicoupled models. The most prominent components were described in Trenberth (1992). This book gives a comprehensive overview of climate modelling.

The individual components, e.g., subsystems (ocean, atmosphere) or parts thereof (glaciers, seaice, icesheets, snow cover) or a specific module (aerosols), are designed to be coupled together. Interfaces between them exchange information, control the mass flow and replace properties from one part to another. The climate state over a specified period is computer-simulated with the whole Earth system, and the result usually reflects the response to the external forcings. Typical simulation periods are 30 years, or 1000 years, or even much longer, depending on the experimental setting. The climate state is described using the 4-dimensional field variables (space and time) and from that characteristic statistics (mean and variance in time) are calculated.

For the atmosphere, the most common variables are temperature, pressure, density, winds and water in all different phases. The most important oceanic variables are salinity, temperature, density and the flows. For the analysis of the

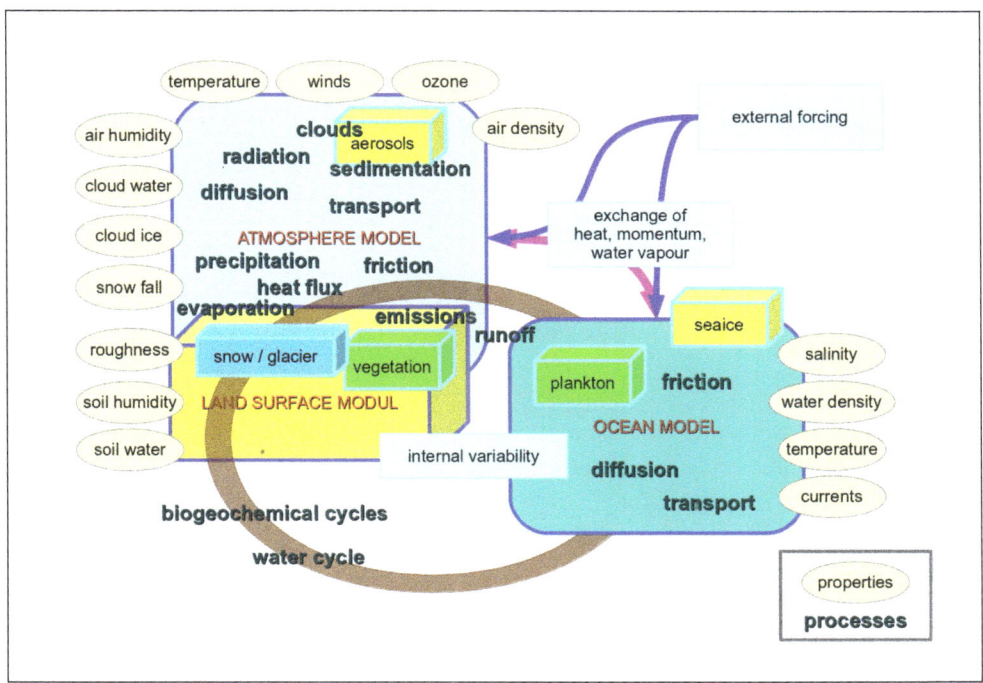

Fig. 2. Essential ESM components, typical properties and important processes.

processes many more variables are calculated. The calculation of mass and energy budgets is a typical part of the diagnostics and necessary for a consistent modelling of biogeochemical cycles (e.g., the carbon cycle or the water cycle).

3   Basic laws and method of solution

The ESM core generally consists of two general circulation models (GCM), the atmosphere and the ocean GCM. Both are useful tools to simulate the global circulation (Washington and Parkinson, 1986). The fundamentals of such models are formulated by partial differential equations that describe the conservation of mass, momentum, and energy, and the law of ideal gases in the atmosphere (Figure 3). The equations are valid for a fluid, for which no simple analytical solution exists so that they must be solved by numerical approximation. The global space is divided into many boxes where the properties are constant and the field variables have to be discretised on the model grid. On the model grid, the value of each single grid cell is coupled with all their neighbours. The equations have to be solved simultaneously. Most of the GCMs use the spectral method (Mote and O'Neill, 1998) as solver. Other techniques are briefly described in Washington and Parkinson (1986). The dynamical core of the model is well separated from the calculation of the physical processes e.g., radiation. Here, the non-linear interactions in the system take place.

The processes in a fluid system are primarily governed by waves. Each physical phenomenon (e.g., cyclones) is related to a characteristic space and time scale (Holton, 1992). The discretisation of the matter and the transport of information in the atmosphere or ocean cause a dependency between the resolution and the maximum possible velocity. The discretisation order of the ESM component determines the dividing line between the resolved and unresolved scales. On the model grid, the local change of a property results from the sum of the substantial change of the property and the change from advection. Due to the advection, the time step Δt has to be small enough to comply the so-called Courant-Friedrichs-Lewy (CFL) stability criterion (Courant et al., 1928).

$$U\frac{\Delta t}{\Delta x} < 1 \qquad (3.1)$$

A stable solution exists only if the CFL criterion holds. Therefore, a smaller grid size Δx requires also a smaller time step Δt. The proportionality is given e.g., by the maximum wind speed U for the atmosphere. A typical value of a time step is 30 minutes for a 100-km grid. Therefore, the resolution of the ESM determines which processes are explicitly resolved. The unresolved processes are approximated by empirical parameterisation. The contribution of the subgrid-scale processes to the tendencies are calculated as a function of the large scale field variables. As described in Chapter 3.6, through the years, more and more components were added to the climate models, parallel to the development of the computer technology. Consequently, the coupling of submodels has become one center of the development.

Fundamental equations of a circulation model (e.g., atmosphere)
(following (Holton, 1992))

momentum balance
$$\tfrac{\partial}{\partial t}\vec{U} = -\vec{U}\bullet\nabla\vec{U} - \omega\tfrac{\partial}{\partial p}\vec{U} + f\vec{k}\times\vec{U} - \nabla\phi + D_M + P_M$$

thermodynamic energy equation
$$\tfrac{\partial}{\partial t}\mathbf{T} = -\vec{U}\bullet\nabla\mathbf{T} + \omega\left(\tfrac{\kappa\mathbf{T}}{p} - \tfrac{\partial\mathbf{T}}{\partial p}\right) + \tfrac{1}{c_p}(\tilde{Q}_{rad} + \tilde{Q}_{con}) + D_H + P_H$$

mass conservation of water vapour
$$\tfrac{\partial}{\partial t}\mathbf{q} = -\vec{U}\bullet\nabla\mathbf{q} - \omega\tfrac{\partial\mathbf{q}}{\partial p} + Eva - Cond + D_q + P_q$$

mass conservation of dry air
$$\tfrac{\partial}{\partial p}\omega = -\nabla\bullet\vec{U}$$

ideal gas law
$$\tfrac{\partial}{\partial p}\phi = -\tfrac{R\mathbf{T}}{p}$$

Steps of solution
- → discretisation on a spherical global grid and truncation
- → implementation of subgrid scale processes as parameterisations
- → coupling of subsystems
- → description of sources and sinks of tracers and chemical compounds
- → time integration scheme

Meaning of the symbols:
$\vec{U}$ horizontal wind vector, ω vertical velocity on pressure coordiantes, f Coriolis force, ϕ geopotential, $D_{M,H,q}$ dissipation of momentum, heat and water vapour, $\mathbf{T}$ temperature, κ Rayleight friction coefficient, p pressure, c_p specific heat at constant pressure, $\tilde{Q}_{rad,con}$ heating due to radiation and condensation, $\mathbf{q}$ specific humidity, $Eva, Cond$ evaporation and condensation, R gas constant, $P_{M,H,q}$ optional physical parameterization terms

Fig. 3. Overview about the model equations of the dynamical part of the ESM and how they are solved, for more details see also infobox 3.2

4 Performance and the big data problem

In addition to the numerical solver, the description of boundary conditions, forcing scenarios and the initial state are necessary to conduct an ESM experiment. Two distinct applications of ESMs are possible, the numerical weather prediction (NWP) and a climate simulation. The dynamical core of the model systems were used for both and any experiment is initialised with a single snapshot of the observed state $X(t_0)$. The temporal evolution of the key variables is calculated step by step using Δt and the tendency $\tfrac{\partial X}{\partial t}$ (see Figure 3), e.g., with the formula:

$$X(t_{new}) = X(t_{old}) + \frac{\partial X}{\partial t}\Delta t \tag{4.1}$$

Due to the uncertainty in the estimate of the initial state, the deterministic forecast is limited to one or two weeks, after which the boundary conditions gain control over the behaviour of the system state. In the operational mode of the NWP, the simulation will be stopped after ten simulation days or a few simulation weeks. For climate scenarios, the integration procedure continues, and the prognostic property X is calculated for many simulation years or centuries. In contrast to NWP, in a climate simulation, the mass balance must be calculated more precisely. The perfect adjustment to the observed last 30 years, called tuning, is one fundamental procedure when applying ESMs.

In a perfect ESM, all climate-relevant details are described. The application of such a model is only possible with an infinitely big and fast computer. Therefore, the development of the ESM has to take into consideration the balance between the complexity of the approximation, the efficient usage of the resources, and the applied scenario. The turn-around time (lapse of time used for execution of the model, e.g., weeks to months) of a single climate experiment has to be manageable during a reasonable project time. There is a strong connection between the complexity of ESMs and the development of computer technology (see Chapter 3.6).

The exponential growth in performance and the increasing parallel methodology create new possibilities of the ESM development, but it also encourages data explosion. Due to the big data problem, the improvements are following different paths. Not all development directions can be combined without bursting the limits of the resources. One of the most important scientific objectives is the explicit simulation of clouds. The decadal forecast will fill the gap between the NWP time horizon of a few weeks and the century-scale climate projections. Ensemble methods are increasingly better in representing the statistical properties of the climate system. The ESMs have the potential for resolving all relevant climate processes and their interactions through their genuine scales of space and time.

Futher reading

McGuffie, K., and Henderson-Sellers, A., A climate modelling primer. John Wiley & Sons, 2005 *is a good starting point for an introduction to climate feedbacks and the levels of complexity of climate models.*

Trenberth, K.E., Climate System Modeling. Cambridge University Press, ISBN-0-521-43231-6, 1992 *presents the components of an Earth system model on an intermediate level.*

Peixoto, J.P., and A.H. Oort, Physics of climate. American Institute of Physics, New York, ISBN-0-88318-712-4, 1992 *gives the theoretical background in depth of the climate system.*

Mote, P., and A. O'Neill, Numerical modeling of the global atmosphere in the climate system. NATO Science Series, Series C, **Vol. 550**, ISBN-0-7923-6301-7,

1998, *presents the modelling aspects with a stronger focus on the details of the implementation.*

Washington, W. M., and C. L. Parkinson, An introduction to three-dimensional climate modeling. Second edition, University science books, Sausalito, California, ISBN-1-891389-35-1, 2005, *as the preceding, presents the modelling aspects with a stronger focus on the details of the implementation.*

Stocker, T., Qin, D., Plattner, G.-K., Tignor, M., Allen, S., Boschung, J., Nauels, A., Xia, Y., Bex, V., and Midgley, P., IPCC Climate Change 2013: The physical science basis. Contribution of Working Group 1 to the Fifth Assessment Report of the Intergovernmental Panel on Climate Change Cambridge University Press, 2013 *discuss the major questions and problems of present and future climate, climate change and climate modelling.*

CHAPTER 3.8

UNCERTAINTIES IN THE MODELING OF THE SOLAR INFLUENCE ON CLIMATE

Eugene Rozanov[1], Timofei Sukhodolov[2] and Kleareti Tourpali[3]

1 Introduction

The response of the Earth's atmosphere and climate to the variability of the Sun obtained from the statistical analysis of the observational data should be explained in terms of the underlying physical, chemical and dynamical processes. It can be done by the analysis of the results of numerical models describing the climate system. The structure of a model aimed at the study of the solar influence on climate is depicted in Figure 1. The core of any model (red circle) consists of the modules to treat the internal processes in the climate system. The set of these processes can be defined differently and depends on the aims of the study. In the case of a 1-D radiative-convective model, mostly the radiation transfer is considered, while in the state-of-the-art Earth System Model (ESM), the treatment of radiation, chemistry, dynamics, physics, biological processes and interaction between them must be included. The model core is responsible for the generation of the climate state.

To take into account the solar influence, the model should be able to treat all solar forcings in the proper modules and provide their contribution to the model core. For example, the increase of the solar irradiance should be treated by the model's radiation code and converted to the additional heating of the atmosphere and land/ocean surface. Accordingly all uncertainties in the models can be related to the: (i) forcing; (ii) interface modules responsible for the forcing treatment;

---

[1] IAC ETH, Zurich, Switzerland and Physikalisch-Meteorologisches Observatorium Davos/World Radiation Center, Davos, Switzerland
[2] Physikalisch-Meteorologisches Observatorium Davos/World Radiation Center, Davos, Switzerland
[3] Aristotle University of Thessaloniki, Lab. of Atmospheric Physics, 54 124 Thessaloniki, Greece

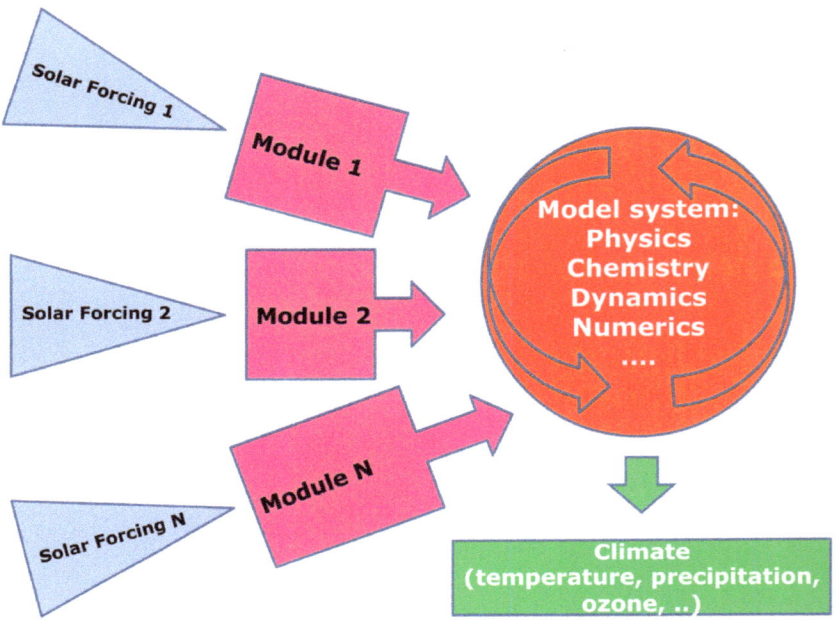

Fig. 1. Typical structure of a numerical model aimed at the study of solar influence on climate.

and (iii) propagation of the initial perturbation by the model core. More detailed discussion of the model uncertainties can be found in the latest IPCC report (IPCC, 2014).

The forcing mechanisms relevant to the solar activity are solar spectral irradiance and different precipitating energetic particles. The uncertainties in the forcing are addressed elsewhere in the book and will not be considered here. The model uncertainties can be related to either not perfect knowledge of the processes (e.g., absorption cross-sections, reaction rates and other parameters) or to the use of too simplified algorithms. These uncertainties can be quantified from the comparison against so-called reference models, which provide the highest accuracy and reflect the state-of-the-art level of knowledge or observational data.

2 Uncertainties in the modeling of the solar irradiance effects

The solar spectral irradiance (SSI) influences the heating rates affecting thermal balance of the atmosphere/surface and photolysis rates affecting the chemical balance. Photolysis is a chemical process by which molecules are broken down into smaller units through the absorption of a photon. The uncertainties in the SSI

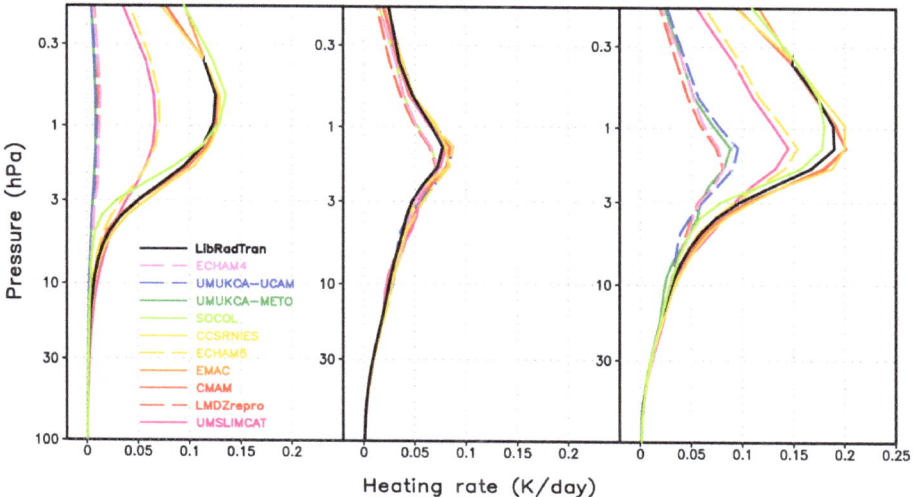

Fig. 2. Global mean, shortwave heating rate differences between minimum and maximum of the 11-year solar cycle in January (K/d), calculated off-line in CCM radiation schemes and reference LBL model (libRadtran). Left panel: Radiative response to prescribed solar irradiance change. Middle panel: Radiative response to solar induced ozone change. Right panel: Total radiative response. The figure is from Forster et al. (2011).

treatment by models depend on the quality of the modules responsible for the radiation and photolysis rate calculations.

Figure 2 illustrates the radiative heating rate response to the increase of the solar UV radiation from the minimum to maximum of the 11-year solar activity cycle simulated by different models. The results of the reference (the most accurate) model are shown in black. The results show that while the influence of the ozone increase is well captured by the models, the radiative response to the increase of solar UV can be correctly simulated only by 3 out of 9 models. Some models perform badly because they were not designed for the study of the effects of SSI variability. Obviously, substantial underestimation of the direct radiative effects leads to uncertainty in the representation of the atmospheric response to the solar variability. For example, if the solar variability induced warming at the tropical stratopause is dramatically reduced, it may lead to much weaker or negligible response of the entire climate system.

Figure 3 illustrates the uncertainties in the simulation of the oxygen photolysis rates to the solar UV increase from the minimum to the maximum of the solar activity. The oxygen photolysis is one of the most important chemical reactions in the atmosphere, because it constitutes a unique source of ozone. The figure shows the relative deviation in percent of different model results from the reference calculations. Different spectral resolution of the models and other simplifications lead to the some uncertainties in the simulated oxygen photolysis and subsequent

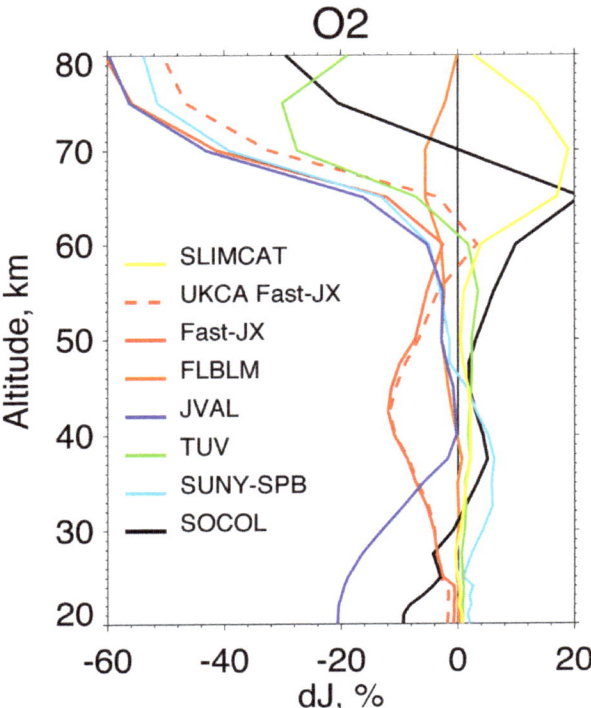

Fig. 3. The deviation (%) in the oxygen photolysis rate response to the increase of solar UV from solar activity minimum to maximum simulated by several models from the reference results obtained with libRadtran.

inaccuracies of the simulated ozone response to the UV increase. In the middle stratosphere, the model errors are about 10%, while more pronounced problems appear in the mesosphere and lower stratosphere where the error can reach 70%.

3 Uncertainties in the modeling of energetic particles effects

Precipitating energetic particles can be divided into galactic cosmic rays (GCR), solar protons (SP), and low- and high-energy electrons (LEE and HEE). The distribution of ionisation rates in space and time depends on the particle type because of their interactions with the variable heliomagnetic and geomagnetic fields, but all these particles are able to ionise neutral molecules (e.g., N_2 and O_2) in the Earth's atmosphere producing reactive hydrogen and nitrogen oxides. These species are able to destroy ozone via a number of catalytic cycles with implications for the temperature and circulation regimes in the atmosphere. The detailed description of ion chemistry is too complicated and computer time consuming for the majority of the chemistry-climate models; therefore the production of reactive

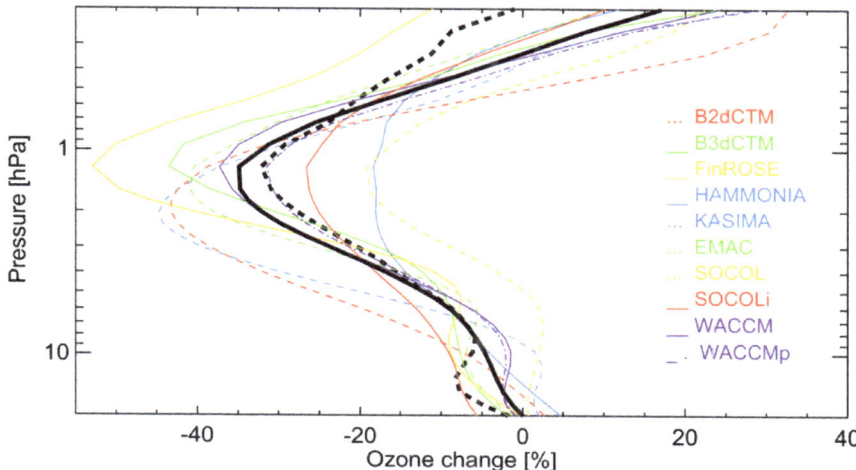

Fig. 4. Observed and modeled relative O_3 changes over 70–90° N during 29.10–4.11.2003 with respect to quiet period. Thick solid and dashed lines represent multimodel mean average and MIPAS observations, respectively. The figure is from Funke et al. (2011).

hydrogen and nitrogen oxides by energetic particles inside the model domain is parameterised as a function of ionisation rates. For the low energy (or auroral) electrons which deposit their energy around 110 km, such an approach is possible only for the models covering the thermosphere while for the low top models the influx of the nitrogen oxides is usually parameterised as a function of geomagnetic activity. More information about precipitating particles and their effects on the atmosphere and climate can be found in Chapters 4.5 and 4.6.

The accuracy of these modules was evaluated by Funke et al. (2011) using the comparison of the simulated and observed atmospheric chemistry response to the strong solar proton event (SPE). The simulated and observed response of the nitrogen oxides to October 2003 SPE is shown in Figure 4. Funke et al. (2011) concluded that the applied methods of energetic particles treatment demonstrate reasonably good performance but some disagreement between models and observations still exists illustrating the uncertainties in modeling. The causes of the disagreement are most likely related to different treatment of important atmospheric processes in model cores, because all models used the same forcing data while the deviations from the observation results differs among the models.

4 Uncertainties in the modeling of the climate system response

The above-described variability of the solar forcing may affect the climate system via different chains of the processes. The weakly absorbed radiation with wavelength exceeding 300 nm is absorbed in the troposphere or at the surface leading to the alteration of the energy balance, hydrological cycle and global climate. The simulated global mean and regional features of the climate response to this particular forcing resembles the response to the greenhouse gases. The recent IPCC report ((IPCC, 2014), Box TS.4, SPM D.1) stated stated that the model uncertainties in the simulation of the continental scale surface temperature are rather small and most features of the observed long-term temperature trend are well reproduced. However, model uncertainties in the surface temperature simulations on the regional scale and short time scales still remain rather high.

The shortwave part of the solar UV irradiance and energetic particles mostly affect the middle atmosphere. Their influence on the tropospheric climate can be achieved through the modulation of the atmospheric circulation and wave patterns. The chain of physical processes responsible for the downward propagation of the stratospheric perturbations is called "top-down" mechanism ((IPCC, 2014), 109 Box 10.2). More intensive solar ultraviolet radiation during the solar maximum years warms up the stratosphere due to enhanced heating rates and strengthens the equator-to-pole temperature gradient. This perturbation propagates downward to enhance the tropospheric mid-latitude westerlies and invoke a positive phase of the North Atlantic Arctic Oscillation (NAO), leading finally to a warmer and dryer climate over the Northern Europe. This mechanism then has small direct influence on global mean climate properties, but can strongly influence the regional temperature pattern during the winter time and mostly over the Northern hemisphere. As it was mentioned earlier, all current models have large uncertainties in the simulation of the surface temperature on regional and short-term scale. Therefore, the uncertainties of the "top-down" mechanism modeling are rather high. The efficiency of this mechanism is model-dependent and at the moment, there is no clear understanding which processes are responsible for its operations. One of the hypotheses is that the efficiency of the "top-down" mechanism depends on the state of the polar vortex, which in turn is regulated by a multitude of processes, including poorly known and represented gravity wave drag. More detailed description of "top-down" mechanism can be found in Chapter 4.2.

5 Conclusions

The uncertainties in the simulation of the climate response to the solar variability can be divided into two groups: representation of the direct effects of solar forcing and transformation of the solar forcing to climate by model core (see Figure 1). The uncertainties of the first kind are better known and can be established using extensive intercomparison campaigns similar to presented by Forster et al. (2011) and Funke et al. (2011). The uncertainties of the second kind are difficult to qualify because they depend on many processes in the system, intensive feedbacks between them and their implementation in models.

Further reading

SPARC CCMVal: SPARC Report on the Evaluation of Chemistry- Climate Models, edited by: Eyring, V., Shepherd, T. G., and Waugh, D. W., SPARC Rep. 5, Univ. of Toronto, Toronto, Ont., Canada, 2010, available at: http://www.sparc-climate.org/publications/sparc-reports/sparc-report-no5/.

CHAPTER 3.9

DETECTION AND ATTRIBUTION: HOW IS THE SOLAR SIGNAL IDENTIFIED AND DISTINGUISHED FROM THE RESPONSE TO OTHER FORCINGS?

Kristoffer Rypdal[1], Martin Rypdal[1] and Sverre Holm[2]

1 Introduction

There will always be variability in the Earth's climate, even in the absence of external forcing, like variation in solar irradiance, volcanic eruptions, or human-induced changes. The nature of internal climate variability is analogous to the change of weather, just extrapolated to longer spatial and temporal scales. This "song of Nature" is comprised of a cacophony of frequencies corresponding to the natural modes of the climate system and forms a background spectrum with a pink-noise character. This means that the power spectral density of global temperature to a crude approximation has the form $S(f) \sim 1/f$, for frequencies f corresponding to periods from months to millennia. The shape of this spectrum implies that internal variability on low frequencies (long time scales) is strong. Another way to put this is that the climate exhibits correlations on long time scales; sometimes described as long-range memory. The variability created by these correlations constitutes a problem when we want to detect climate signals and trends with external causes.

Signal detection means to establish the statistical significance of a trend, an oscillation, or a spatiotemporal pattern. This is successfully done if we can establish that it is very unlikely that the pattern of interest has arisen by chance from the internal background noise. Once a pattern has been successfully detected, the next issue is to identify a cause, or more general, to identify and assess the relative weight of a number of causes. This process is what we call attribution.

---

[1] Department of Mathematics and Statistics, UiT - The Arctic University of Norway, N-9037 Tromsø, Norway
[2] Department of Informatics, University of Oslo, Norway

© EDP Sciences 2015
DOI: 10.1051/978-2-7598-1733-7.c124

2 Detection

It is often claimed that the 11-year solar-cycle signal is detectable in global temperature. However, this statement is not precise, since the term "solar-cycle signal" refers to attribution, not to detection. Here, we shall illustrate how detection works by testing the hypothesis of a multidecadal oscillation in the global land temperature record. Numerous observational and some theoretical studies suggest that such an oscillation exists with main period around 70 years. The period is not sharp and fixed, but as a crude model, we can represent is as a sinusoid with this period. The hypothesis is then that the temperature time series can be written in the form;

$$T(t) = \delta + A_1 t + A_2 \sin(2\pi f t + \varphi) + \sigma w(t), \qquad (2.1)$$

where f has been fixed to about 70 years and $w(t)$ is a random process with standard deviation σ. The parameter δ represents the background temperature level, ϕ the phase of the oscillation, while the interesting parameters are the trend coefficients A_1 and A_2 which represent the strengths of the linear and oscillatory trends. The parameters can be estimated to give the best least-square fit to the observed record. We shall assume here for simplicity that the linear trend is a physical reality and known. The method for testing whether the oscillation is real is to assume that the converse is true, i.e., to hypothesise that it can be described as a natural fluctuation of the internal noise. This *null hypothesis* takes the form $\delta + A_1 t + \varepsilon(\theta; t)$, where $\varepsilon(\theta; t)$ represents a model for the internal noise depending on a set of parameters θ, which will be estimated by fitting to the observed data. When θ has been estimated, we can generate an ensemble of synthetic samples satisfying the null hypothesis and then fit the trend model Eq. (2.1) to each sample series. For each sample, we find estimates $(\hat{A}_1, \hat{A}_2)$ satisfying the null hypothesis, and we can build a probability density distribution (PDF) $P(\hat{A}_1, \hat{A}_2)$. In Figure 1a, we have plotted a contour of this joint PDF inside which the probability is 95%. Since (A_1, A_2) estimated from the observation data (red dot) is outside this contour, we may conclude that it is unlikely that it can occur by chance from the null hypothesis. In Figure 1b we have plotted the cumulative probability function (CDF) over the oscillation amplitude A_2. The dotted line marks the 95% cumulative probability and the full vertical line is A_2 estimated from the observed data. This means that the observed amplitude is outside the 95% confidence interval of the null hypothesis, and hence that the oscillation is statistically significant. We have *detected* an oscillatory trend, but we have not attributed it to any particular external cause.

3 Attribution by multiple regression

A standard method in attribution studies is that of multiple linear regression. The idea is to separate the climate signal into a number of components assumed to represent the climate response to individual forcings. Each of these components have a certain characteristic shape (or fingerprint). In order to determine these

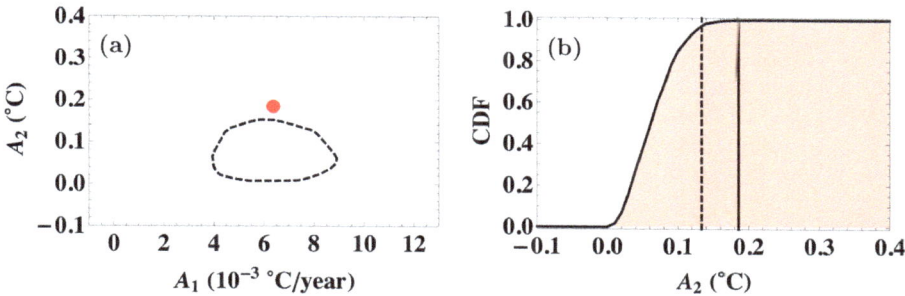

Fig. 1. (a): The 95% confidence contour (dashed) of the distribution $P(\hat{A}_1, \hat{A}_2)$ for global land temperature obtained by the null model where $\varepsilon(\theta;t)$ is assumed to be a pink-noise stochastic process. Observed trend coefficients are indicated by the red dot. (b): The CDF derived from $P(\hat{A}_2)$ for this null model, with upper 95% confidence limit as dotted vertical line and the A_2 estimated from observation as full vertical line.

fingerprints, we need models of some sort. Full-scale general circulation climate models can be used, but often also simpler, conceptual models are useful. The rationale for attribution studies is that even the most advanced climate models may estimate wrongly the magnitude of individual responses. even though they have got the fingerprints right. Hence, we may write the total climate signal $T(t)$ as a linear combination of the fingerprints. Consider, for instance, the global temperature $T(t)$ and the fingerprints of various forcings and an internal mode. Then we can write,

$$T(t) = c_S S(t) + c_V V(t) + c_A A(t) + c_I I(t) + \sigma w(t), \tag{3.1}$$

where $S(t)$, $V(t)$, and $A(t)$ are the fingerprints of solar, volcanic, and anthropogenic forcing, respectively, and $I(t)$ is the fingerprint of the internal mode. The fitting parameters c_S, c_V, c_A, and c_I translate each fingerprint into a temperature response, and can be estimated by minimizing the least square error with respect to the observed data. A measure of how successfully the method attributes variability to the various forcing components is to compute how much of the observed variance that is explained by the model.

One common problem with this approach is that if there are many causal factors to consider, and hence many parameters to fit, there is a risk of overfitting. This means that a good fit can be obtained even though the result is unphysical. Another problem is that the fingerprints of forcing in general are distorted and delayed by inertia in the climate response caused by slow heat exchange between the ocean surface layer and the deep ocean, sea ice, and ice sheets. This inertia may, for instance, lead to a small response to the relatively fast solar cycle forcing, while the response to slow trends in solar irradiance may be stronger, but considerably delayed. Delay effects are not accounted for in the regression model (3.1) if the model defining the fingerprint does not involve a dynamic response to forcing. A conceptual model of such a dynamic response is described in Box 4.1, and in Figure 1c, there one can observe a 0.1 °C increase in temperature in the period 1960–2010 attributed to solar irradiation, although the solar forcing did not grow in this period. This increase is a delay effect of the increasing trend in solar irradiance during the first half of the 20th century. Note also that the 11-year solar-cycle signal in that figure is very weak. This is because of the inertia in the temperature response.

Sometimes crude approaches, such as studying the effect of removing a particular forcing component from a model, can provide convincing evidence that this forcing must be required to explain observed climate changes. Fig 1 in Box 4.1 demonstrates that natural forcing only (solar + volcanic) cannot explain the warming trend in global temperature over the last 50 years.

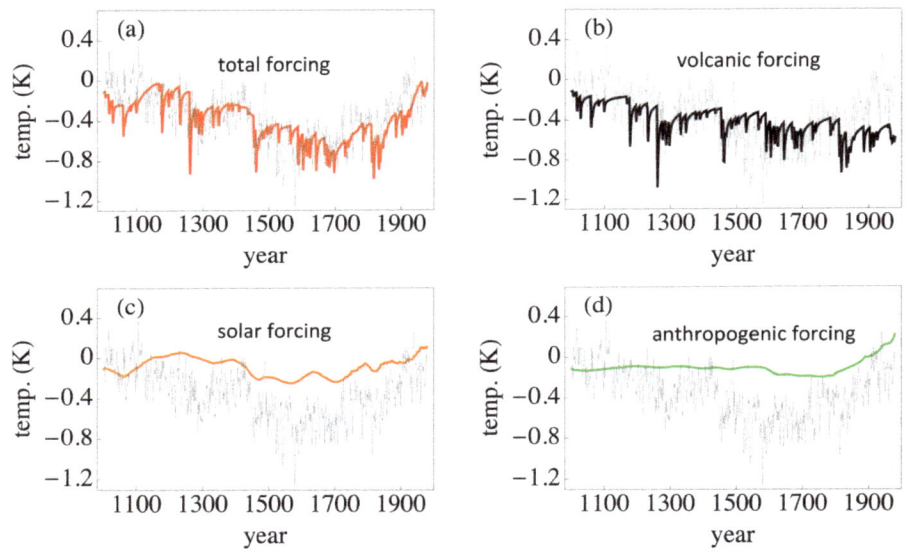

Fig. 2. Forced temperature change 1000–1979 AD according to the conceptual response model. Grey curve is the observed temperature record as reconstructed from paleoproxies. (a) From total forcing. (b) From volcanic forcing. (c) From solar forcing. (d) From anthropogenic forcing.

4 Attribution example using a conceptual model

Even though the natural forcing, including solar, seems unable to explain recent global warming, it could have played a significant role in the near pre-industrial past. In Figure 1 we have applied the same conceptual model as in Box 4.1, with the same model parameters as computed for the instrumental 20th century data, to the forcing reconstructed for the period 1000–1979 AD. We observe that the response to the total forcing reproduces the large-scale features of the Northern-hemisphere temperature reconstruction represented by the grey curve. It also appears that most of the temperature decline between the medieval warm period peaking around 1100 AD to the "little ice age" peaking around 1650 AD is caused by volcanic activity, which was very low in the medieval period. It seems that only one third of this temperature drop was caused by the reduction in solar irradiance associated with the grand solar minimum called the Maunder minimum.

5 Attribution of solar signal in regional data

The solar signal in global temperature appears to be weak throughout the 20th century, but it may be stronger in regional and seasonal climate variability. Such variability, associated with the 11-year cycle has been studied in some models, and several physical mechanisms of amplification of the solar forcing have been proposed. For such signals, the attribution issue is mathematically complicated since the fingerprints now depend both on space and time coordinates. One technique used for such studies is to decompose the spatial temperature field into patterns called empirical orthogonal functions (EOFs), and these patterns are ranked according to how much of the total variance they explain. Each EOF is multiplied by a time-dependent coefficient, which is called the principal component of that EOF. Often a few of these EOFs with their principal components explain most of the variability and in some cases a particular principal component has a temporal fingerprint that matches the solar forcing signal. In such cases, the principal component time series can be analysed by a multiple, linear regression model and one can investigate how much of the variability of the climate mode corresponding to the particular EOF can be attributed to solar forcing.

6 Attribution of cycles

It has been suggested that the multidecadal oscillation, whose significance was studied in Section 2, can be attributed to cycles in the motion of the sun caused by the giant planets. The rationale for such a proposal is that some spectral analyses of the climate time series and time series for planetary motions show some spectral lines at the same frequencies. Two properties can be used to test such a hypothesis. The first is to compare the stationarity of the spectral lines in question, in particular those at around 20 and 60 years. Lack of stationarity means that what appears as a sharp frequency and phase of an oscillation really changes with time, and can be investigated by different techniques classified as time-frequency analysis. Another technique is test for coherence using the well established magnitude squared coherence function. Application of these kinds of tests demonstrate very clearly that the climate signal lacks the stationarity and the coherence required to explain the spectral lines as oscillation driven by the planetary motions. There are some more substantiated studies based on spectral analysis that suggest a connection between planetary motions and solar activity reconstructed from radio-isotope proxies over the last ten millennia. These results are also controversial, however, and they do not imply that the corresponding spectral lines detected in solar activity have a discernible effect on the climate.

In general, the detection and attribution of cycles remains a topic of much controversy. It is known that there are cycles in the solar forcing due to variations in the orbit of the Earth around the Sun, and in the tilt and precession of the Earth's rotation axis. These so-called Milanković cycles have periods of thousands of years and are thought to act as pacemakers of ice ages. The attribution of ice ages and variability within ice ages to this orbital forcing, however, is not trivial, and a number of unsolved issues remain.

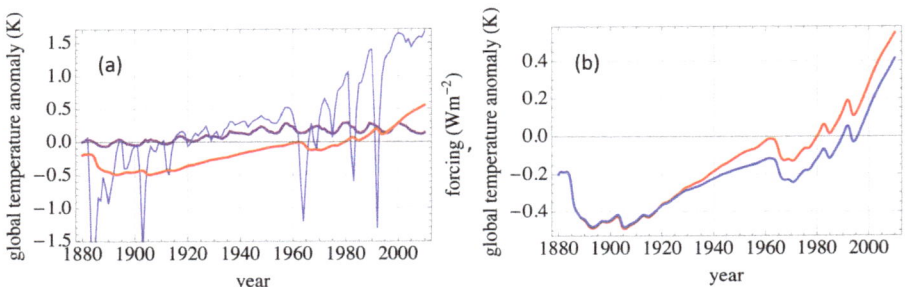

Fig. 3. (a): shows total forcing (blue), where the negative dips are due to volcanic eruptions, and forcing from total solar irradiance (magenta). The red curve is the temperature response to the total forcing. Note that the three last dips due to eruptions nearly coincide with solar minima. (b): shows the response to the total forcing (red) and the response to the forcing where the solar forcing has been subtracted.

What about the temperature response to the 11-year solar cycle? As discussed in Section 3, the inertia in the climate response gives reason to believe that this response is rather weak, and it is very difficult to detect such a signal in the noisy temperature signal. Nevertheless, claims have been made that a solar cycle is visible in temperature records over the last 4-5 cycles. We find such oscillatory structures with peak-to-peak amplitude of almost 0.1 K in the temperature response to total forcing and they seem to be in phase with the solar cycle. However, their real cause becomes apparent when looking at the total and solar forcing signals in Figure 3a, and at the response to forcing with, and without, the solar component, as shown in Figure 3b. Surprisingly, the oscillatory response remains unaltered after the solar forcing has been eliminated. It turns out that this strong oscillation is caused exclusively by a few volcanic eruptions incidentally taking place just prior to solar minima (Figure 3a), and hence demonstrates the necessity of careful attribution studies.

Further reading

Abreu, J. A., Beer, J., Ferriz-Mas, A. , McCracken, K. G., and Steinhilber F., 2012, Is there a planetary influence on solar activity?, Astronomy and Astrophys., A88, 548-557, doi:10.1051/0004-6361/201219997.

Bindoff, N. L., P. A. Stott, K. M. AchutaRao, M. R. Allen, N. Gillett, D. Gutzler, K. Hansingo, G. Hegerl, Y. Hu, S. Jain, I.I. Mokhov, J. Overland, J. Perlwitz, R. Sebbari and X. Zhang, 2013: Detection and Attribution of Climate Change: from Global to Regional. In: Climate Change 2013: The Physical Science Basis. Contribution of Working Group I to the Fifth Assessment Report of the Intergovernmental Panel on Climate Change [Stocker, T. F., D. Qin, G. -K. Plattner, M. Tignor, S. K. Allen, J. Boschung, A. Nauels, Y. Xia, V. Bex and P. M. Midgley (eds.)]. Cambridge University Press, Cambridge, United Kingdom and New York, NY, USA.

Holm, S., 2014, On the alleged coherence between the global temperature and the sun's movement, J. Atmos. Solar-Terrestr. Phys., 110-111, 23-27.

Lean, J. L. and Rind, D. H., 2009, How will Earth's surface temperature change in future decades? Geophys. Res. Lett., 36, L15708, doi:10.1029/2009GL038932.

Muller, R. A., and MacDonald, G. J., 2000, Ice Ages and Astronomical Causes – Data, spectral analysis and mechanisms, Springer.

Rypdal, K., 2012, Global temperature response to radiative forcing: Solar cycle versus volcanic eruptions, J. Geophys. Res., 117, D06115, doi:10.1029/2011JD017283.

Rypdal, K., Østvand, L., and Rypdal, M., 2013, Long-range memory in Earth's surface temperature on time scales from months to centuries, J. Geophys. Res., 118, doi:10.1002/jgrd.50399.

Rypdal, M., and Rypdal, K., 2014, Long-memory effects in linear-response models of Earth's temperature and implications for future global warming, J. Climate, 27, 5240-5258.

Scafetta, N., 2012, Testing an astronomically based decadal-scale empirical harmonic climate model versus the IPCC (2007) general circulation models, Journal of Atmospheric and Solar-Terrestrial Physics, 80, 124-137, doi:10.1016/j.jastp.2011.12.005.

von Storch, H. and Zwiers, F. W., 1999, Statistical Analysis in Climate Research, Cambridge University Press.

Østvand, L., Rypdal, K., and Rypdal, M., 2014, Statistical significance of rising and oscillatory trends in global ocean and land temperature in the past 160 years, Earth Syst. Dynam. Discuss., 5, 327-362. doi:10.5194/esdd-5-327-2014

INFOBOX 3.1

WHY ARE MODELS NEEDED IN THE FIRST PLACE, AND CAN THEY BE TRUSTED?

Dann Mitchell[1]

    The Earth's coupled ocean-atmosphere system is highly complex and can vary due to many different natural and anthropogenic processes. These are known as 'forcings' and in addition to the solar forcing, include volcanic gases, well mixed greenhouse gases, aerosol effects and various other internal climate variations. At any particular point, both temporally and spatially, our observed climate may vary due to any combination of these forcings. Unfortunately we do not have a long enough time series of good-quality observations to be able to clearly pick apart the different influences that different forcings have on the climate, so that estimates of these influences are statistically uncertain. The advantage of using models is that we can isolate individual forcing components because we can control the model inputs. If, for instance, we would like to see how the climate might vary with enhanced solar forcing, we can change this parameter in the model and compare the subsequent model output with a control simulation of normal solar forcing. We might also wish to see how the climate might vary if humans never existed, and therefore we could simply exclude the anthropogenic forcings from our model simulations. The adaptation of climate models therefore allows us to perform hypothetical experiments on the Earth's climate that are impossible with observations alone.

In addition to the increased mechanistic understanding of solar variability and its effect on climate, models are essential for forecasting future climate conditions. The 11-year solar cycle component of incoming solar irradiance is quasi-periodic, and therefore if we know how climate is changed during different phases of the solar cycle, our skill in predicting seasonal-decadal forecasts can be improved.

One potential drawback of using models is that if they do not represent climatic processes well, then performing experiments with them may lead to misleading results. The aforementioned drawback naturally leads to the question 'can models be trusted?'. Indeed, there is a famous quote by the British statistician George Box; "essentially, all models are wrong, but some are useful". This statement is

---

[1] National Centre for Atmospheric Science, University of Oxford, Oxford, United Kingdom

of course correct because if models were perfect, and we knew exactly the initial conditions of the simulations along with the precise forcing components, our hindcasts of past climate would always be correct. This is not the case, and so one must ask 'to what degree are our models wrong and therefore to what degree can they be used to understand climate?'. The answer is of course subjective, and dependent on the type of scientific question we are addressing, but modelling groups from around the world will rigorously assess many aspects of their models in order to test the bounds to which they can be used. Any science based on model experiments is interpreted in terms of our physical understanding of the climate system, and crucially, in terms of what our observations of the historical climate system tell us. For instance, global-mean temperatures near the stratopause (about 50 km) are observed to vary with a cycle of 11 years (as well as other superimposed cycles of varying length). This is coincident with the 11-year cycle of incoming total solar irradiance, and physically we understand this because we know that incoming UV radiation (a part of the total solar irradiance) interacts with the ozone at this level, to cause the 11-year cycle in temperature. Models that are able to capture this behaviour may therefore be used to gain further insight into our understanding of this particular effect. This example can be generalised to broader aspects of the climate system leading to our hypothesis that models can be 'trusted', if they capture the observed behaviour and statistics of climate, for the right reasons, and on the timescales of interest.

INFOBOX 3.2

MODEL EQUATIONS AND HOW THEY ARE SOLVED

Ulrike Langematz[1]

1 The fundamental equations

The state of Earth's atmosphere and its temporal and spatial behaviour are governed by the fundamental laws of fluid mechanics and thermodynamics, which are the conservation of momentum, energy, and mass. For an infinitesimal volume of fluid particles in the Lagrangian framework, these can be expressed in terms of partial differential equations involving the physical quantities wind speed and direction, temperature, pressure, and humidity, for example, as functions of space and time. The evolution of the three-dimensional wind vector $\vec{v}$ is governed by Newton's second law and given by the Navier-Stokes momentum equation:

$$\frac{d\vec{v}}{dt} = -\frac{1}{\rho}\nabla p - 2\vec{\Omega} \times \vec{v} + \vec{g} + \vec{F_R} \qquad (1.1)$$

Equation (1.1) states that the acceleration $d\vec{v}/dt$ of an infinitesimal volume of air particles is determined by the sum of the acting forces, i.e., the pressure (p) gradient force (divided by air density ρ), the Coriolis force due to Earth's rotation with the angular rotation rate $\vec{\Omega}$, the gravitational force $\vec{g}$, and the frictional force $\vec{F_R}$. Note that the terms on the right side of the equation represent forces per unit mass. Vectors are marked by arrows on top.

The conservation of energy is represented by the first law of thermodynamics

$$c_p \frac{dT}{dt} - \frac{1}{\rho}\frac{dp}{dt} = Q, \qquad (1.2)$$

where c_p stands for the specific heat capacity of air at constant pressure, and Q for the diabatic net heating rate per unit mass. The equation involves changes of temperature T due to energy conserving adiabatic processes, such as expansion cooling or compression heating, as well as heating or cooling by diabatic processes,

---

[1] Institut für Meteorologie, Freie Universität Berlin, Carl-Heinrich-Becker-Weg 6-10, 12165 Berlin, Germany

for example by radiative transfer or the release of latent heat associated with condensation processes.

Finally, the conservation of mass is expressed by the continuity equation

$$\frac{1}{\rho}\frac{d\rho}{dt} + \nabla \cdot \vec{v} = 0 \qquad (1.3)$$

Equation (1.3) states that the fractional change of air density following the motion of an air parcel is equal to the negative of the velocity divergence. In other words, locally air density can only change if a convergent or divergent mass flux exists.

Equations (1.1) to (1.3) describe the state of the atmosphere on all spatial and temporal scales. For meteorological applications, a number of transformations and approximations of the basic equations are introduced, which facilitate the solution of the equations. These include:

- the replacement of the vectorial Navier-Stokes momentum equation by three scalar equations for the wind components u (zonal wind), v (meridional wind) and w (vertical wind),

- the replacement of the Cartesian coordinate system (with the x-axis pointing from west to east, the y-axis pointing from south to north, and the z-axis pointing upward) by spherical coordinates with λ the geographical longitude in the easterly direction, ϕ the geographical latitude in the northerly direction, and z pointing upward,

- the replacement of the total (or material) time derivative d/dt using the Euler operator $d/dt = \partial/\partial t + \vec{v} \cdot \nabla$ The total derivative, which describes the total time rate of change of a fluid parcel in a fixed Eulerian framework, can be split into the local time rate of change and the advection of the property by the wind field,

- using an isobaric (constant pressure) vertical coordinate instead of the geometric height coordinate. Log-pressure altitude is then derived by

$$z_p = -H \ln p/p_0, \qquad (1.4)$$

where p_0 is a reference pressure. H is a scale height that is determined by the temperature profile of the considered height range. In stratospheric applications for example is $H = 7$ km, representative for a mean stratospheric temperature of 240 K.

In addition, three approximations are commonly applied to the basic system of equations, i.e.,

- the assumption of hydrostatic balance. Except for small-scale motions, the acceleration of the vertical wind component w can be neglected ($dw/dt = 0$)

and the vertical component of the momentum equation reduces to the hydrostatic equation:

$$\partial p/\partial z = -g\rho \qquad (1.5)$$

- the approximation of the distance between an altitude r in the atmosphere and Earth's center ($r = a + z$) by Earth's radius, $r \approx a$, and

- the neglect of the Coriolis acceleration due to the horizontal component of Earth's rotational vector.

Equations including the above three simplifications are called the primitive equations. The nature of the simplifications still allows to apply the equations to a wide range of circulations, except for small-scale, non-hydrostatic motions. With the introduction of the geopotential Φ ($d\Phi = gdz$), the following system of primitive equations forms the basis for global prognostic models of the horizontal wind components and temperature on a sphere:

$$\frac{du}{dt} - \frac{1}{a}\tan\phi\, uv - 2\Omega\sin\phi\, v + \frac{1}{a\cos\phi}\frac{\partial\Phi}{\partial\lambda} = F_\lambda \qquad (1.6)$$

$$\frac{dv}{dt} + \frac{1}{a}\tan\phi\, u^2 + 2\Omega\sin\phi\, u + \frac{1}{a}\frac{\partial\Phi}{\partial\phi} = F_\phi \qquad (1.7)$$

$$\frac{\partial\Phi}{\partial z} - \frac{R_d}{H} = 0 \qquad (1.8)$$

$$\frac{dT}{dt} + w\frac{\kappa T}{H} = \frac{Q}{c_p} \qquad (1.9)$$

$$\frac{1}{a\cos\phi}\frac{\partial u}{\partial\lambda} + \frac{1}{a\cos\phi}\frac{\partial(v\cos\phi)}{\partial\phi} + \frac{1}{\rho_0}\frac{\partial\rho_0 w}{\partial z} = 0, \qquad (1.10)$$

where $\kappa = R/c_p$ and R is the specific gas constant for air. ρ_0 is a reference air density. When integrated with suitable initial and boundary conditions, the integration of the primitive equations allows to derive the spatial and temporal evolution of atmospheric dynamical and thermal processes. This basic set of primitive equations can be extended to determine other prognostic variables, such as specific humidity. Moreover, depending on the considered spatial scales approximated sets of equations may be used which neglect individual terms of negligible magnitude. More detailed information on the conventional systems of equations can be found for example in Holton (1992) and Andrews et al. (1987).

2 Solution of the model equations

The above system of equations provides the basis for the analysis and prediction of the atmospheric state in numerical weather prediction (NWP) or climate models. As the primitive equations are too complex to allow exact solutions, they are solved by applying numerical integration techniques, such as finite-difference, series expansion, or finite-volume methods. For example, by applying the finite-difference approximation, each continuous differential operator d in a differential equation is replaced with a discrete difference operator Δ. This means, a meteorological variable, which continuously varies in space and time like the zonal wind component u, is mapped to a finite number of values at each spatial node of a grid at discrete time steps. Figure 1 shows as an example the zonal wind component u in a west-east grid with i grid cells at a fixed time step. The distance from the western edge of the grid to the western edge of cell i is x_i. In such a grid, the differential zonal velocity du at point x_i can be approximated with $\Delta u_i = u_{i+1} - u_{i-1}$ (central difference) or $\Delta u_i = u_{i+1} - u_i$ (forward difference) or $\Delta u_i = u_i - u_{i-1}$ (backward difference), respectively. Applying the corresponding discretisation to x_i, the partial derivative of u with x at point x_i can be approximated by:

$$\frac{\partial u}{\partial x} \approx \frac{\Delta u_i}{\Delta x_i} = \frac{u_{i+1} - u_{i-1}}{x_{i+1} - x_{i-1}}. \tag{2.1}$$

An alternative method to horizontal discretisation is to expand the prognostic variables in terms of truncated series of spherical harmonics (spectral transform models). A variable $X(\lambda, \mu, \eta, t)$ (with $\mu = \sin\phi$ and η the vertical coordinate) is represented by:

$$X(\lambda, \mu, \eta, t) = \sum_{m=-M}^{M} \sum_{n=m-1}^{N(m)} X_n^m(\eta, t)\, P_n^m(\mu)\, e^{im\lambda} \tag{2.2}$$

with $P_n^m(\mu)$ the latitude-dependent associated Legendre functions and $X_n^m(\eta, t)$ the spectral, complex coefficients of the array X. The horizontal resolution of a model is then determined by the maximum zonal wavenumber M considered in the expansion. Depending on the application of the model, horizontal resolutions of spectral models vary between T21 (i.e., 21 zonal wavenumbers, corresponding to about 625 km at the equator) in early climate models[2] and T1297 (i.e., 1297 zonal wavenumbers, corresponding to approximately 17 km) in the operational European Centre for Medium-Range Weather Forecast (ECMWF) model (Diamantakis and Flemming, 2014).

---

[2] Roeckner, E. *et al.* 1992, Simulation of the present-day climate with the ECHAM model: Impact of model physics and resolution. MPI für Meteorologie, Report No. 93.

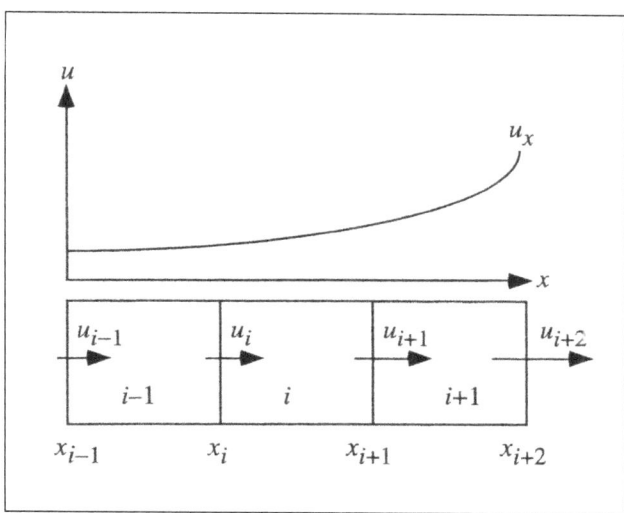

Fig. 1. Discretisation of a continuous west-east zonal scalar velocity u. The grid is broken into discrete cells, and u-values are mapped from the continuous function to the edge of each cell. The arrows in the cells represent magnitudes of the wind speeds. (After Jacobson (2005)).

In the vertical domain, the atmosphere is discretised at a prescribed number of layers. An example of a vertical coordinate system is shown in Figure 2.[3] While the vertical coordinate p is given on the half levels (dashed lines), the prognostic variables vorticity ζ, divergence D, temperature T and specific humidity q are derived on the intermediate full levels (solid lines). The displayed system is called a hybrid coordinate system as it is a combination of terrain following levels in the lower part of the vertical domain and constant pressure levels in the upper domain, with a transition region normally located in the lower stratosphere. While conventional climate models usually have their highest prognostic level in the middle stratosphere at about 30 km ($p = 10$ hPa), state-of-the-art NWP models and chemistry-climate models (CCMs) resolve the full stratosphere and most of the mesosphere up to 60 or 80 km ($p = 0.1$ to 0.01 hPa).

Similarly, different approximations are used for temporal derivatives, such as the Forward Euler, the Implicit, Leapfrog or Runge-Kutta schemes. As these schemes have different characteristics with respect to the stability of the approximate solution of the equation, the optimal integration timestep depends on the choice of the integration scheme. More detailed information on methods in numerical modelling is given e.g., in Jacobson (2005).

---

[3] Deutsches Klimarechenzentrum (DKRZ) Modellbetreuungsgruppe 1992, The ECHAM3 atmospheric general circulation model. DKRZ Tech. Report No. 6, ISSN 0940-9237, Deutsches Klimarechenzentrum, Hamburg, Germany, 184 pp.

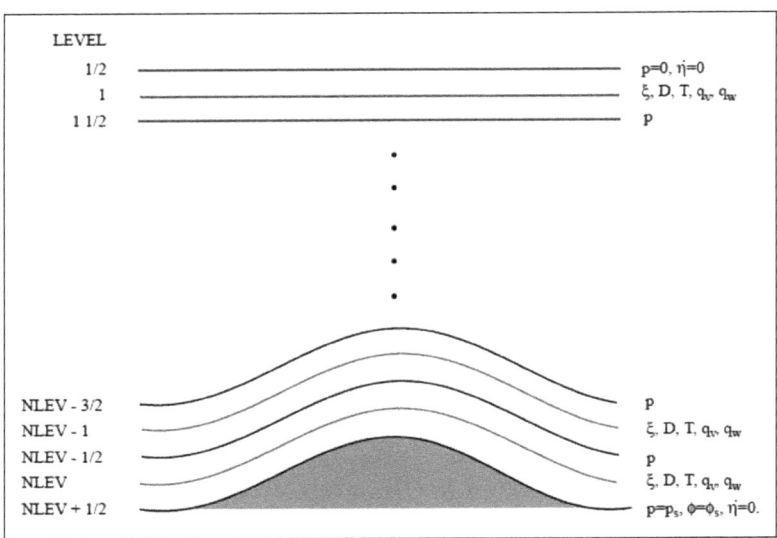

Fig. 2. Hybrid vertical coordinate system of the ECMWF Forecast Model (DKRZ, 1992).

3 Physical parameterisations

Depending on the horizontal and vertical resolution of the model, small-scale processes that are not resolved by the model need to be parameterised, i.e., their total effect in a given grid box needs to be expressed in terms of the resolved model variables. Small-scale horizontal turbulent fluxes of momentum, dry static energy and specific humidity are usually parameterised using empirical horizontal diffusion methods, such as $K_x = -k_x \nabla^4 X$, with K_x the tendency due to horizontal diffusion of the prognostic variable X and k_x the appropriate horizontal diffusion coefficient. To avoid energy accumulation in the smallest resolved scales, scale-selective horizontal diffusion parameterisations are used. Diabatic processes that are not resolved by the model, such as small-scale vertical fluxes of momentum, dry static energy and specific humidity, are parameterised using physical models. These include radiation schemes for the radiative heating by absorption and emission of solar and terrestrial radiation, turbulent vertical diffusion in the planetary boundary layer and the free atmosphere, large scale and convective precipitation, as well as surface processes. In particular for studying the impact of solar variability on climate, short-wave (SW) radiation schemes with appropriate spectral resolution are required to capture the stratospheric effects of short- and long-term variations in the solar ultra-violet (UV) and visible (VIS) spectral ranges (Forster et al., 2011). The temperature tendency (i.e., the heating rate) due to the absorption of solar radiation at a given solar zenith angle χ is determined by the spectrally integrated vertical divergence of the solar irradiance $F_{s,\nu}$

$$\frac{dT}{dt} = -\frac{1}{\rho\, c_p} \int_\nu \frac{dF_{s,\nu}}{dz}\, d\nu \qquad (3.1)$$

with ν: frequency. Considering only the effects of absorption by ozone and molecular oxygen, the spectral solar irradiance at an atmospheric altitude z is calculated by:

$$F_{s,\nu}(z,\chi) = F_{s,\nu}(\infty)\, e^{-(\sigma(O_3) \int_z^\infty n(O_3)\, dz' + \sigma(O_2) \int_z^\infty n(O_2)\, dz')\sec\chi} \qquad (3.2)$$

For more information please see Brasseur and Solomon (2005). According to Equation (3.2), the spectral solar flux at the top of the atmosphere $F_{s,\nu}(\infty)$ is exponentially weakened on its downward path to an altitude z depending on the spectral absorption cross sections of ozone and molecular oxygen $\sigma(O_3)$ and $\sigma(O_2)$ and the amount of the absorbing gases between the top of the atmosphere and z (number densities $n(O_3)$ and $n(O_2)$). In addition, scattering effects by air molecules and aerosols need to be considered. As the spectral integration requests enormous computer resources, solar radiation codes in global models use parameterisations, i.e., solar fluxes that are derived from mean absorption coefficients and mean solar fluxes incoming at the top of the atmosphere, which are integrated over a limited number of spectral intervals. Nissen et al. (2007) present a SW radiation scheme for the middle atmosphere that considers absorption by ozone and molecular oxygen in 49 UV/VIS spectral intervals between 121.56 and 683 nm, and is thus able to capture the enhanced radiative heating during the maximum of the 11-year solar activity cycle. As in the stratosphere, the radiative solar effect is enhanced by ozone photochemistry, the full solar signal can only be reproduced by models that include both SW radiation codes with sufficient spectral resolution as well as interactive chemical models.

With increasing computer capacities, global models include more and better parameterisations of the relevant physical processes. However, with rising complexity uncertainties in the projections may also increase due to non-linear interactions between the included processes. An area with potential for further improvement is the parameterisation of the effects of gravity wave dissipation on the circulation in the stratosphere and mesosphere.

Further reading

Andrews, D. G., J. R. Holton, and C. B. Leovy 1987, Middle Atmosphere Dynamics, Academic Press, San Diego, Calif., Vol. 40

Brasseur, G. P., and S. Solomon 2005, Aeronomy of the Middle Atmosphere, Springer, 3rd rev. and enlarged ed. 2005, XII

Holton, J. R. 1992, An Introduction to Dynamic Meteorology, Academic Press, San Diego, Calif., third edition, Vol. 48

Jacobson, M. Z. 2005, Fundamentals of Atmospheric modeling, Cambridge University Press, second edition

References of Part III

Andrews, D. G., J. R. Holton, and C. B. Leovy. *Middle atmosphere dynamics.* Academic Press, 1987.

Baumgarten, G. Doppler Rayleigh/Mie/Raman lidar for wind and temperature measurements in the middle atmosphere up to 80 km. *Atmospheric Measurement Techniques*, **3**, 1509–1518, 2010, DOI: 10.5194/amt-3-1509-2010.

Bernath, P. F., C. T. McElroy, M. C. Abrams, C. D. Boone, M. Butler, et al. Atmospheric Chemistry Experiment (ACE): Mission overview. *Geophysical Research Letters*, **32**, L15S01, 2005, DOI: 10.1029/2005GL022386.

Bony, S., and B. Stevens, I. Held, J. Dufresne, E. K.A., P. Friedlingstein, G. S., and C. Senior. Carbon Dioxide and Climate: Perspectives on a Scientific Assessment, in: G.R. Asrar and J.W. Hurrell (eds.), *Climate Science for Serving Society: Research, Modeling and Prediction Priorities.* Springer, 2013.

Brasseur, G. P., and S. Solomon. *Aeronomy of the Middle Atmosphere: Chemistry and Physics of the Stratosphere and Mesosphere.* Springer, 2005.

Bremer, J., and U. Berger. Mesospheric temperature trends derived from ground-based LF phase-height observations at mid-latitudes: comparison with model simulations. *Journal of Atmospheric and Solar-Terrestrial Physics*, **64**, 805–816, 2002, DOI: 10.1016/S1364-6826(02)00073-1.

Brönnimann, S., T. Ewen, T. Griesser, and R. Jenne. Multidecadal Signal of Solar Variability in the Upper Troposphere During the 20th Century. *Space Science Reviews*, **125(1-4)**, 305–317, 2007, DOI: 10.1007/s11214-006-9065-2, http://adslabs.org/adsabs/abs/2006SSRv..125..305B/.

Chiodo, G., D. R. Marsh, R. Garcia-Herrera, N. Calvo, and J. A. García. On the detection of the solar signal in the tropical stratosphere. *Atmospheric Chemistry & Physics*, **14**, 5251–5269, 2014, DOI: 10.5194/acp-14-5251-2014.

Courant, K., R., F. Friedrichs, and L. H. On the Partial Difference Equations of Mathematical Physics. *Mathematische Annalen*, **100**, 32–74, 1928. Republished 1967.

Crooks, S. A., and L. J. Gray. Characterization of the 11-Year Solar Signal Using a Multiple Regression Analysis of the ERA-40 Dataset. *Journal of Climate*, **18**, 996–1015, 2005, DOI: 10.1175/JCLI-3308.1.

Damadeo, R. P., J. M. Zawodny, L. W. Thomason, and N. Iyer. SAGE version 7.0 algorithm: application to SAGE II. *Atmospheric Measurement Techniques*, **6**, 3539–3561, 2013, DOI: 10.5194/amt-6-3539-2013.

Dee, D. P., S. M. Uppala, A. J. Simmons, P. Berrisford, P. Poli, et al. The ERA-Interim reanalysis: configuration and performance of the data assimilation system. *Quarterly Journal of the Royal Meteorological Society*, **137(656)**, 553–597, 2011, DOI: 10.1002/qj.828, http://doi.wiley.com/10.1002/qj.828.

Diamantakis, M., and J. Flemming. Global mass fixer algorithms for conservative tracer transport in the ECMWF model. *Geoscientific Model Development*, **7**, 965–979, 2014, DOI: 10.5194/gmd-7-965-2014.

Duchon, C. E., and M. S. O'Malley. Estimating Cloud Type from Pyranometer Observations. *Journal of Applied Meteorology*, **38**, 132–141, 1999, DOI: 10.1175/1520-0450(1999)038¡0132:ECTFPO¿2.0.CO;2.

Fiedler, J., G. Baumgarten, and F.-J. Lübken. NLC observations during one solar cycle above ALOMAR. *Journal of Atmospheric and Solar-Terrestrial Physics*, **71**, 424–433, 2009, DOI: 10.1016/j.jastp.2008.11.010.

Forster, P. M., V. I. Fomichev, E. Rozanov, C. Cagnazzo, A. I. Jonsson, et al. Evaluation of radiation scheme performance within chemistry climate models. *Journal of Geophysical Research (Atmospheres)*, **116**, D10302, 2011, DOI: 10.1029/2010JD015361.

Funke, B., A. Baumgaertner, M. Calisto, T. Egorova, C. H. Jackman, et al. Composition changes after the "Halloween" solar proton event: the High Energy Particle Precipitation in the Atmosphere (HEPPA) model versus MIPAS data intercomparison study. *Atmospheric Chemistry & Physics*, **11**, 9089–9139, 2011, DOI: 10.5194/acp-11-9089-2011.

Gleisner, H., and P. Thejll. *Patterns of tropospheric response to solar variability*. grl, **30**, 1711, 2003, DOI: 10.1029/2003GL017129.

Gray, L. J., J. Beer, M. Geller, J. D. Haigh, M. Lockwood, et al. Solar Influences on Climate. *Reviews of Geophysics*, **48**, RG4001, 2010, DOI: 10.1029/2009RG000282.

Gray, L. J., J. Beer, M. Geller, J. D. Haigh, M. Lockwood, et al. Correction to Solar influences on climate. *Reviews of Geophysics*, **50**, RG1006, 2012, DOI: 10.1029/2011RG000387.

Haigh, J. D. The role of stratospheric ozone in modulating the solar radiative forcing of climate. *Nature*, **370**, 544–546, 1994, DOI: 10.1038/370544a0.

Haigh, J. D., M. Blackburn, and R. Day. The Response of Tropospheric Circulation to Perturbations in Lower-Stratospheric Temperature. *Journal of Climate*, **18**, 3672–3685, 2005, DOI: 10.1175/JCLI3472.1.

Holton, J. R. *An introduction to dynamic meteorology*. Academic Press, 1992.

Jacobson, M. *Fundamentals of atmospheric modeling*. Cambridge University Press, 1999.

Jacobson, M. Z. *Fundamentals of Atmospheric Modeling.* Cambridge University Press, 2005.

Janowiak, J. E., and P. Xie. CAMS-OPI: A Global Satellite-Rain Gauge Merged Product for Real-Time Precipitation Monitoring Applications. *Journal of Climate,* **12**, 3335–3342, 1999, DOI: 10.1175/1520-0442(1999)012¡3335:COAGSR¿ 2.0.CO;2.

Kalnay, E., M. Kanamitsu, R. Kistler, W. Collins, D. Deaven, et al. The NCEP/ NCAR 40-year reanalysis project. *Bull. Amer. Meteor. Soc.,* **77**, 437–471, 1996.

Keckhut, P., J. D. Wild, M. Gelman, A. J. Miller, and A. Hauchecorne. Investigations on long-term temperature changes in the upper stratosphere using lidar data and NCEP analyses. *Journal of Geophysical Research (Atmospheres),* **106**, 7937–7944, 2001, DOI: 10.1029/2000JD900845.

Lindsay, R., M. Wensnahan, A. Schweiger, and J. Zhang. Evaluation of Seven Different Atmospheric Reanalysis Products in the Arctic*. *Journal of Climate,* **27**, 2588–2606, 2014, DOI: 10.1175/JCLI-D-13-00014.1.

Lübken, F.-J., U. Berger, and G. Baumgarten. Temperature trends in the midlatitude summer mesosphere. *Journal of Geophysical Research (Atmospheres),* **118**, 13,347, 2013, DOI: 10.1002/2013JD020576.

Mayr, H. G., J. G. Mengel, F. T. Huang, and E. R. Nash. Equatorial annual oscillation with QBO-driven 5-year modulation in NCEP data. *Annales Geophysicae,* **25**, 37–45, 2007, DOI: 10.5194/angeo-25-37-2007.

Mayr, H. G., J. G. Mengel, F. T. Huang, and E. R. Nash. Solar cycle signatures in the NCEP equatorial annual oscillation. *Annales Geophysicae,* **27**, 3225–3235, 2009, DOI: 10.5194/angeo-27-3225-2009.

Mc Guffie, K., and A. Henderson-Sellers. *A Climate Modelling Primer,* 1987.

McLandress, C., D. A. Plummer, and T. G. Shepherd. Technical Note: A simple procedure for removing temporal discontinuities in ERA-Interim upper stratospheric temperatures for use in nudged chemistry-climate model simulations. *Atmospheric Chemistry & Physics,* **14**, 1547–1555, 2014, DOI: 10.5194/acp-14-1547-2014.

Mitchell, D. M., L. J. Gray, M. Fujiwara, T. Hibino, J. A. Anstey, et al. Signatures of naturally induced variability in the atmosphere using multiple reanalysis datasets. *Quarterly Journal of the Royal Meteorological Society,* n/a–n/a, 2014, DOI: 10.1002/qj.2492, http://dx.doi.org/10.1002/qj.2492.

Mlynczak, M. G., B. T. Marshall, F. J. Martin-Torres, J. M. Russell, R. E. Thompson, E. E. Remsberg, and L. L. Gordley. Sounding of the Atmosphere using Broadband Emission Radiometry observations of daytime mesospheric $O_2(^1\Delta)$ 1.27 μm emission and derivation of ozone, atomic oxygen, and solar and chemical energy deposition rates. *Journal of Geophysical Research (Atmospheres)*, **112**, D15306, 2007, DOI: 10.1029/2006JD008355.

Moberg, A., D. M. Sonechkin, K. Holmgren, N. M. Datsenko, and W. Karlén. Highly variable Northern Hemisphere temperatures reconstructed from low- and high-resolution proxy data. *Nature*, **433**, 613–617, 2005, DOI: 10.1038/nature03265.

Mote, P., and A. O'Neill. *Numerical Modeling of the Global Atmosphere in the Climate System, Series C: Mathematical and Physical Sciences*, Vol. 550. Kluwer Academic Publisher, 1998. ISBM 0-7923-6301-9 (HB) 0-7923-6302-7 (PB).

Murtagh, D., U. Frisk, F. Merino, M. Ridal, A. Jonsson, et al. Review: An overview of the Odin atmospheric mission. *Canadian Journal of Physics*, **80**, 309, 2002, DOI: 110.1139/p01-157.

Nissen, K. M., K. Matthes, U. Langematz, and B. Mayer. Towards a better representation of the solar cycle in general circulation models. *Atmospheric Chemistry & Physics*, **7**, 5391–5400, 2007.

Offermann, D., P. Hoffmann, P. Knieling, R. Koppmann, J. Oberheide, and W. Steinbrecht. Long-term trends and solar cycle variations of mesospheric temperature and dynamics. *Journal of Geophysical Research (Atmospheres)*, **115**, D18127, 2010, DOI: 10.1029/2009JD013363.

Onogi, K., J. Tsutsui, and H. Koide. The JRA-25 reanalysis. *Journal of Meteorological Society*, **85**(3), 369–432, 2007. http://ecco2.jpl.nasa.gov/data2/data/atmos/jra25/Onogi_2007_JRA25.pdf.

Østvand, L., T. Nilsen, K. Rypdal, D. Divine, and M. Rypdal. Long-range memory in internal and forced dynamics of millennium-long climate model simulations. *Earth System Dynamics*, **5**, 295–308, 2014, DOI: 10.5194/esd-5-295-2014.

Peixoto, J., and A. Oort. *Physics of climate*. American Institute of Physics, 1992.

Rodger, C. J., J. B. Brundell, R. H. Holzworth, and E. H. Lay. Growing Detection Efficiency of the World Wide Lightning Location Network. In: *American Institute of Physics Conference Series*, vol. 1118 of *American Institute of Physics Conference Series*, 15–20, 2009, DOI: 10.1063/1.3137706.

Rodgers, C. D. *Inverse Methods for Atmospheric Sounding: Theory and Practice*, vol. 2 of *Series on Atmospheric, Oceanic and Planetary Physics*, F. W. Taylor, ed. World Scientific Publishing Co. Pte. Ltd, Singapore, 2000.

Saha, S., S. Moorthi, H.-L. Pan, X. Wu, J. Wang, et al. The NCEP Climate Forecast System Reanalysis. *Bulletin of the American Meteorological Society*, **91**(8), 1015–1057, 2010, DOI: 10.1175/2010BAMS3001.1, http://journals.ametsoc.org/doi/abs/10.1175/2010BAMS3001.1.

Schneider, S. *Introduction to climate modeling, Chapter 1 in Climate system modeling*. Cambridge University Press, 1992.

Solanki, S. K., I. G. Usoskin, B. Kromer, M. Schüssler, and J. Beer. Unusual activity of the Sun during recent decades compared to the previous 11,000 years. *Nature*, **431**, 1084–1087, 2004.

Soon, W., K. Dutta, D. R. Legates, V. Velasco, and W. Zhang. Variation in surface air temperature of China during the 20th century. *Journal of Atmospheric and Solar-Terrestrial Physics*, **73**, 2331–2344, 2011, DOI: 10.1016/j.jastp.2011.07.007.

Steinhilber, F., J. Beer, and C. Fröhlich. Total solar irradiance during the Holocene. *Geophysical Research Letters*, **36** (**L19704**), 2009, DOI: 10.1029/2009GL040142.

Stocker, T. *Introduction to Climate Modelling*. Springer, 2011.

Stocker, T., D. Qin, G.-K. Plattner, M. Tignor, S. Allen, J. Boschung, A. Nauels, Y. Xia, V. Bex, and P. Midgley. IPCC, 2014: *Climate Change 2013: The physical science basis*. Contribution of Working Group 1 to the Fifth Assessment Report of the Intergovernmental Panel on Climate Change. Cambridge University Press, 2013.

Toohey, M., M. I. Hegglin, S. Tegtmeier, J. Anderson, J. A. Añel, et al. Characterizing sampling biases in the trace gas climatologies of the SPARC Data Initiative. *Journal of Geophysical Research (Atmospheres)*, **118**, 11,847, 2013, DOI: 10.1002/jgrd.50874.

Trenberth, K. Climate System Modeling. Cambridge University Press, ISBN 0-521-43231-6 (HB), 1992.

Uppala, S. M., P. W. Kållberg, A. J. Simmons, U. Andrae, V. D. C. Bechtold, et al. The ERA-40 re-analysis. *Quarterly Journal of the Royal Meteorological Society*, **131**, 2961–3012, 2005, DOI: 10.1256/qj.04.176.

van Loon, H., G. A. Meehl, and D. J. Shea. Coupled air-sea response to solar forcing in the Pacific region during northern winter. *Journal of Geophysical Research (Atmospheres)*, **112**, D02108, 2007, DOI: 10.1029/2006JD007378.

Washington, W., and C. Parkinson. *An Introduction to Three-dimensional Climate Modelling*. Oxford University Press, Oxford, 1986. ISBN 0.935702-52-0.

Waters, J. W., L. Froidevaux, R. S. Harwood, R. F. Jarnot, H. M. Pickett, et al. The Earth Observing System Microwave Limb Sounder (EOS MLS) on the Aura Satellite. *IEEE Transactions on Geoscience and Remote Sensing*, **44**, 1075–1092, 2006, DOI: 10.1109/TGRS.2006.873771.

WMO. *Guide to Meteorological Instruments and Methods of Observation*. Tech. Rep. 8, World Meteorological Organisation, Geneva, 1988.

Part IV

IMPACTS ON THE EARTH SYSTEM

CHAPTER 4.1

DIRECT IMPACT OF SOLAR IRRADIANCE VARIABILITY

Jean Lilensten[1] and Kleareti Tourpali[2]

1 Introduction

Chapter 1.2 shows that the solar irradiance is the main contributor to the climatic engine and Chapter 2.2 gives the values of the Total Solar Irradiance, its variability and how it is measured. The question addressed in this Chapter is what do we know or do we not know on the influence of this variability on climate. The Chapter itself will focus on the impacts on the atmosphere. Infobox 4.1 will give more hints on understanding this impact on the Earth surface.

2 Different timescales of solar variability

Solar energy is produced by thermonuclear reactions. It is mainly emitted in the form of electromagnetic radiation. Each second, a surface of 1 square meter on the Sun produces an average energy of 6.40×10^7 J (see Chapter 2.2). This energy travels throughout the solar system, and at the level of the Earth orbit, one receives an average of about 1361 Joules per second and square meter. These values are variable. Solar variability on timescales of more than 100,000 years is constrained observationally mainly by studies of solar particles implanted in lunar samples and meteorites. These studies have yielded evidence for an early active phase (referred to as the T-Tauri phase of a star) and for a long-term decline in solar wind flux, for example. The Sun also experiences secular variations, of which the exact origin is still poorly understood. The exploration of solar variability on time scales of up to 100,000 years is now made possible through the study of radioactive isotopes such as ^{14}C and ^{10}Be, whose production rate is modulated by the solar magnetic field. These isotopic records are especially useful for times

---

[1] Institut de Planétologie et d'Astrophysique de Grenoble (IPAG), 38041 Grenoble cedex 9, France
[2] Aristotle University of Thessaloniki, Lab. of Atmospheric Physics, 54 124 Thessaloniki, Greece

up to 10,000 years ago (the Holocene) and reveal possible periodicities of about 11, 90, 200, and 2300 years. The last major minimum in the latter cycle may have coincided with the observed Maunder Minimum in solar activity during the 15th and 16th centuries. At the scale of a 10,000 years and below, the magnetic field maintained by the solar dynamo is the main driver of solar variability. The history of solar magnetic activity has been recorded, for example, by the number of sunspots on the solar disk since the 15th century and by measurements of solar radio flux at 10.7 cm wavelength (F10.7) since the late 1940's. It is responsible for the Hale magnetic cycle, lasting about 18 to 24 years, which is divided into two 9 to 13 years parts - the Schwabe cycle, also known simply as the solar cycle - each with a reverse orientation of the solar magnetic dipole. Smaller time scales (month to second) involve other mechanisms such as the solar rotation and the complex dynamics of the magnetic field in the solar atmosphere. The Carrington rotational cycle (about 27 day) is due to a common tendency for active regions to be concentrated on one side of the Sun. It is the most important short-term solar spectral irradiance quasi-periodicity with amplitudes that can be comparable to those occurring on the solar cycle timescale. Sporadic events such as flares may extract up to 10^6 Wm^{-2} on a limited area (see Chapter 2.4) and have a strong short-term impact on the Earth's upper atmosphere. Their effect on the climate has often been disregarded due to their short duration. But the energy released by solar eruptions, like that of many natural phenomena, is distributed according to a power law: many low-energy events could eventually impact the climate. Thus, all time scales must be considered in the climate system. Our poor knowledge of the history of the evolution and variability of the solar wind irradiance is today one of the major obstacles to understanding the mechanisms of climate forcing.

3 Effects in the thermosphere / ionosphere

Different communities have different definitions of the "upper atmosphere". Here, we will define it as the layer where the solar energy produces ions, creating the ionosphere (i.e., the ionized part of the atmosphere) and the thermosphere (i.e., its neutral counterpart). The lower altitude of this upper atmosphere varies with the solar activity, with a lowest value for the ionospheric D-region (see the glossary) at about 70 km. The extreme-UV (EUV, 10-120 nm) and soft X-ray (XUV, 0.1-10 nm) components of solar irradiance variability influence the upper atmospheric layers, where they are absorbed essentially by O_2 and N_2 (below 150 km) and O (above) through heating, excitation, dissociation and ionization, resulting in a mixture of neutrals (the thermosphere), and electrons and ions (the ionosphere). Therefore, the electron and ion concentrations and temperatures strongly depend on solar activity. Between the minimum and maximum of the Schwabe solar activity cycle, the EUV flux varies by up to 100%. As a result, the temperature of the neutral gas and ions at 400 km can be multiplied by 2 (about 800 K at solar min to about 1500 K at solar max) and the electron temperature by up to 4 times (about 1000 K at solar min to about 4000 K at solar max). Figure 1 shows a modeling of

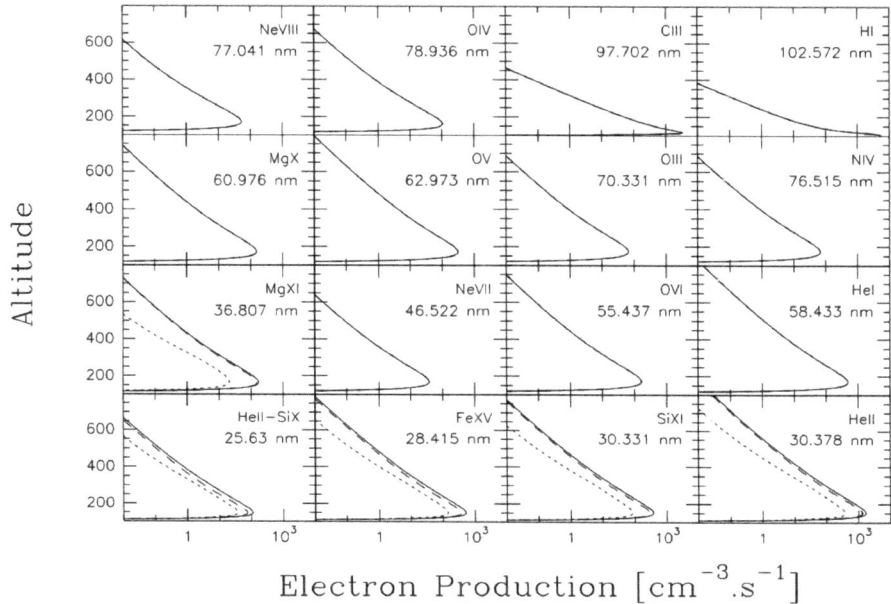

Fig. 1. This plots shows the altitude where different solar emission lines deposit their energy in the upper atmosphere through ionization. The dashed line is through direct photoionization and the dotted line through secondary electron impact. The full line is the total production.

the energy deposition through ionization for 16 intense lines of the solar EUV spectrum during mean solar conditions (details may be found in Lilensten et al. (2007)).

Chapters 4.2, 4.4 and 4.5 show that the solar wind effects in the upper atmosphere are quite similar to those of EUV radiation (excitation, dissociation, ionization and heating) with a yet stronger and more complex variability, and furthermore generating auroras. In the frame of the present Chapter, it is sufficient to mention that the total energy released in the thermosphere is small: the total solar emittance (the total energy released by the sun through radiation) is 6.40×10^7 Wm^{-2} of which the UV/EUV/XUV represents 10^5 Wm^{-2}. The regular solar wind extracts about 10^4 Wm^{-2} of solar surface, but the Earth receives about 5×10^{-4} Wm^{-2} and the flares can extract sporadically up to 10^6 Wm^{-2}, from the Sun but in small areas. To link the climate change with such small amounts request amplification mechanisms. If such mechanisms exist, they are still unknown.

4 Temperature effects in the mesosphere / stratosphere

In the upper part of the mesosphere, approximately 75 km. the enhanced radiation at solar maximum conditions leads to a substantial increase in solar heating due to a combination of radiative heating and heating due to chemical reactions (absorption of O_2 and O_3 and exothermic chemical recombination), with the largest amount of heating due to chemical reactions. This is the release of chemical potential energy which has been added to the system before through dissociation, i.e., absorption of solar radiation (Brasseur and Solomon, 2005). Overall, the heating due to exothermic reactions exceeds the direct heating by solar radiation in the altitude range 70 - 95 km. The increase of chemical heating in the upper mesosphere is mainly a result of odd oxygen chemical reactions (primary: O + O + M; secondary: O + O_2 + M), which is consistent with the increase of atomic oxygen due to stronger solar irradiance at this altitude region. This increased heating is balanced largely by enhanced infrared cooling. Above about 100 km, the increase in solar heating becomes larger as altitude becomes higher (Beig et al., 2008). Below about 70 km, the temperature difference between solar minimum and maximum becomes smaller and insignificant (following the ozone variations; ozone is decreasing at solar maximum; see paragraph below). A secondary maximum is found just above the stratopause and solar variability induced temperature effects are positive from there on and below into the stratosphere.

In the upper stratosphere, the main response to the solar irradiance changes is a warming at the solar maximum (Gray et al., 2010). In the equatorial region, this warming can be as large as about 1 to 2 K for the annual mean and for an 'average' solar cycle. On the whole, this upper stratospheric temperature response is a combined result of an increase in ozone concentration and of the larger amount of solar flux available for absorption by ozone and oxygen during solar maximum conditions.

Below these altitudes, the warming signal drops and, although it remains positive (a combination of positive response due to solar irradiance changes and a much smaller, even negative response due to ozone changes) it becomes much smaller with decreasing altitude, reaching a minimum in the middle stratosphere. Moving further below into the lower stratosphere, a local temperature maximum exists between about 30 and 70 hPa (20 - 25 km), which peaks in the subtropics at either side of the equator. This lower stratospheric temperature response is not directly linked to radiative heating, but rather indicates a large-scale response in the dynamics of this region.

The annual mean response of temperature to the Schwabe solar cycle averaged over the equatorial region (30°S 30°N) in the stratosphere is presented in Figure 1 (left panel).

5 Solar induced effects on ozone in the mesosphere / stratosphere

Effects of solar variability on ozone in the middle atmosphere are primarily due to enhanced solar irradiance (mainly in the UV part of the spectrum), to changes in the precipitation rate of energetic particles producing NOx and HOx which cause additional ozone destruction, as well as to changes in the transport of ozone, occurring from indirect effects on circulation (induced by the same solar cycle related ozone and temperature changes). In this section we focus on the role of increases in solar irradiance from the minimum to the maximum phase of the Schwabe solar cycle. The mechanism involving changes in particle precipitation will be discussed in detail in Chapter 4.6.

Variations of the solar flux at the Lyman-alpha line at 121.6 nm are up to 30% over a 27-day rotation and about a factor of 2 over the Schwabe solar cycle. They play a very important role in the mesosphere, affecting the photolysis of water vapour and oxygen in the mesosphere. Both these species are involved in ozone chemistry, and the net effect on ozone depends on the balance between competing effects (i.e., loss and production rates).

However, Lyman-alpha radiation penetrates the atmosphere only down to about 65 km. Longer wavelengths are controlling the solar interactions at lower altitudes. In the lower mesosphere and upper stratosphere, changes in the wavelength range in the Schumann-Runge bands and Herzberg continuum (from about 170 to 240 nm) can affect photolysis that controls production of ozone through the dominant oxygen photolysis process at these altitudes (Brasseur and Solomon, 2005). This acts to increase ozone during periods of high solar activity as the photochemical reactions that lead to ozone production and destruction are enhanced.

In the upper mesosphere, the largest changes in ozone and atomic oxygen occur in the vicinity of the mesopause. These changes result from the enhanced photolysis of O_2 (ozone increases up to 25% and atomic oxygen more than 50%). Note that although mesospheric ozone exhibits a large diurnal variation (nighttime mixing ratios about an order of magnitude larger than daytime values above 0.01 hPa), the relative solar responses of day and nighttime ozone have very similar patterns.

In the lower regions in the mesosphere, ozone concentration is less at solar maximum. Near 70 km, the large increase in solar flux at the Lyman alpha line induces a general decrease in ozone. This is the result of increased photodissociation of water vapor, a source for hydroxyl and peroxy radicals (OH and HO_2), and the subsequent catalytic destruction of ozone through interactions with them.

In the upper stratosphere where solar UV variations directly affect ozone production rates, an increase of 2% to 4% is found in the ozone abundance during solar maximum (Figure 2, right panel). Positive responses are also present at middle and higher latitudes in the middle stratosphere. In the tropical middle stratosphere observations indicate a weak response in the ozone amount, with its concentration increasing again in the lower stratosphere.

Thus the most characteristic features of the equatorial stratospheric ozone response to the increase in solar irradiance during a Schwabe solar cycle are a maximum in the upper stratosphere (the chemically - controlled altitudes), replaced by an area with lower and insignificant response in the middle stratosphere, and a secondary maximum in the equatorial lower stratosphere. In this last region photochemistry is slow due to the attenuation of the solar radiation field that is able to drive photochemical reactions; Solar UV radiation in the spectral interval 120 to 300 nm is almost completely absorbed in the Earth's middle atmosphere.

Therefore the ozone lifetime exceeds dynamical transport time scales. This implies that the ozone change is an indirect effect, due to changes in lower stratospheric dynamics and transport.

The "total column ozone" is a variable equal to the height integration of ozone at any location (defined by latitude and longitude). A clear decadal oscillation in phase with the Schwabe solar cycle is evident in both satellite data and ground-based measurements, the main cause of which is the ozone response in the lower stratosphere (because of the high number densities).

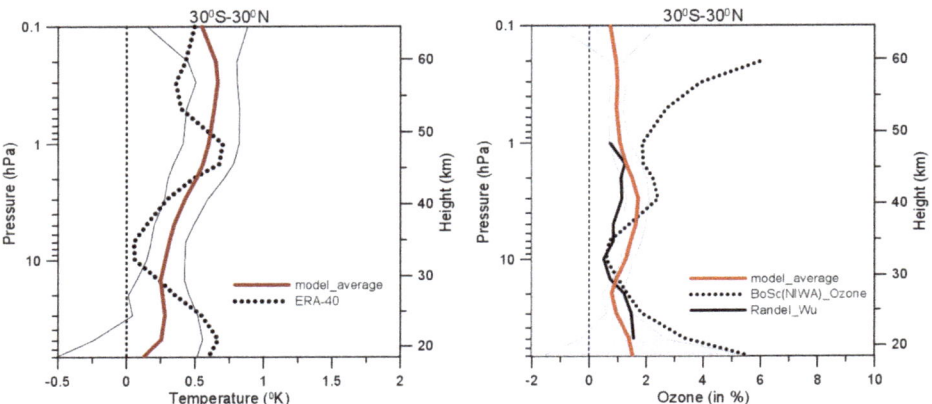

Fig. 2. Schwabe solar cycle effects on stratospheric temperature (left panel) and stratospheric ozone (right panel) as derived from observations and a set of Chemistry-Climate Models simulating the solar cycle. Results are shown as temperature change (in K) and ozone change (in %) from solar minimum to maximum. Figure adapted from the SPARC CCMVal2 Report (CCMVAL, 2010), using CCMVal-2 CCMs (model average; 1960-2004) and observations (Bodeker Scientific (NIWA)-3D ozone data set, 1979-2004; Randel and Wu ozone (1979-2005), and ERA-40 Reanalysis (1979-2001) from 100 to 0.1 hPa.

6 Effects in the troposphere

Over a solar cycle, much larger changes take place at shorter wavelengths in the ultraviolet than in the visible and IR bands of the solar spectrum, thus the direct impact of solar irradiance variability is larger in the middle and upper atmosphere than it is at lower altitudes. Already at the level of the tropopause, almost all radiation below 300 nm has been absorbed at upper levels, so that only part of UV-B radiation along with UV-A, visible and IR radiation penetrate through the troposphere and reach lower levels and the surface, where the UV radiation plays an important role for tropospheric photochemistry.

In the visible and IR there is not much absorption in the atmosphere. The main absorbers are tropospheric O_3 at the Chappuis bands (400-850 nm), some types of aerosols, and water vapor (clouds), oxygen, and carbon dioxide, which absorb in certain IR bands, all resulting to a warming in the lower troposphere.

As radiation propagates through the troposphere, part of it is involved in various scattering processes by clouds, aerosols, atmospheric molecules and finally by the surface and is directed back to space (Haigh, 2007). Clouds play the most important role over areas that are not covered by ice or snow. The remaining part, particularly in the visible and IR bands, where the bulk of solar flux resides, reaches the surface and heats it directly.

The visible and IR bands, which have the largest contribution to the TSI, exhibit only small variations over the solar cycle (only up to 0.5%). However, due to the large amounts of energy carried by the solar flux at these bands, even these small differences may have important consequences for the Earth system.

The direct impact of the solar variability at these wavelengths on the Earth's climate might be relatively small, but it involves amplification mechanisms (e.g. the bottom-up mechanism, see Chapter 4.2).

Recent satellite measurements and model reconstructions indicate the possibility of an out-of-phase relationship at the longer wavelengths of solar output with respect to solar variability, indicating that that the radiative forcing of surface climate by the Sun is out of phase with solar activity.

7 What we know, what we do not know, what to take home

A series of assessments are certain, and may be considered as facts:

- The sun radiates energy at all wavelengths, from the radio to the XUV range. Most of the energy is radiated in the visible range and is relatively stable in time (Chapter 2.1). The radio and energetic parts (UV, EUV, XUV) are very variable both in amplitude and time but constitute a very small amount of the total solar irradiance in term of energy (see Chapter 2.2).

- Each wavelength deposits its energy at specific altitudes, in different layers of the atmosphere, through different mechanisms ranging from ionisation in the upper atmosphere, to heating, dissociation and excitation of molecules at all altitudes.

- The XUV / EUV spectral ranges deposit their energy above 70 km while the less energetic, less variable, but more abundant UV / Visible ranges deposit their energy below, down to the troposphere.

- The variability of the upper atmosphere due to that of the solar flux is well documented and well estimated. The neutral temperature varies by a factor up to 3, and the electron temperature varies by a factor up to 10. However, this thermospheric layer is a vacuum where the notion of temperature does not have the same meaning as in the troposphere.

- Because of the thermal structure of the atmosphere from ground to the upper thermosphere, convection between the upper layers and the lower layers is impossible.

Some points are less clear though

- The variability of the atmospheric temperature below the thermosphere due to that of the solar flux is not fully known. It may be as high as 3 K in the upper stratosphere but drops drastically in the mesosphere. Its exact value is poorly known. It increases again in the upper mesosphere, where we estimate a signal of 3 to 50 K.

- Although the amount of the TSI energy deposited in the thermosphere is very small, it cannot be excluded that a still unknown amplification mechanism exists. The scientific community focuses its attention toward the role of ozone (Chapters 3.2 and 3.3). Other minor atmospheric constituents (NO, ...) may play a role (Chapter 4.6). Exploring how the solar irradiance variability affects these constituents will be a major challenge in the future years.

Further reading

J. Lilensten et J. Bornarel, Space weather, environment and societies, Ed. Springer, 2006, ISBN-10 1-4020-4331-7

G. Brasseur and S. Solomon, Aeronomy of the middle atmosphere, Ed. Springer, 2005, Chemistry and Physics of the Stratosphere and Mesosphere, Series: Atmospheric and Oceanographic Sciences Library, Vol. 32, ISBN 978-1-4020-3824-2

Charles P. Sonett, Mark S. Giampapa, and Mildred S. Matthews, editors, The Sun in Time, University of Arizona Press, Tucson, Arizona 990 pages, 1991.

J.D. Haigh., The Sun and the Earths Climate. Living Reviews in Solar Physics, 4:265, 2007

CHAPTER 4.2

'TOP-DOWN' VERSUS 'BOTTOM-UP' MECHANISMS FOR SOLAR-CLIMATE COUPLING

Amanda C. Maycock[1] and Stergios Misios[2]

1 Introduction

The expected change in global mean surface temperature between the maximum and minimum phases of the solar cycle of about 11 years can be estimated from simple radiative arguments. Between solar maximum and minimum, the typical increase in total solar irradiance (TSI) at the Sun-Earth distance is $\sim$1 W m^{-2} (see Chapter 2.2). Around 30% of this excess energy is reflected back to space by clouds and the 'brighter' parts of the Earth's surface (e.g., snow covered regions). The remaining absorbed energy must be distributed over the entire globe, an area 4 times the size of the intercepting area, resulting in a radiative imbalance at the top of the atmosphere of $\sim$0.175 W m^{-2}. For comparison, this is around 20 times smaller than the radiative perturbation resulting from a doubling in the abundance of atmospheric CO_2. If we assume that the equilibrium change in global mean surface temperature is $\sim$0.8 K for every 1 W m^{-2} of radiative imbalance, this implies a meagre increase in global surface temperature of up to $\sim$0.14 K between solar maximum and minimum.

Despite the relatively small effect on global mean surface temperature, a number of mechanisms have been proposed to explain locally amplified responses to solar forcing as a result of dynamical and thermodynamical feedback processes in the climate system. These can be broadly categorised as 'bottom-up' and 'top-down' mechanisms. Bottom-up mechanisms focus on the effects of changes in visible and near-infrared radiation on surface temperature, while top-down mechanisms focus on changes in solar ultraviolet (UV) radiation and associated effects on stratospheric ozone, temperatures and winds. This Chapter summarises our current understanding of these mechanisms and highlights some open questions around their relative importance for the observed response to solar variability.

---

[1] Centre for Atmospheric Science, University of Cambridge, UK
[2] Laboratory of Atmospheric Physics, Aristotle University of Thessaloniki, Greece

2 Bottom-up mechanisms

The literature around bottom-up mechanisms is less mature than that for the top-down pathways, since the oceans were traditionally considered as a passive layer that would damp the response to the solar cycle. Indeed, the simple estimate of an increase in global surface temperature of $\sim$0.14 K between solar maximum and minimum discussed in the preceding Section 1 assumes that the incoming and outgoing radiative fluxes at the top of the atmosphere are balanced. However, the large thermal inertia of the Earth's oceans means that the climate system takes many years to fully adjust to an imbalance in the global energy budget. Because of that the global surface temperature response to a time-varying perturbation, such as the solar cycle of about 11 years, will be smaller than if a change in TSI were instantaneously 'switched-on' and the climate system allowed to adjust to a new equilibrium state.

The thermal inertia of the oceans also means that the maximum surface warming will occur after the peak in solar forcing, and this likely explains the observational evidence for a warming of the global upper ocean of 0.08$\pm$0.02 K which lags the solar forcing by one to two years. Despite the traditional view of the oceans having a passive role, recent observational and modelling studies have provided evidence for a possible regional amplification of surface solar cycle signals owing to coupled air-sea processes.

As noted in the first part of this handbook, analysis of surface temperature observations dating back to the late 19th century has shown a cooling of $\sim$1 K in the eastern tropical Pacific Ocean in composites of peak sunspot years (see Figure 1). Such a large cooling is certainly unexpected in radiative terms (see above discussion), and highlights the potential for dynamical and thermodynamical processes to amplify the surface response to solar forcing.

The starting point to explain such a strong cooling in the east Pacific is the surface absorption of solar radiation that occurs primarily in cloud free regions (see top panel Figure 2). Model simulations indicate that the amplitude of the anomalous surface heating could be as much as 2 W m^{-2} locally as a result of reductions in cloud cover. In response to the increase in surface heating, the oceans release more water vapour into the atmosphere, which is converged into precipitation zones by the prevailing east-west trade winds (middle panel Figure 2). Water vapour stores energy in the form of latent heat that is released upon condensation to water droplets; in the tropics, this mostly happens within deep convective clouds. The release of energy from latent heating fuels the tropical meridional (Hadley) and zonal (Walker) overturning circulations. When the Walker circulation is accelerated, the surface trade winds are enhanced, which drives more upwelling of cold water in the eastern Pacific leading to a decrease in surface temperature. Amplified subsidence or air and relatively cooler sea surface temperatures in this region would be expected to further reduce cloud cover (bottom panel Figure 2), allowing stronger surface heating.

The cycle described above forms a positive feedback loop that is thought to result in relatively cold water in the tropical east Pacific in years of high solar

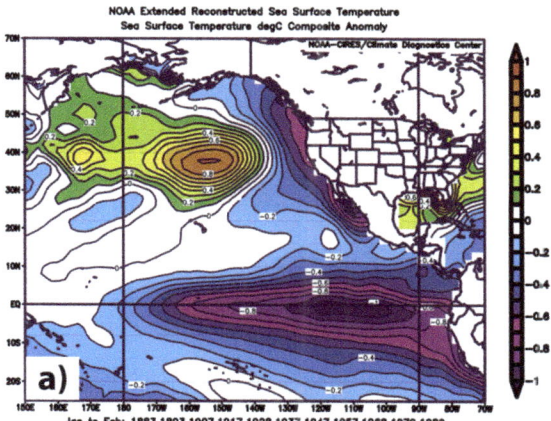

Fig. 1. January-February averaged Pacific sea surface temperature anomalies (°C) during 11-year solar cycle peak years (1883, 1893, 1907, 1917, 1928, 1937, 1947, 1957, 1968, 1979, 1989) from the NOAA Extended Reconstructed Sea Surface Temperature data set. Reproduced with permission from van Loon et al. (2007).

activity. It is thought that this bottom-up mechanism does not work in isolation, but in symphony with the 'tropical top-down' mechanism (see Section 3.2), and that these two distinct pathways act to reinforce each other. Global climate models are capable of simulating aspects of this coupled air-sea mechanism and, if it is confirmed in future observations, it could be an important bottom-up pathway in the solar-climate connection.

Nevertheless, the observational evidence for relatively cooler east Pacific sea surface temperatures under solar maximum conditions is somewhat complicated by the occurrence of the El Niño-Southern Oscillation (ENSO), which is the dominant source of interannual variability in the tropical Pacific. Every three to five years, the sea in the central and eastern Pacific warms extensively, initiating an El Niño event, which typically lasts for 9 to 12 months. The strongest El Niño event of the 20th century occurred in the winter 1997/98, when the eastern equatorial Pacific warmed by more than 5 K. In the opposite phase of ENSO, known as La Niña, the eastern Pacific is characterised by cooler than average ocean temperatures.

It has been suggested that periods of increased solar irradiance over the geological past driven by changes in the Earth's orbit, such as during the Holocene (geological epoch of the last 12 000 years), could have been associated with more La Niña conditions. This cooling is thought to be the result of a dynamical thermostat mechanism in which a uniform heating of the Pacific is stabilised by ocean dynamics, leading to negative temperature anomalies in the eastern sector. Although such mechanisms suggest solar irradiance changes could affect ENSO on centennial and/or millennial time scales, it is not yet clear whether the solar cycle of about 11 years could drive similar changes.

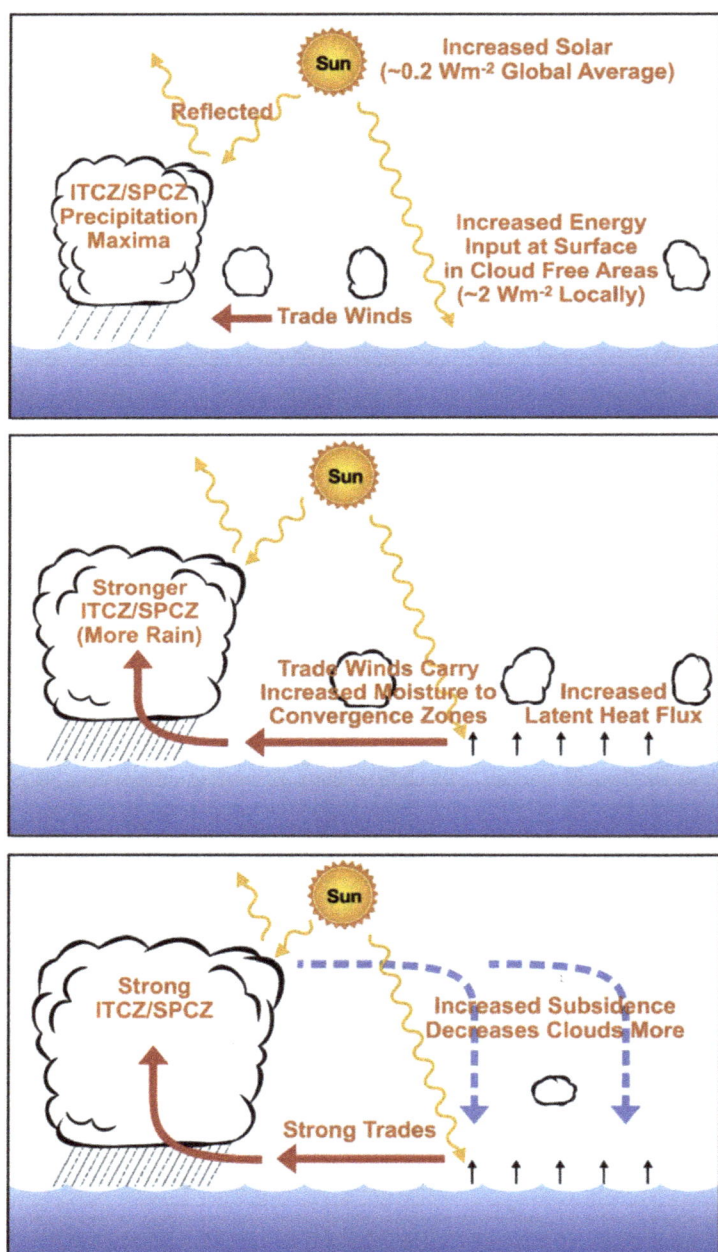

Fig. 2. Schematic showing the 'bottom-up' mechanism for an amplified surface temperature response over the tropical Pacific. Reproduced from Meehl et al. (2009). © American Meteorological Society. Used with permission.

ENSO is not the only quasi-periodic phenomenon active in the tropical Pacific. Historical reconstructions of sea surface temperatures back to 1880 indicate a 10- to 12-year oscillation, with ENSO-like spatial characteristics. Diagnosis of the ocean thermal budget has shown that this quasi-decadal oscillation can be explained by a similar mechanism that generates ENSO variability. Although such oscillations could be generated naturally through air-sea coupling, simulations with conceptual and comprehensive models suggest that the solar cycle can trigger a resonant excitation of this decadal variability. This would mean that the tropical Pacific ocean-atmosphere system exhibits a natural mode of variability on quasi-decadal time scales that resonates when forced by the approximately 11-year cycle in TSI. This bottom-up mechanism could also explain the observational evidence for a basin-wide warming in years after solar maximum as indicated by analysis of surface ocean temperature observations (see Chapter 3.3 and 3.6).

If proven to be of significant strength, the surface changes described above could impact on the overlying atmosphere via radiative-convective coupling. Recent work has suggested that the bulk of the energy absorbed by the oceans goes into evaporating water, which is then transported to the upper tropical troposphere and deposited as latent heat. This evaporation feedback reduces the net surface warming, but leads to a warmer troposphere aloft. The water vapour feedback compounds the warming in the upper tropical troposphere as more terrestrial longwave radiation is absorbed by the moistened troposphere. Such a synergy of bottom-up mechanisms could induce changes in the large-scale circulation, forcing the subtropical tropospheric jets to shift toward the poles. An intensification of the poleward heat transport by the atmosphere would warm the air over the polar regions, which would subsequently emit more longwave radiation heating the polar surface.

3 Top-down mechanisms

One of the most well understood aspects of the atmospheric response to solar variability is a relative warming of the upper stratosphere ($\sim$30–50 km) between the maximum and minimum phases of the approximately 11-year solar cycle. This amounts to an increase in temperature near the tropical stratopause ($\sim$50 km) of $\sim$1.5 K (upper panel Figure 3). This warming comes from two main effects: (1) an increase in incoming solar UV (wavelengths $\sim$200–300 nm) radiation, which is absorbed by ozone in the stratosphere; (2) an increase in shorter wavelength UV radiation (wavelengths less than 242 nm), which leads to more chemical production of ozone in the mid and upper stratosphere through the photolysis of oxygen. The latter effect leads to an increase in mid and upper stratospheric ozone of a few percent between solar maximum and minimum conditions (bottom panel Figure 3).

The increases in UV radiation and ozone both cause additional shortwave heating in the mid and upper stratosphere leading to an increase in temperature. The relative importance of each effect for the temperature response is estimated to be around 50/50. In addition to the mid and upper stratospheric temperature

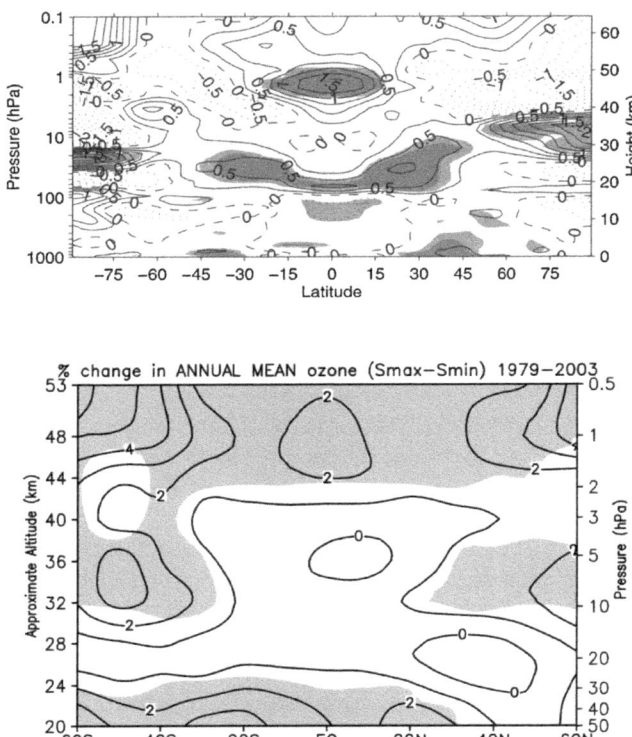

Fig. 3. (Top panel) The annual-mean temperature (°C) difference between the 11-year solar maximum and minimum conditions derived from European Centre meteorological re-analysis data covering 1978–2008. Reproduced from Frame and Gray (2010) (© American Meteorological Society). Used with permission. (Bottom panel) As in top panel, but showing the percentage changes in ozone (%) derived from SBUV satellite observations covering 1979–2003. Reproduced with permission from Soukharev and Hood (2006). Both estimates have been derived using a multiple linear regression analysis. The grey shading denotes regions where the differences are found to be statistically significant at the 95% confidence level using a two-sided Student's t-test.

changes, which are well understood from direct radiative and chemical arguments, a secondary peak temperature response has also been identified in the tropical lower stratosphere; this will be discussed further in Section 3.2.

It has been suggested that the stratospheric response to solar variability described above could impact on weather and climate in the underlying troposphere. Two main pathways have been proposed that are known as the 'polar route' and the 'tropical route'; the next sections describe these in turn. Whilst individual studies have demonstrated possible influences of both pathways in isolation, the relative importance of each route, and the potential interactions between them, is not currently well understood.

> **Box 4.2.1: The high latitude stratosphere during winter-time**
>
> The horizontal temperature gradient between the tropics and the winter high latitudes sets up a polar vortex each winter which extends throughout the stratosphere. This vortex consists of a region of relatively isolated air over the pole surrounded by a strong westerly jet that acts as a barrier for mixing of air between the poles and the mid-latitudes.
>
> The mean strength of the polar vortex from year-to-year is more variable in the Northern hemisphere than in the Southern hemisphere. This is primarily due to the effects of planetary scale Rossby waves, which are formed in the troposphere from flow disturbances over topography and land/sea temperature contrast, that propagate upwards into the stratosphere and break causing a disruption of the stratospheric flow: this wave breaking results in a transient warming of the polar region and a weakening of the polar vortex. On average, more planetary waves are generated in the Northern hemisphere because of the greater land mass there, and thus the polar vortex is more disturbed. In around two out of every three winters, the Northern hemisphere stratospheric flow undergoes a period of large excursion from its climatological state in events known as major sudden stratospheric warmings (SSWs). During these events, the polar stratospheric temperature can increase by up to ~60 K over a period of a few days and the stratospheric winds temporarily reverse from westerly to easterly.
>
> It has been shown that changes in winds and temperature in the lower stratosphere can impact on the evolution of weather systems in the underlying troposphere. In particular, following SSWs, there is on average an increase in mean sea level pressure over the polar region and a decrease in mid-latitudes, which corresponds to a negative phase of the North Atlantic Oscillation (NAO). This pressure pattern is associated with a southward shift in the storm track over the North Atlantic, which impacts on weather over western Europe and North America. Therefore processes in the climate system that alter the high latitude stratospheric flow can influence regional weather patterns via so-called 'stratosphere-troposphere dynamical coupling' (see e.g., Gerber et al. (2012)).

3.1 The 'polar route'

The so-called polar route describes the proposed influence of the solar cycle on the high latitude stratospheric flow during winter. To understand this mechanism, some background on the characteristics of the stratospheric flow in winter is given in Box 4.2.1.

As described above, the direct stratospheric response to the solar cycle consists of a temperature change in the mid and upper stratosphere. The structure of this temperature response is not homogeneous, but rather the relative warming under solar maximum conditions tracks the background seasonal cycle in temperature, with the largest changes occurring at the summer stratopause. This means that a change in solar forcing alters the horizontal temperature gradient near the subtropical stratopause.

On sufficiently large spatial scales, the horizontal temperature gradient is strongly coupled to the vertical gradient in the east-west (zonal) component of the wind (i.e., thermal wind balance is maintained). This means that the increase in stratospheric heating at solar maximum would lead to a westerly wind anomaly in the subtropical upper stratosphere. On its own, this change might not be expected to have a strong impact on the rest of the atmosphere. However, the background stratospheric winds can affect the propagation of the planetary scale waves that drive variations in the strength of the polar vortex (see Box 4.2.1). It has been hypothesised that such interactions between planetary waves and the background flow can act to propagate and amplify the initial westerly wind anomaly from the subtropical stratopause polewards and downwards through the winter season (see Figure 4(a)).

Such a change in the strength of the polar vortex can then influence regional surface climate at high latitudes through stratosphere-troposphere dynamical coupling (see Box 4.2.1). This 'polar route' for solar-climate coupling has been highlighted as a mechanism for driving a more positive North Atlantic Oscillation (NAO) index under solar maximum conditions, as has been identified in modelling and observational studies.

Some very recent studies have also suggested that atmosphere-ocean coupling over the North Atlantic could introduce a lagged component (i.e., delayed by about 1–3 years) to the NAO response to solar variations; this implies a combined role for top-down and bottom-up mechanisms in determining regional responses to solar cycle variations in the Northern hemisphere. Owing to the dependence of the 'polar route' on the presence of the stratospheric polar vortex, these mechanisms can only be active during winter-time or at the spring transition in each hemisphere.

3.2 The 'tropical route'

The second top-down route for solar-climate coupling focuses on the impacts of the tropical lower stratospheric secondary temperature maximum on the troposphere (see lower panel of Figure 3). It has been hypothesised that the modulation of planetary waves at high latitudes described in Section 3.1 would also affect the strength of the upward transport of air in the tropical lower stratosphere, which forms the ascending branch of the global stratospheric overturning circulation. This modulation would correspond to a decrease in upward motion under solar maximum conditions leading to a relative warming in the tropical lower stratosphere, since air expands and cools as it rises (see Figure 4(b)).

Tropical upwelling also affects the transport of ozone, such that a decrease in the circulation leads to an increase in ozone in the tropical lower stratosphere under solar maximum conditions, resulting in further warming because ozone absorbs

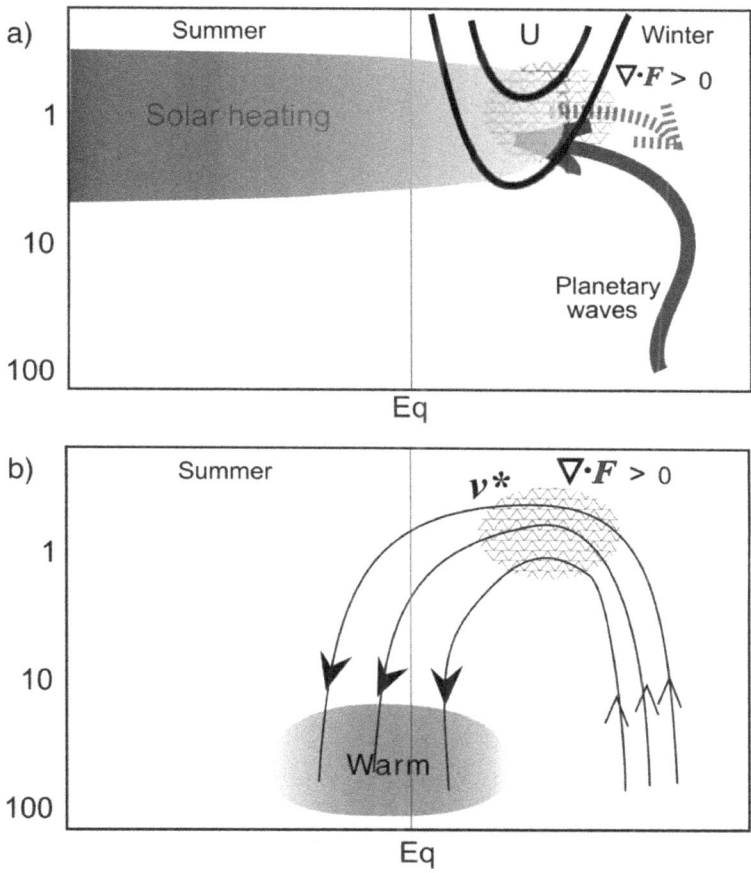

Fig. 4. Schematic showing the top-down 'polar route' for solar-climate coupling. $\nabla \cdot F$ is a measure of the zonal torque due to breaking of planetary scale waves, U is the east-west component of the wind and v^* is the north-south component of the wind forced by the wave breaking. Reproduced with permission from Kodera and Kuroda (2002).

solar and terrestrial radiation. Together these mechanisms could explain a relative warming of the tropical lower stratosphere between solar maximum and minimum conditions.

Whilst some global atmospheric models simulate a temperature increase in the tropical lower stratosphere that is comparable to that found in meteorological reanalysis data (see Figure 3), there have been questions raised around the reliability of the signals diagnosed in reanalyses. Due to the lack of global observations of the stratosphere before the satellite era began (pre-1979), reliable stratospheric reanalyses only exist for the past few solar cycles. During this period, there were two large tropical volcanic eruptions that coincided with solar maximum years: Mount Pinatubo in June 1991 and El Chichón in April 1982. These eruptions injected large quantities of sulphur dioxide into the tropical lower stratosphere, which is oxidised to form sulphate aerosols. These particles absorb radiation and warm the tropical lower stratosphere for a few years after an eruption. It has therefore been suggested that part of the apparent warming in the tropical lower stratosphere under solar maximum conditions identified in reanalysis data may be the result of aliasing between solar and volcanic effects.

A number of studies, mostly using more idealised numerical models of the atmosphere, have shown that heating in the tropical lower stratosphere could have a direct impact on surface climate by modulating the propagation of synoptic weather systems in the troposphere. This has been shown to lead to a poleward shift in the storm tracks under solar maximum conditions and an increase in the width of the tropical Hadley circulation. This mechanism could operate in both hemispheres and, in contrast to the 'polar route', may also act outside of the winter season.

It has also been suggested that the tropical route may act in concert with the bottom-up mechanism in the tropics (see Section 2). However, since there is still debate around the robustness of the lower stratospheric temperature maximum in response to the solar cycle, it remains an open question as to the relative importance of the 'polar route' and 'tropical route' for top-down solar-climate coupling.

Further reading

Gray, L. J., Beer, J., Geller, M., Haigh, J. D., Lockwood, M., Matthes, K., Cubasch, U., Fleitmann, D., Harrison, G., Hood, L., Luterbacher, J., Meehl, G. A., Shindell, D., van Geel, B., and White, W.: 2010, *Rev. Geophys.* **48**(4), 1

Haigh, J. D.: 2007, *Living Rev. Solar Phys.* **4**, 2. http://www.livingreviews.org/lrsp-2007-2

Pap, J. M., and Fox, P. [eds.]: 2004, *Solar Variability and Its Effect on Climate*, Geophysical Monograph Series Volume **141**, American Geophysical Union, Washington

Rind, D.: 2002, *Science* **296**, 673-677

CHAPTER 4.3

INTERACTIONS OF DIFFERENT SOURCES OF VARIABILITY

Katja Matthes[1], Rémi Thiéblemont[1], Markus Kunze[2] and Matthew Toohey[1]

The two previous Chapters in part 4 (Chapters 4.1 and 4.2) discussed direct impacts of solar irradiance variability as well as existing mechanisms for solar-climate coupling. This Chapter gives an overview on other variability sources that complicate the identification of pure solar signals in the atmosphere.

1 Introduction

The identification of solar irradiance signals in the atmosphere and the ocean is further complicated by the interaction with other sources of variability which potentially overshadow the solar signal and are schematically presented in Figure 1. As discussed previously (Chapters 4.1 and 4.2), solar irradiance variations in the UV produce higher temperatures during solar maximum years in the tropical upper stratosphere (pink shaded area in Figure 1). This direct radiative effect leads through complex dynamical interactions between atmospheric planetary waves (black arrows in Figure 1) and the mean flow to indirect effects such as a strengthening of the polar vortex and the westerly winds over the winter pole as well as to a warming in the tropical lower stratosphere (pink shaded areas in Figure 1). In addition, there is a small warming of the surface, in particular over the tropical oceans caused by increased absorption of visible and near infrared radiation (VIS/NIR) during solar maxima at the ocean surface which is further enhanced through atmosphere-ocean coupling (Meehl and Arblaster, 2009). These solar variability induced signals in the tropics and at the winter poles compete with signals from other sources of natural and man-made climate variability.

The difficulty to disentangle these signals in observations is partly due to the shortness of the available observational record and also depends on the analysis

---

[1] GEOMAR Helmholtz Centre for Ocean Research Kiel, Düsternbrooker Weg 20, 24105 Kiel, Germany; and Christian-Albrechts Universität zu Kiel, Kiel, Germany
[2] Institut für Meteorologie, Freie Universität Berlin, Carl-Heinrich-Becker Weg 6-10, 12165 Berlin, Germany

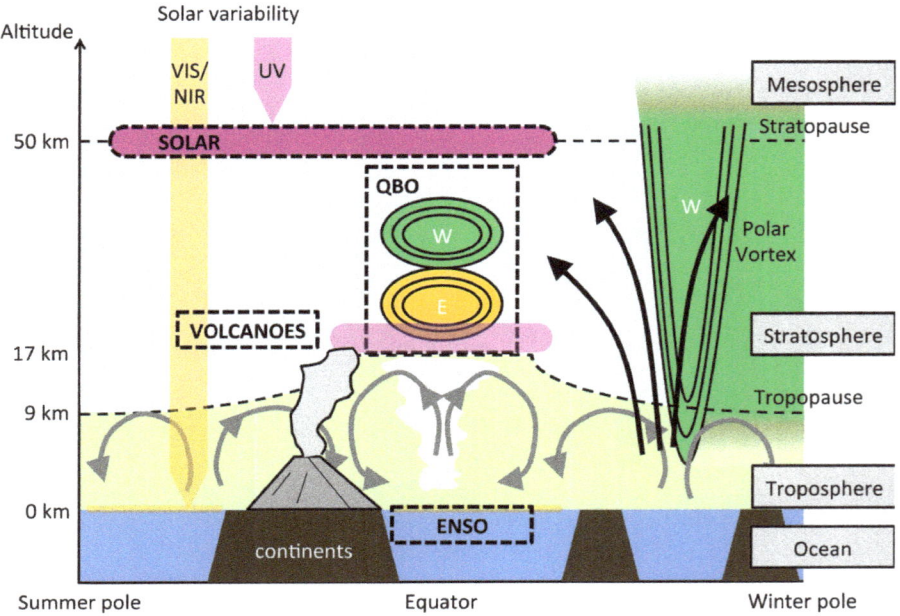

Fig. 1. Schematic representation of possible natural variability sources that interact with the solar irradiance signal (SOLAR) and complicate a solar signal detection in the atmosphere-ocean system. Sources of variability (solar, QBO, ENSO, and volcanoes) are surrounded by dashed rectangles. Higher temperatures during solar maximum are shown in pink, westerly winds (W) are green, and easterly winds (E) yellow, see text for details.

technique. Sufficient observations covering the atmosphere from the surface to the stratosphere and mesosphere only exist for the last three solar cycles. However, for a reliable solar signal detection, at least 100 years would be needed. Another difficulty to disentangle signals arises from the fact that the variability factors such as ENSO, QBO and volcanoes interact with each other. Thus, the resulting solar signal is not necessarily a linear superposition of the individual factor contributions. The commonly used mathematical method of multiple linear regression might therefore not be appropriate to study these non-linear effects. In contrast, a composite analysis requires a long enough record of observational data to make firm conclusions.

Due to the shortness of observational records, climate modeling studies are ideal to investigate the observed findings. Unlike in nature, in a climate model, it is easy to switch on and off different climate variability factors to understand their potential contributions to the solar signal as well as the interaction between the variability factors themselves. The climate model can be run for a sufficient period of time (more than the currently observed three solar cycles) multiple times, so the number of realisations is larger than the only one we have in reality and allows for a robust detection and attribution of solar signals. Nevertheless, also models have uncertainties and do not necessarily represent the observed climate state perfectly. Therefore, a combined analysis of observational and model experiments is needed to tackle questions like the contribution of different natural and man-made variability factors.

One prominent natural variability factor in the stratosphere is the Quasi-Biennial Oscillation (**QBO**) of equatorial zonal mean winds (Baldwin et al., 2001). Approximately every two years, the zonal mean winds in the tropical lower stratosphere change their direction from easterly (QBO east) to westerly (QBO west) blowing winds which in turn alter the circulation in the polar atmosphere through interaction of the mean flow with planetary waves: QBO east conditions in the tropical lower stratosphere favor a more disturbed and warmer polar vortex at the stratospheric winter pole whereas QBO west conditions favor a colder and undisturbed polar vortex (Holton and Tan, 1980).

Another prominent natural variability factor is the El Niño Southern Oscillation (**ENSO**) phenomenon in the tropical Pacific, a coupled atmosphere-ocean oscillation. Every two to seven years, surface temperatures in the tropical eastern Pacific and surface pressure in the western Pacific are higher than normal (El Niño, ENSO warm phase). La Niña (ENSO cold phase) is characterised by lower surface temperatures in the eastern Pacific and low surface pressure in the western Pacific. ENSO warm phase conditions in the tropical Pacific favor a more disturbed and warmer polar vortex at the stratospheric winter pole as compared to ENSO cold phase conditions (van Loon and Labitzke, 1987). The ENSO warm signal propagates from the tropical Pacific into the stratosphere, increases planetary wave activity there which in turn disturbs the polar vortex Ineson and Scaife (2009); Butler and Polvani (2011).

A third prominent natural variability factor is volcanic eruptions (volcanoes) which occur sporadically. The injection of sulfur species into the stratosphere by major volcanic eruptions leads to the formation of sulfate aerosols (H_2SO_4) on a timescale of weeks, which can strongly affect the Earth's radiative balance. These aerosols reflect solar visible radiation, causing a net cooling at the surface, and absorb solar near-infrared and terrestrial infrared radiation, causing a warming of the lower stratosphere. The impact of volcanic eruptions on the stratosphere has been observed for three major eruptions: Mount Agung (1963), El Chichón (1982), and Mount Pinatubo (1991). During winters after major volcanic eruptions, the polar vortex tends to be colder and less disturbed (Robock, 2000). This stronger vortex occurs because of a stronger temperature difference between equator and pole resulting from aerosol heating in the tropical lower stratosphere (Stenchikov et al., 2002; Toohey et al., 2014) and has an effect similar to that seen during solar maxima (Figure 1).

Since all variability factors occur at the same time, it is very difficult to separate their effects. On top of the natural climate signals shown in Figure 1, there is also a possible contribution of solar energetic particles (see Chapter 4.6) as well as a human-induced warming in the troposphere and cooling in the stratosphere due to the man-made increases in greenhouse gases.

2 Contributions in the polar stratosphere

QBO east phase conditions as well as ENSO warm events favor a disturbed and warm polar stratopheric vortex during winter. Prominent manifestations of disturbed vortex events are sudden stratospheric warmings (SSWs), which occur every other year on average and are characterised by a strong temperature increase within a few days and a reversal of the normal circulation from westerly to easterly winds. Labitzke and van Loon (1990) discovered the dependence of SSWs not only on the phase of the QBO and ENSO, but also on the phase of the solar cycle. Figure 2 shows an update of their famous original figure, which shows a clear dependence of the North Pole 30 hPa (24 km) geopotential heights[3] on the solar cycle when the data are grouped according to the phase of the QBO. SSWs (filled circles and squares in Figure 2) tend to occur during QBO west and solar maximum as well as during QBO east and solar minimum conditions. Under QBO east conditions (left-hand Figure 2), the 30 hPa geopotential height decreases with increasing solar activity (negative correlation coefficient r), whereas under QBO west conditions (right-hand Figure 2), it increases with increasing solar activity (positive correlation coefficient r). The impact of ENSO (red and blue years in Figure 2) on the frequency of SSWs is rather complex and recent work suggests a stronger occurrence of SSWs during both ENSO phases (Butler and Polvani, 2011). Figure 1 gives a flavor of the complex Interactions of the different variability factors on the polar atmosphere.

Parts of these observed findings can be reproduced in climate model experiments (Calvo et al. (2009); Gray et al. (2010)). It has been shown for example that by removing ENSO and QBO variability from the model atmosphere, the amount of SSWs is significantly reduced (Richter et al., 2011). The preferred occurrence of SSWs during certain QBO and solar phases seems to be related to a QBO modulation of the mean state which in turn modifies the solar signal transfer (Matthes et al., 2013). However, details of the solar-QBO and other non-linear interactions with ENSO and volcanoes are less clear and understood and are still under investigation (e.g., Camp and Tung (2007); Calvo and Marsh (2011); Matthes et al. (2013), Chiodo et al. (2014)).

3 Contributions in the tropical stratosphere

The solar induced warming in the tropical lower stratosphere (see Figure 1 and Chapter 4.2) has been subject of a number of studies to understand whether it is related to non-linear interactions of different variability factors or arises from contamination by or aliasing with other signals, such as the QBO, ENSO, and volcanoes depending on the respective analysis method (e.g., Gray et al. (2010), Chiodo et al. (2014)). Figure 3 shows the annual mean solar induced temperature

---

[3] Gravity-adjusted height of the pressure level, meaning valleys and mountains of height surfaces in the stratosphere.

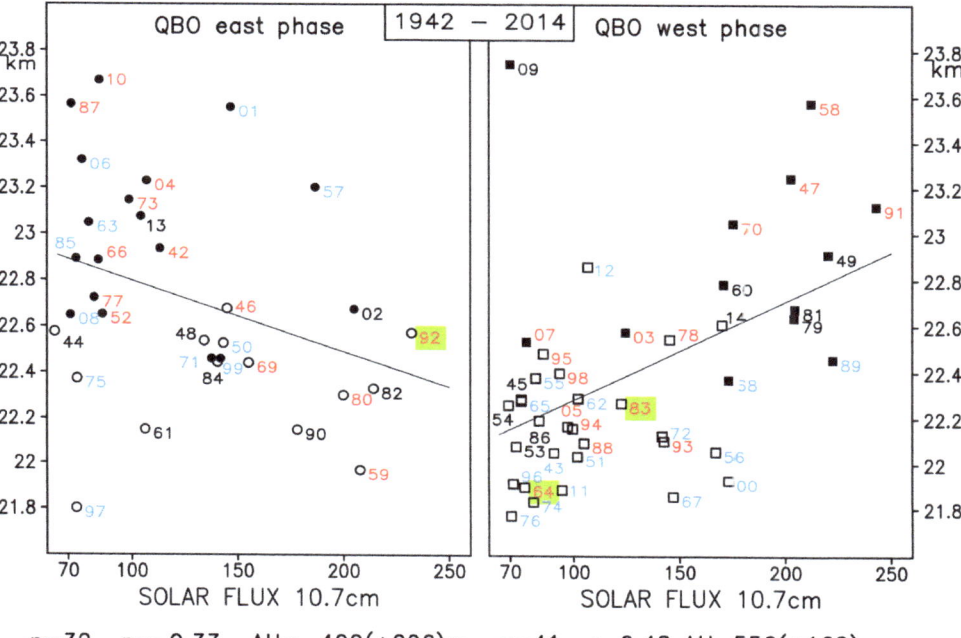

Fig. 2. Scatter diagrams of the monthly mean 30-hPa geopotential heights (geopot. km) in February at the North Pole (1942–2014), plotted against the 10.7 cm solar flux, a proxy for solar activity, in solar flux units (1 sfu = 10−22 W m^{-2} Hz^{-1}). Left: Circles: years in the east phase of the QBO ($n = 32$). Right: Squares: years in the west phase of the QBO ($n = 41$). The numbers indicate the respective years, with ENSO warm events in red and ENSO cold events in blue; r=correlation coefficient; dH gives the mean difference of the heights (geopotential m) between solar maxima and minima (minima are defined by solar flux values below 100). Filled squares and filled circles denote Sudden Stratospheric Warmings (SSWs), i.e., a winter with a reversal of the zonal wind over the Arctic at the 10-hPa level. Years in which a volcanic eruption affected the stratosphere are highlighted with a green square. (Reconstructions 1942 till 1947; NCEP/NCAR re-analyses: 1948 till 2014) (Labitzke and van Loon (1990), updated from Labitzke et al. (2006)).

signals in a number of different sensitivity experiments by switching on and off the above described factors, such as QBO, ENSO, and VOLCANOES with a state-of-the-art climate Model including atmospheric chemistry (Hansen et al., 2014). All figures depict a similar warming signal during solar maximum in the tropical upper stratosphere of about 1 K at 1 hPa (48 km) height. However, the shape and strength of the solar induced warming in the tropical lower stratosphere depends on the presence of volcanic, QBO and ENSO effects. The warming signal in the tropical lower stratosphere is particularly different when volcanoes and the QBO are removed from the model experiments. Since in nature all variability

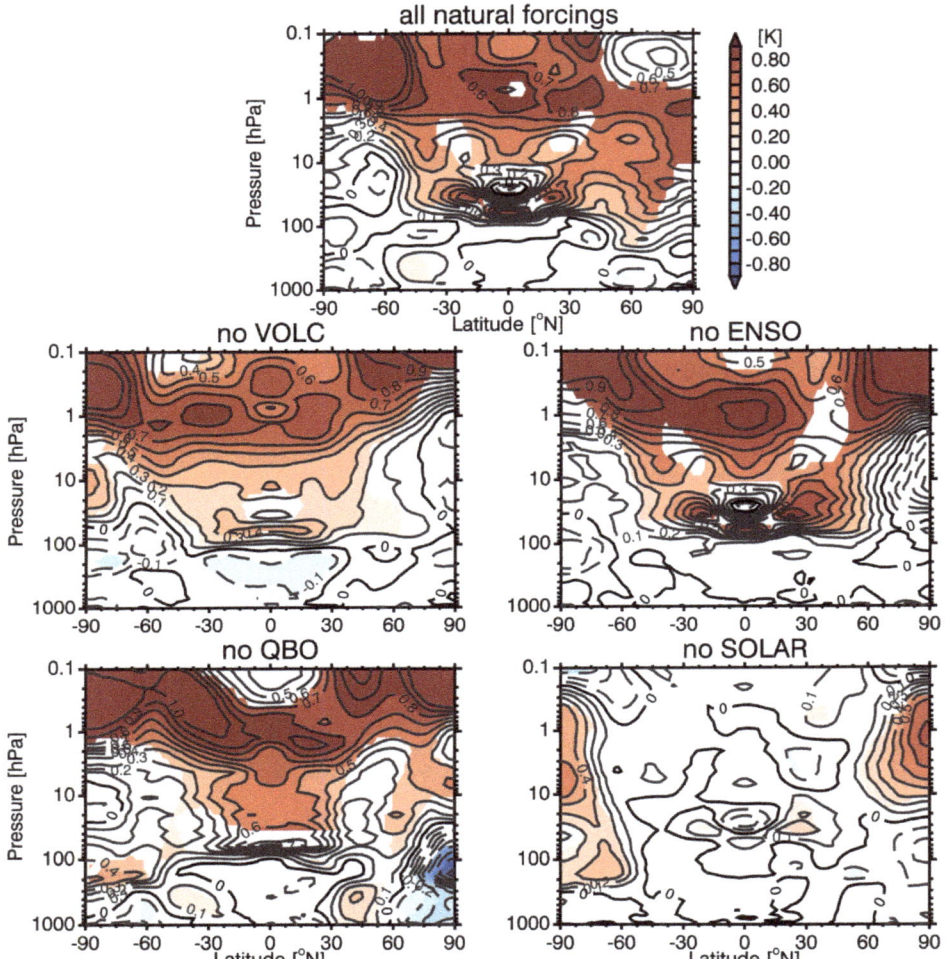

Fig. 3. Differences in annual mean temperatures between solar maximum and solar minimum years for a number of different multi-year model simulations with different sets of variability factors in a fully coupled atmosphere-ocean climate model with atmospheric chemistry (CESM-WACCM). a) Model experiments using all natural forcings (SOLAR, volcanoes, ENSO, and QBO); b) model experiments removing volcanoes; c) model experiments removing ENSO; d) model experiments removing the QBO, and e) model experiments removing the solar cycle (sampling in this case is done for the same years as in the runs with solar cycle). Statistically significant temperature signals are shaded.

factors occur at the same time, it is very difficult to separate their effects and attribute the warming in the tropical lower stratosphere to the solar cycle alone (e.g., Chiodo et al. (2014)). There are additional factors to be considered, such as the man-made GHG increase, and particle precipitation, which will be discussed later on (Chapter 4.6).

Take home message

Besides solar cycle variability there are a number of other natural variability factors, such as ENSO, QBO and volcanoes that influence climate variability. These factors act at the same time and therefore interact with each other and overshadow or modulate solar signals. It is therefore difficult to detect pure solar signals, in particular in the troposphere where synoptic-scale, i.e., small-scale, eddies contribute to a large variability. A major problem is the relatively short length of observational records, which makes it extremely difficult to detect robust solar signals throughout the atmosphere and in the ocean. Recent progress has been made with improved and more complex climate models that can be used as simulators of past, present and future climate to produce a sufficiently long record under idealised conditions to investigate solar signals. However, also these models have uncertainties and it is therefore important on the one hand to improve climate models and on the other hand to continue the production of spatially and temporally well-resolved records of the climate system.

Further reading

Baldwin, M. P. et al.: 2001, The quasi-biennial oscillation, Rev. Geophys., 39, 179229, doi:10.1029/1999RG000073.

Brönnimann, S. et al.: 2006, Impact of El Niño-Southern Oscillation on European climate, Rev. Geophys., 45, RG3003, doi:10.1029/2006RG000199.

Gray, L.J., et al.: 2010, Solar Influences on Climate, Rev. Geophys., 48, RG4001, doi:10.1029/2009RG000282.

Labitzke, K., and H. van Loon (1999), The Stratosphere: History, Phenomena, Relevance, Springer.

Robock, A.: 2000, Volcanic Eruptions and Climate, Rev. Geophys., 38(2), 191219, doi:10.1029/1998RG000054.

CHAPTER 4.4

IMPACT OF SOLAR VARIABILITY ON THE MAGNETOSPHERE

Eija I. Tanskanen[1] and Patrizia Francia[2]

Solar storms shake the magnetic environment of the Earth on time scales from seconds to solar cycles and beyond. Effects of solar storms are transferred via solar wind to the magnetosphere (the near-Earth space dominated by the terrestrial magnetic field instead of the solar magnetic field). Solar storms vary in strength, length and type, as described in Part 2, and thus vary their impact on the magnetosphere as well. In this chapter, we describe how the solar storms affect to the near-Earth space.

1 Solar wind - magnetosphere interaction

The two main manifestations of magnetic activity at the magnetosphere are geomagnetic storms and auroral substorms. The other solar storm induced geomagnetic variations include geomagnetic pulsations, steady magnetospheric convection and continuous magnetospheric dissipation. Geomagnetic storms were discovered in the 1700s and they are the strongest geomagnetic variations caused by the solar originated disturbances, producing strong currents and changes to the Earth's magnetic field at the equator of several hundred nano Tesla (nT). Auroral substorm-like features were reported a century later, in the 1800s (Birkeland, 1908). The important link between the solar wind and the magnetosphere was discovered in the mid-1900s (Fairfield and Cahill, 1966). It was noticed that the magnetospheric storm occurrence depends strongly on the interplanetary magnetic field direction such that southward oriented interplanetary magnetic field opens the "magnetosphere's door" to the solar wind plasma to enter to the near-Earth space. This process is called magnetic reconnection (see Figure 1). The solar wind plasma (electrons and protons) is transported into magnetosphere, ionosphere and

---

[1] Eija Tanskanen, Finnish Meteorological Institute, Erik Palmenin aukio 1, FI-00100 Helsinki, Finland
[2] Department of Physical and Chemical Sciences University of L'Aquila, via Vetoio 1, 67100 L'Aquila, Italy

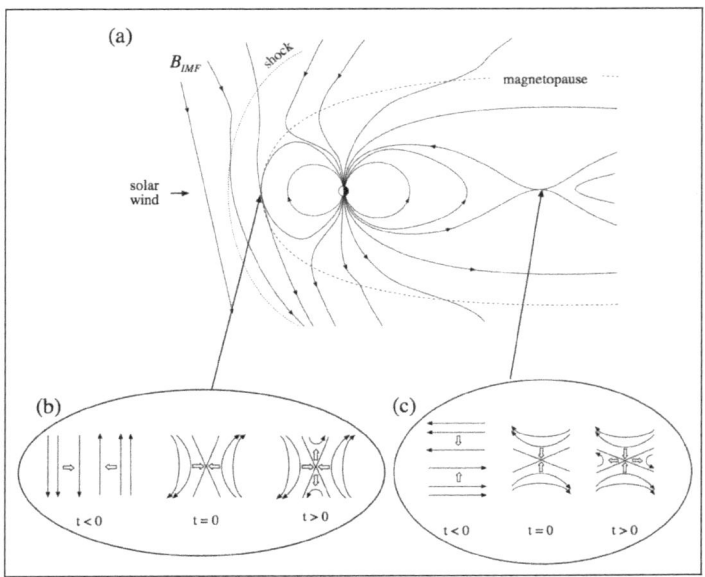

Fig. 1. A schematic picture is showing how the "magnetospheric door" is opened for the solar wind particles and magnetic disturbances to enter the magnetosphere during solar storms.

atmosphere. On its way from the Sun to the terrestrial atmosphere the particles and magnetic field cause variations in magnetospheric particle density, temperature, magnetic field and pressure. This variation affects technological systems and human beings in space and on ground, as well as conditions in the atmosphere and on ground. Space weather disturbances triggered by solar storms cause harm for example for satellite communication, ship and airplane navigation close the magnetic poles, power transmission and oil drilling.

Geomagnetic storms and substorms play an important role in transferring the solar energy into the magnetosphere, the previous one to the equatorial region and the latter one to the high-latitudes. Substorms are able to transfer the energy closer to the Earth than magnetic storms. Substorms occur several times per day, on average, while storms occur only 1-2 times per month. Furthermore, substorms transfer energy closer to the Earth than magnetic storms. Solar wind coupling efficiency varies over the course of a day, month, season, year and solar cycle. Most often used coupling functions include vB^2, v^2B, E_y and ϵ-parameter, where v is solar wind velocity, B interplanetary magnetic field, E_y solar wind electric field, and ϵ-parameter is a combination of velocity, magnetic field and solar wind flow angle. Solar wind disturbances "shake" and disturb the terrestrial environment in many different ways. Interplanetary plasma clouds and coronal mass ejections (ICME) affect the Earth's magnetic environment strongly for intervals of few hours, while high-speed streams (HSS) enhance magnetospheric activity over long

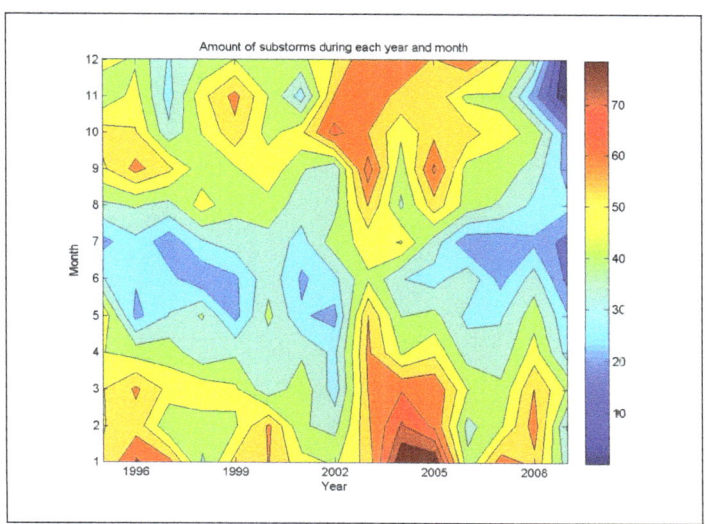

Fig. 2. Seasonal and yearly variation of auroral substorm occurrence rate. Substorms are identified from ground-based magnetometer observations by an algorithmical identification method.

intervals of several days and weeks. ICMEs cause abrupt and dramatic changes in the magnetosphere, atmosphere and at ground, while HSS disturb magnetosphere continuous, but more smooth than ICMEs. The high-speed streams cause slow degradation of technological systems rather than rapid burning of transformers like ICMEs. The abrupt changes in the magnetosphere caused by ICME-type disturbances are easier to detect compared to the slow changes caused over the years. ICME occurrence peaks at the solar activity maxima and the high-speed stream occurrence maximises in the declining phase of the solar cycle (Tanskanen et al., 2005). Recent studies also show that most substorms occur in winter (see Figure 2) while most intense auroral substorms occur in spring and fall (Tanskanen, 2009).

2 Magnetospheric ULF waves and their effects

Magnetospheric Ultra-Low-Frequency (ULF) waves are short period (0.2–1000 s) and small amplitude oscillations of the Earth's magnetic field (Liu, 2011). Amplitudes of waves range from less than 0.1 nT to hundreds of nT. Ground signatures of ULF waves are named geomagnetic pulsations. Pulsations with quasi-sinusoidal waveform are called continuous (Pc1-Pc5, according to their period). Their amplitude is found to increase with the latitude of the observing point and thus the wave amplitude is larger on the geomagnetic field in more external regions of the magnetosphere. Like longer period magnetospheric disturbances, such as magnetic storms and substorms, waves are mostly produced by the interaction of the magnetosphere with geoeffective solar wind structures, such as HSSs and ICMEs.

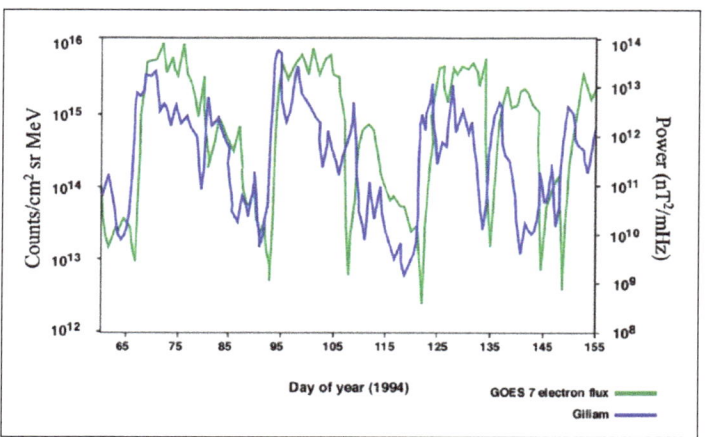

Fig. 3. ULF wave power, observed by a ground magnetometer at Gillam in Canada, together with energetic electron fluxes observed at geosynchronous orbit (Courtesy: (Rostoker et al., 1998)).

They represent an essential element in the magnetospheric physics, in that they provide a means for the magnetospheric field and plasma to release the energy transmitted from the solar wind.

Waves can be related to different interaction processes depending on their period. Waves with periods of few minutes (corresponding to frequencies of few mHz, Pc5-4) can be generated by: a) the Kelvin-Helmoltz instability, due to the relative motion of the solar wind with respect to the stationary plasma of the magnetosphere along the flanks of the magnetopause; b) a broad band impulse on the magnetopause, such as a solar wind pressure pulse; c) the transmission through the magnetopause of compressional waves already existing in the solar wind. Waves with periods of tens of seconds (or frequencies of tens of mHz, Pc3) are related to the penetration in the magnetosphere of upstream waves generated in the Earth's foreshock region through a wave-particle interaction by solar wind protons reflected off the bow shock and backward propagating along the interplanetary magnetic field lines.

Magnetospheric waves are useful in probing the magnetospheric environment, because they carry information on the regions through which they propagate. In recent years, magnetospheric waves have received considerable attention regarding their interaction with magnetospheric particles. In particular, long period ULF waves, such as Pc5 waves seem to have a role in accelerating radiation belt particles to relativistic energies (MeV). The increases in energetic electron fluxes at geosynchronous orbit are known to be preceded by increases in Pc5 wave activity at high latitude by a few days, which are associated with high-speed solar wind streams (Figure 3; (Baker et al., 1998; Rostoker et al., 1998). The highest correlation between solar wind speed, ULF wave power and MeV electron flux was found in the declining phase of the solar cycle (Mann et al., 2004). The acceleration process is believed to involve resonant interaction of the Pc5 electric and magnetic field oscillations with the electron drift motion outside geosynchronous orbit, leading to radial inward transport and relativistic electron flux build up between 4.5 and 6 RE in the outer radiation belt (Takahashi, 2006). The precipitation of 1 MeV electrons into the atmosphere, along dipolar magnetic field lines located at about 4-5 Earth's radii at the equator, has been associated with electromagnetic ion-cyclotron waves (EMIC waves) observed by ground-based magnetometers during periods of moderate geomagnetic activity (Rodger et al., 2008). EMIC waves occur in the ULF highest frequency range (Pc 1-2, f about 0.1–5 Hz) and are generated near the magnetic equator by unstable distributions of ring current ions during geomagnetic storms. Theoretical studies have demonstrated that such waves should be an effective mechanism for loss of larger than 1 MeV electrons from the radiation belts through a resonant interaction and scattering into the loss cone. Both electron acceleration and loss processes need to be active in order to generate significant level of energetic electron precipitation during geomagnetic disturbances associated to HSSs and ICMEs, which has potentially a significant influence on the Earth's upper and middle atmosphere, particularly at high latitudes (Clilverd et al., 2010).

Further reading

Akasofu, Syun-Ichi, Exploring the secret of the auroral, Kluwer Academic Publishers, 2002.

Carlowicz, Michael and Ramon E. Lopez, Storms from the Sun, National Academies of Sciences, US, 2002.

Substorm Zoo, data and browser-based analysis tools, link: `http://www.substormzoo.org`

CHAPTER 4.5

ATMOSPHERIC IONISATION BY SOLAR ENERGETIC PARTICLE PRECIPITATION

Pekka T. Verronen[1] and Craig J. Rodger[2]

1 Solar particles arrive to Earth with the solar wind

Earth's atmosphere is affected by both continuous and short-lived impulsive solar particle precipitation. The Sun continuously emits a stream of high-energy particles, mostly consisting of protons and electrons. This stream is known as the solar wind and it affects the whole heliosphere, including the Earth and its nearby space (see Chapters 2.3 and 2.4). The intensity of the Solar wind varies slowly across the 11-year solar activity cycle, but also on shorter time scales due to violent eruptions on the Sun's corona, which lead to increases in solar wind density and energy.

2 Earth's magnetic field shields the atmosphere from particles, but not from all of them

Earth's magnetic field shields most of the atmosphere from solar charged particles. Due to their relatively low energy, the solar wind protons and electrons are not able to penetrate the Earth's atmosphere directly. As an exception, during big solar storms, the protons gain enough energy to do this, but even then they can enter the atmosphere only at polar latitudes where the magnetic shielding is weakest. Solar electrons are captured by the Earth's magnetic field and stored near the Earth, for example in the so-called Van Allen radiation belts. From their storage, the electrons are eventually lost either to outer space or to the polar atmosphere. The latter is especially important during solar-wind-driven magnetic storms, which also accelerate the electrons to high-enough energies, so that they can precipitate into the atmosphere.

---

[1] Earth Observation Unit, Finnish Meteorological Institute, PO Box 503, 00101 Helsinki, Finland
[2] Department of Physics, University of Otago, PO Box 56, Dunedin 9016, New Zealand

3 Atmospheric effects of particle precipitation

Except for galactic cosmic rays, which are not of solar origin and are not discussed here (they are discussed in detail in Chapters 2.3, 4.7, and 4.8), the energy of the precipitating particles is not high enough to cause nuclear reactions. Instead, most of their energy ionises atmospheric neutral molecules, mainly the most abundant ones N_2 and O_2. The ionisation process creates an ion-electron pair, and thus adds to the existing ionisation produced by solar extreme ultraviolet and X-ray radiation, contributing significantly to the characteristics of the Earth's ionosphere in the polar regions.

In addition to ionisation, a part of the particle energy dissociates atmospheric neutral species. Further, ionisation initiates ion chemical reaction chains. Together, these lead to productions of ozone-depleting constituents in the atmosphere, such as odd hydrogen (HO_x) from water vapour and odd nitrogen (NO_x) from N_2. For further details on particle effects on atmosphere and climate, see Chapter 4.6.

4 Main types of particle precipitation and their characteristics

There are basically three main types of solar particle precipitation: solar proton events, auroral electron precipitation, and radiation belt electron events (which includes medium and relativistic energy electrons). These differ from each other by energy of the particles, duration and rate of the events, as well as the distribution of events over the solar cycle. The energy of the particles is important because it determines the atmospheric altitudes affected (the larger the energy, the deeper the penetration). Table 1 lists some of the event characteristics. Figure 1 shows the range of altitudes and ionisation rates for the different types, together with a range of ionisation rates to which the magnitude of atmospheric effects correspond to. Figure 2 demonstrates the frequency of different types of particle precipitation and their distribution during the solar cycle. Note, however, that both solar proton and radiation belt electron events are sporadic, so their distribution is not the same for any two solar cycles.

Solar proton events (SPE) are caused by large solar explosions in solar flares and so-called coronal mass ejections (CME). When a CME hits the Earth, a larger number of high-energy protons penetrate into a wider range of latitudes across the polar cap areas, i.e., at magnetic latitudes higher than about 60°, in both hemispheres. SPEs can lead to very strong ionisation in the upper stratosphere and mesosphere. At these altitudes, the ionisation due to the slowly varying solar radiation and galactic cosmic rays is weak, and SPEs can increase ionisation rates by a factor of 100 – 10 000. SPEs are infrequent and sporadic, however, although they are more probable during the maximum of a solar cycle at which time a few of them occur per year. A high-ionisation period caused by an SPE typically lasts up to 10 days.

Table 1. Characteristics of different types of particle precipitation events.

Type	Solar proton events	Auroral electron precipitation	Radiation belt electron events
Typical energies	1 – 200 MeV	< 30 keV	30 keV – 1 MeV
Main atmospheric effect altitudes	30 – 90 km	>90 km	50 – 90 km
Precipitation duration and rate	< 10 days per event Sporadic, a few very large events during a solar cycle	Essentially continuous	A few days, about 60 very large events during the last 11 years cycle
Solar-cycle distribution	Most likely during sunspot maximum	About 2 years after sunspot maximum	About 2 years after sunspot maximum
Ionisation rates (ion pairs/cm^3/s)	Up to about 10 000	Up to about 10 000	Up to about 1000

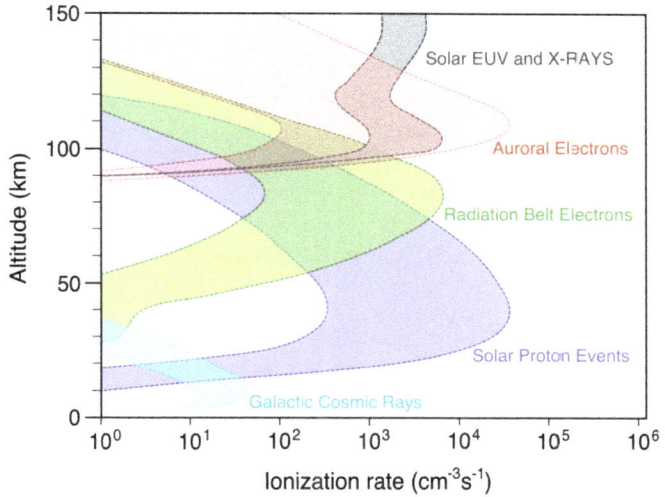

Fig. 1. A schematic demonstration of the magnitude and altitude extent of different atmospheric ionisation sources (courtesy of ME Andersson, Finnish Meteorological Institute). The Particle forcing exceeds that of solar irradiance, and is the main source of ionisation, at altitudes below about 90 km. Note that the figure is a modified version of the one presented by D.N.Baker, EOS Transactions, American Geophysical Union, Vol. 93, No. 34, 21 August 2012.

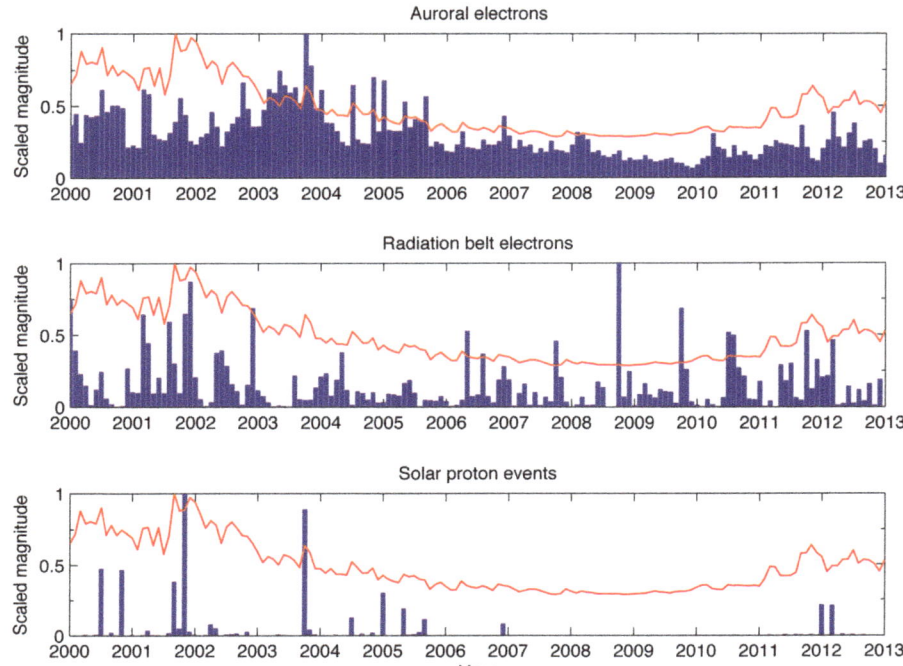

Fig. 2. A demonstration of the frequency of different types of particle ionisation events, and also the 2-3 year lag of the electron ionisation maximum (compared to the solar radio flux). Shown are the scaled magnitudes of: 1) auroral electrons (magnetic Ap index as a proxy), 2) precipitating radiation belt electrons (flux observations of electrons with energy larger than 30 keV), and 3) solar proton events (peak proton flux units, energy larger than 10 MeV) from 2000 to 2012. The red line shows the scaled solar radio flux (indicator of the solar activity cycle) and is the same in all panels.

Auroral electrons (AE) are of relatively low energy and ionise a latitude band around the magnetic poles, i.e., the so-called auroral ovals. The peak of the oval is typically located at 67–77° of magnetic latitude. AE mostly affect the lower thermosphere, with peak ionisation rates occurring at about 110 km altitude. In addition to ionisation, the auroral electrons are responsible for creating aurora (often referred to as the Northern and Southern lights) when interacting with atmospheric molecules. The ionisation produced by AE is more or less a continuous process but there is still significant day-to-day variability with increased ionisation taking place during magnetic storms.

Higher-energy electron precipitation (EEP) ionises a latitude band principally in the range of 55–75° of magnetic latitude, partly overlapping with the auroral oval. These latitudes are the atmospheric footprints of the Van Allen radiation belts. During magnetic storms (and in the days after), the number of electrons increases across all energies, such that EEP-produced ionisation increases in the upper mesosphere but also penetrates to lower altitudes, down to about 50 km. The ionisation rates from EEP in storms are higher than those from AE but lower than SPE. Thus, ionisation caused by these electrons can significantly affect mesospheric but not stratospheric altitudes. Like SPEs, EEP is strongly an event type phenomenon although this source is still much more frequent than SPEs.

Our scientific understanding of the causes and properties differs between the three main types of particle precipitation. This is partially reflected by the experimental observations, which are available to describe each type. Satellite measurements of SPE from geostationary and low-Earth orbit have provided good-quality representations of the varying flux with energy for the range relevant to the atmosphere, and there is a fairly high level of understanding as to the portions of the atmosphere affected by SPE across most of the event. Auroral electrons have been monitored by low-Earth orbiting satellites for many decades. Due to their essentially continuous nature and the long-lived good-quality datasets, there are multiple models available to predict auroral particle precipitation and its variation in space and time. It is EEP for which there has been lowest quality experimental data, and for which there remains the largest uncertainties. There is a growing interest in EEP across radiation belt and atmospheric communities which are, over time, improving our knowledge as to the properties of EEP. As EEP is common and drives fairly strong ionisation rate increases it is important for our understanding of this source to significantly improve.

Further reading

Brasseur, G. P., and S. Solomon: 2005, Aeronomy of the Middle Atmosphere, 3rd ed., Springer, Dordrecht.

Hargreaves, J. K.: 1992, The solar-terrestrial environment, Cambridge Atmospheric and Space Science Series, Cambridge University Press, Cambridge, UK.

CHAPTER 4.6

IMPACT OF ENERGETIC PARTICLE PRECIPITATION ON ATMOSPHERIC CHEMISTRY AND CLIMATE

Annika Seppälä[1], Bernd Funke[2] and Pekka T. Verronen[3]

1 Direct impact of energetic particles

As discussed in the previous chapter "Atmospheric ionisation by energetic particle precipitation", precipitating charged particles increase the amount of ionization in the polar atmosphere. In the atmosphere, between about 30 km and 100 km altitudes, the enhanced ionization results in production of gases known as Odd Hydrogen, or HO_x, and Odd Nitrogen, which is also known as NO_x. These are families of gases, which actively affect the atmospheric ozone balance, particularly via catalytic chemical reaction cycles, where ozone molecules are destroyed, but the catalyst, the NO_x or HO_x, remains. Their role in atmospheric ozone chemistry was initially discovered in the 1970s, leading to a Nobel price in Chemistry in the 1990s (Crutzen et al., 1975). It was not until the 21st century and a new era of Earth observing satellite instruments, such as those onboard the European environmental Envisat-satellite, that the full scale of energetic particle precipitation driven HO_x and NO_x production and their consequent long lasting effect on mesospheric and stratospheric ozone was understood (e.g., Funke et al. (2011)).

The HO_x family is formed of H, OH, and HO_2 (Brasseur and Solomon, 2005). These gases have a very short chemical lifetime, but they destroy ozone in very rapid chemical reactions, particularly in the mesosphere between 60–80 km. During night-time conditions, Odd Hydrogen is only produced via ionization by energetic particles (see Figure 1) and following large solar storms the abundance of HO_x can increase by thousands of percent. After the ionization subdues, the HO_x amounts return to normal background levels in the matter of hours. The increase in HO_x during energetic particle precipitation events can lead to up to 80–90%

---

[1] Finnish Meteorological Institute, Helsinki, Finland; also at British Antarctic Survey/NERC, Cambridge, United Kingdom
[2] Instituto de Astrofísica de Andalucíca, CSIC, Granada, Spain
[3] Earth Observation Unit, Finnish Meteorological Institute, PO Box 503, 00101 Helsinki, Finland

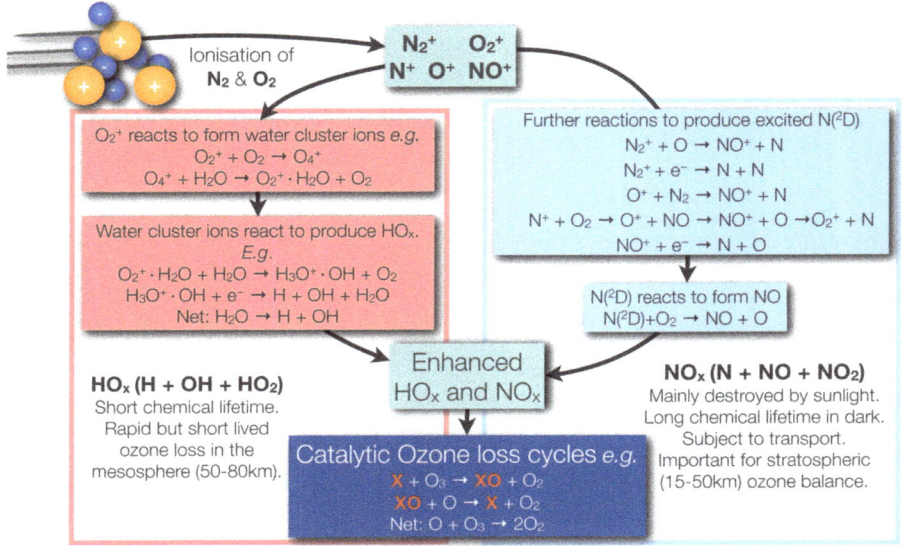

Fig. 1. HO_x and NO_x production by energetic particle precipitation and the subsequent impact on ozone. The catalyst in the example ozone loss cycle, where X can be any of the HO_x or NO_x gases, is shown with bold red font.

ozone loss in the mesosphere, with gradual recovery as the HO_x concentration starts to level off.

In the stratosphere and mesosphere, the NO_x family is formed of N, NO, and NO_2 (see Figure 1). Sometimes the more extensive NO_y family Brasseur and Solomon (2005) can also be used. In addition to the NO_x gases, the NO_y family includes the so-called NO_x reservoir species e.g., N_2O_5, and HNO_3. The term "reservoir species" refers to molecules that have a longer chemical lifetime in the atmosphere and to which the other members of the chemical family are gradually converted to, thus providing a "reservoir". NO_x gases in general have a long chemical lifetime when sunlight is not present. Most NO_x gases are only destroyed in photolysis, needing sunlight. Thus, during polar night conditions (absence of sunlight), energetic particle driven ionisation, which produces NO_x in polar regions, provides ideal conditions for large amounts of NO_x both being produced and maintained in the atmosphere. Because of the long chemical lifetime (months) during polar winter, atmospheric transport becomes very important for the distribution of the NO_x and thus the potential extent of the ozone impact leading to the so-called "Indirect Effect" discussed in the next section.

Both NO_x and HO_x react with atmospheric ozone (O_3) or atomic oxygen (O, needed for ozone production: $O + O_2 \rightarrow O_3$) in several possible catalytic reaction cycles (see Figure 1). An example of such a reaction cycle is:

$$XO + O_3 \rightarrow XO_2 + O_2 \qquad (1.1)$$

$$XO_2 + O \rightarrow XO + O_2 \qquad (1.2)$$

where the X can be either N or H. The sum of these reactions is:

$$O + O_3 \rightarrow 2\,O_2 \qquad (1.3)$$

with the produced molecular oxygen (O_2) requiring the presence of solar UV light before breaking into atomic oxygen ready to form ozone again. During large energetic particle precipitation events such as solar storms, the immediate direct impact on mesospheric and upper-stratospheric ozone levels can reach up to 80–90% ozone loss in the mesosphere and 40–50% in the upper stratosphere.

2 Downward transport of NO_x and the role of sudden stratospheric warming events-the indirect effect

As described above, during winter the small amount of sunlight present in the polar region can result in accumulation of NO_x produced by energetic particle precipitation. In these conditions, the NO_x becomes subject to being transported with winds in the horizontal direction and by large-scale atmospheric downward descent. In the winter hemisphere strong winds are formed around the pole, typically at latitudes of about 60°. These winds, known as the "polar night jet", extend across the stratosphere and mesosphere isolating the air inside (i.e., preventing mixing with air equatorwards of the polar night jet). This phenomenon is called the "polar vortex" (see Chapter 4.2) (Holton, 2004). When energetic particle precipitation produces NO_x during wintertime, the NO_x rich air becomes isolated inside the polar vortex, helping to maintain the elevated NO_x levels. Inside the vortex, large-scale downward descent of air masses takes place transporting the NO_x rich air downwards to lower altitudes. Once transported to lower altitudes, the NO_x can react with atmospheric ozone. This kind of effect from energetic particle precipitation requiring interaction with atmospheric transport is known as the Indirect Effect to separate from the immediate direct effects described in the previous section.

In the Northern hemisphere in particular, atmospheric dynamics show large variability during the wintertime. One manifestation of this variability are events known as Sudden Stratospheric Warming (SSW). A sudden stratospheric warming event begins with a sudden increase of temperature in the lower stratosphere and cooling of temperatures above. This is accompanied by weakening, or, sometimes, even reversal of direction, of the winds in the polar night jet, enabling mixing of the polar air with lower latitude air masses. Following SSW events, the polar vortex is sometimes formed again, and in some cases, the descent of air inside the vortex is accelerated. In cases like these, large amounts of NO_x

are pulled down towards the stratosphere from higher altitudes. As the amount of NO_x at higher altitudes is strongly linked to the amount of energetic particle precipitation, also the amount of the NO_x descending to the stratosphere in these cases depends on particle precipitation levels. Apart from large solar storms, the biggest increases of NO_x reaching the stratosphere in the Northern hemisphere have been linked to the strong, rapid descent following SSW events (Randall et al., 2009).

In the Southern hemisphere, atmospheric dynamics exhibit much less variation and the wintertime polar vortex is much more stable than its Northern counterpart. This enables a steady descent of NO_x throughout the winter, reaching altitudes as low as 25 km by the end of the Southern winter. It has been estimated that, due to the stability of the vortex and the steady descent, the NO_y produced by energetic particle precipitation regularly contributes 10–30% of the total Southern polar stratospheric and lower mesospheric NO_y budget. In contrast, in the Northern hemisphere, this is generally less than 10% (Funke et al., 2014).

3 Indirect effects on ozone

In addition to energetic particle precipitation affecting mesospheric ozone levels, when the NO_x initially produced at mesospheric altitudes is transported to the stratosphere it can provide a long lasting source for ozone loss, particularly in the upper stratosphere. Chemistry-Climate Model simulations examining the long-term impacts on ozone from solar storms alone (i.e., not including energetic electron precipitation) have found that the indirect impact on ozone from the descending NO_x/NO_y between about 25-50km can be on average 10–20% for several months following the initial precipitation event (Jackman et al., 2009). When looking at the average effect over a much longer period of time (several years), the high latitude ozone loss from the sporadic solar storms is estimated to be of the order of few percent.

4 Indirect effects on atmospheric dynamics and potential links to regional climate

Ozone provides an important source for local atmospheric heating and cooling in the mesosphere and stratosphere via absorption of solar UV radiation (leading to heating) and radiative infrared thermal emission (leading to cooling). Any changes in the ozone balance thus have a potential to influence the thermal balance of the atmosphere. If the thermal balance is altered, the atmosphere will seek to reach a new equilibrium state by for example changing winds. Changes in winds can further affect the atmospheric dynamical state via the so-called wave-mean flow interaction (Holton, 2004; Brasseur and Solomon, 2005). The wave-mean flow interaction describes the coupling of zonal mean winds (or mean flow) with atmospheric waves, which provide means for large-scale energy transfer in the atmosphere and are a major source of dynamical variability. Wind speeds

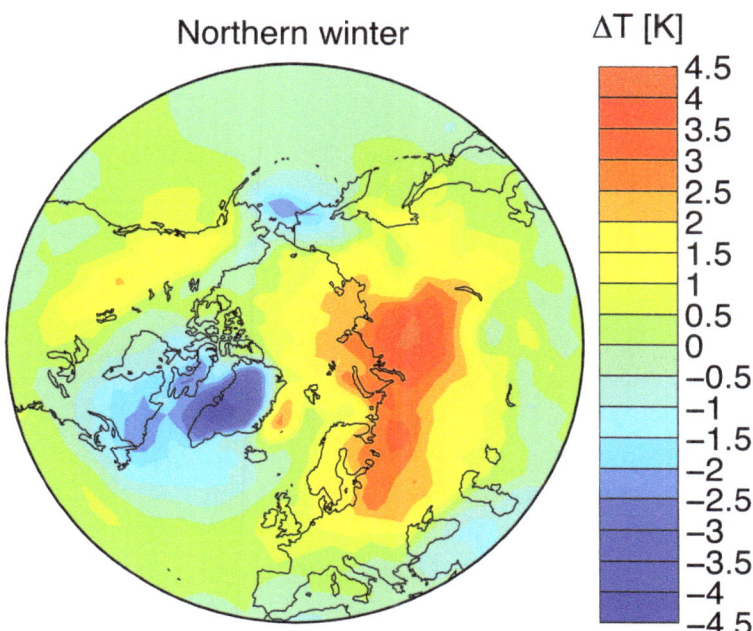

Fig. 2. Average Dec-Feb surface air temperature difference between times of high energetic particle precipitation and low energetic particle precipitation in reanalysis data (see Chapter 3.4 for information on reanalysis data) over a 50 year time period. For further details see Seppälä et al. (2013) and references therein.

are critical for any wave propagation in the atmosphere. On the other hand, modification of atmospheric wave propagation can feed back into the atmospheric winds as the wave energy can accelerate or decelerate the mean flow, highlighting the complex feedback system.

The wave–mean flow interaction and coupling to atmospheric waves also provides means for initially high altitude perturbations to propagate downwards towards the troposphere and the surface. Recent studies using Chemistry-Climate Model simulations and meteorological re-analysis data (Figure 2) have suggested that the changes in atmospheric polar wintertime ozone balance induced by energetic particle precipitation produced NO_x could result in dynamical changes across the stratosphere and propagate towards the surface, where regional polar, but not global, winter climate may be affected (Rozanov et al., 2005; Seppälä et al., 2013). However, the full details of understanding the mechanisms connecting the initially chemical changes and any surface responses (Figure 2) are not yet well understood. In addition, any potential regional climate signals are likely a combination of different sources of variability and therefore the contamination of a specific signal by other sources needs to be carefully assessed (see Chapter 4.3).

Further reading

Brasseur, G. P., and S. Solomon: 2005, Aeronomy of the Middle Atmosphere, 3rd ed., Springer, Dordrecht.

Jackman, C. H., and R. D. McPeters: 2002, The Effect of Solar Proton Events on Ozone and Other Constituents, Rev. Geophys., 128.

Krivolutsky, A., and A. I. Repnev: 2012, Impact of space energetic particles on the Earths atmosphere (a review), Geomagn. Aeron., 52(6), 685716, doi:10.1134/S0016793212060060.

Rozanov, E. V., M. Calisto, T. A. Egorova, T. Peter, and W. Schmutz: 2012, Influence of the Precipitating Energetic Particles on Atmospheric Chemistry and Climate, Surv. Geophys., 33(3), 483501, doi:10.1007/s10712-012-9192-0.

Seppälä, A., K. Matthes, C. E. Randall, and I. A. Mironova: 2014, What is the solar influence on climate? Overview of activities during CAWSES-II, Progress in Earth and Planetary Science, 1(1), 24, doi:10.1186/s40645-014-0024-3.

Sinnhuber, M., H. Nieder, and N. Wieters: 2012, Energetic Particle Precipitation and the Chemistry of the Mesosphere/Lower Thermosphere, Surv. Geophys., 33(6), 12811334, doi:10.1007/s10712-012-9201-3.

CHAPTER 4.7

THE IMPACT OF COSMIC RAYS ON CLOUDS

Benjamin A. Laken[1] and Jaša Čalogović[2]

1 Introduction

Even a small cloud response to changes in the cosmic ray flux could provide a powerful amplification mechanism, enabling energetically minute variations in solar activity to impact the Earth's climate system. Consequently, of all the proposed solar–terrestrial linkages, arguably none has garnered as much controversy, nor been as hotly debated, as the cosmic ray–cloud link. Potentially, this hypothesis could have over-turned conventional wisdom regarding natural climate forcings and the role of anthropogenic emissions in recent observed global warming. But, as we will discuss in this chapter, clear evidence in favor of a cosmic ray–cloud link has yet to be presented.

1.1 Aerosols and clouds

Clouds exist in an astonishing panoply of forms, ranging from small patches of ground-level fog, to giant cumulonimbus clouds. Despite their seemingly endless variety, most clouds form by the same basic process: by cooling air containing both water vapor and sufficient amounts of particles (so-called aerosols) on which droplets of water or ice crystals can form.

Aerosols are droplets or particles suspended in a gas, these can be anything from sulphates, to salt blown into the air from the spray of breaking ocean waves, soil or desert dust entrained by the wind, soot particles (black carbon), secondary organic aerosols, or organic matter, such as pollen, bacteria, or even viruses. For cloud droplets to form on aerosol particles (heterogeneous nucleation), they typically need to be water soluble and have diameters >50 nm: such aerosols are referred to as 'cloud condensation nuclei' (CCN). Without aerosol particles,

---

[1] Department of Geosciences, University of Oslo, Oslo, Norway.
[2] Hvar Observatory, Faculty of Geodesy, University of Zagreb, Kačićeva 26, 1000 Zagreb, Croatia.

supersaturation of >300 % relative to water would be required to form clouds (homogeneous nucleation). However, supersaturation levels in the atmosphere rarely exceed 1–2 %. Consequently, without aerosols clouds simply would not form.

One of the hypothesized links between cosmic rays and clouds depends on the relationship between aerosols and clouds. However, before we discuss this, we will first briefly describe the importance of clouds to the Earth's climate.

1.2 A cooling sunshade and a warming blanket

The Earth's climate system is highly sensitive to perturbations in the radiative balance – the amount of shortwave (SW) radiation the Earth receives from the Sun *vs.* the amount of outgoing longwave (LW) radiation at the top of the atmosphere. If any perturbations occur to the SW or LW radiative fluxes, such as the increasing retention of LW radiation due to increasing human greenhouse gas emissions, global temperatures will change. Even a small change in clouds and water vapor – the dominant greenhouse gas in our atmosphere – will significantly alter the Earth's radiative fluxes. As there are still many remaining uncertainties regarding clouds and how they may respond to global warming, clouds remain one of the largest difficulties in understanding and projecting future climate.

Clouds cover around two thirds of the Earth's surface (70±3 % for clouds with optical depths of >0.1) (Stubenrauch et al., 2013), and exert a strong influence on the radiation balance in two distinct ways. Firstly, the denser a cloud is, the greater its ability to reflect sunlight: low, dense clouds can reflect 20–90% of incoming sunlight. Satellite observations show that over the whole Earth this effect reduces incoming SW radiation by around -50 W m^{-2}, and without clouds, the Earth would absorb ~20% more SW radiation and be ~ 12 °C warmer. Secondly, all clouds (and water vapor) also absorb outgoing LW radiation, emitting energy in all directions, including back towards the Earth's surface. Satellite observations show the global reduction in outgoing radiation flux due to clouds is ~30 W m^{-2}. In particular, high, thin clouds (cirrus) play a key role in this reduction, limiting the rate at which the Earth's surface loses LW energy to space. Without this effect, the Earth would be ~ 7 °C cooler. Considering these impacts, clouds have a net global radiative forcing of around -20 W m^{-2}, and cool the climate by ~ 5 °C. Consequently, even a small change in cloud amount could significantly alter climate (Ramanathan et al., 1989).

The situation becomes more complex at local scales, however, the localized impact of cloud on the radiation balance depend on factors, such as the time of day, latitude, season, surface type, altitude and thickness of the cloud, cloud thermodynamic phase, and various micro-physical properties of the clouds. Hence, the potential impacts of theoretical solar-cloud links on climate differ depending on the exact nature of the link, vis-à-vis the location, extent, and type of cloud properties which may be influenced.

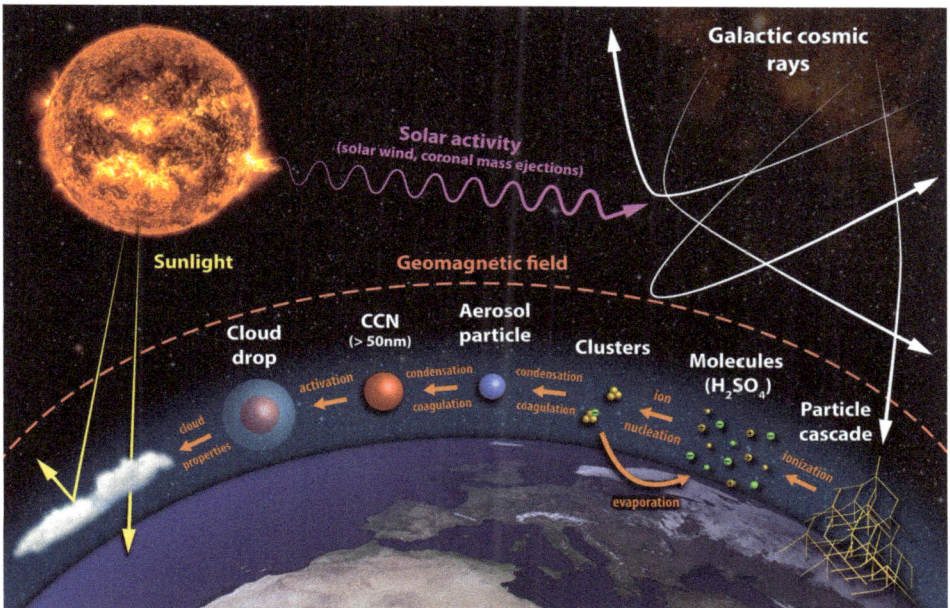

Fig. 1. Illustration of the clear-air mechanism, where galactic cosmic rays (consisting mainly of high-energy protons) hit the nuclei of atoms in the Earth's upper atmosphere with enough force to generate particle cascades through the atmosphere producing ionisation. This increases the formation of new aerosol particles, which may possibly grow to sufficient sizes to influence cloud properties.

2 A cosmic connection?

Atmospheric ions – such as those produced by the solar-modulated flux of cosmic radiation and their secondary particles, ubiquitous throughout the Earth's atmosphere – may alter the nucleation and growth of the aerosols upon which cloud droplets form (Figure 1). This so-called 'ion-mediated nucleation' (or 'clear-air') mechanism, could provide an indirect link between solar activity and the Earth's troposphere, explaining how relativity small changes in solar activity could become amplified into a climatologically significant effect.

Over the past few decades, much work has been done to better understand the clear-air mechanism. This has included studies ranging from the development of a detailed theoretical understanding of ion-mediated nucleation processes (e.g., Yu and Turco, 2000; Yu and Luo, 2014) to the addition of the clear-air mechanism into general circulation models with aerosol microphysics (e.g., Snow-Kropla et al., 2011; Dunne et al., 2012); and most importantly, the creation of sophisticated laboratory experiments, such as the Cosmics Leaving OUtdoor Droplets (CLOUD) at CERN, designed to test the impact of cosmic ray induced ionisation on aerosol nucleation within a specialised chamber where conditions may be carefully controlled (Kirkby et al., 2011; Almeida et al., 2013).

Results of the CLOUD experiment have demonstrated that the clear-air mechanism is important in the nucleation of new aerosol particles under specific laboratory conditions – those conditions being low temperatures characteristic of the upper-troposphere, and low concentrations of amines and organic molecules. The researchers found that ionisation enhances aerosol nucleation by a factor of 2–10 times. This result confirms that ionisation can speed-up the birth of new aerosol particles, although it remains unclear from these results how effectively the process operates in the Earths atmosphere.

Despite the possible existence of the clear-air mechanism, recent model studies have shown that changes in the nucleation rate due to cosmic rays may still fail to significantly alter CCN populations. This is because although the clear-air mechanism causes small aerosol particles to form faster than they would without ionisation, these particles remain at small sizes for relatively long time periods. At small sizes, they are susceptible to scavenging by larger pre-existing aerosols, and so are unlikely to survive and grow to CCN sizes. Consequently, as the majority of the troposphere is rich in aerosols, the clear-air mechanism is unlikely to alter the number of CCN available to sufficiently impact clouds. The ultimate impact on CCN concentrations over the solar cycle of about 11 years (Schwabe cycle) is estimated to be approximately $\sim$0.2% change for aerosols of >80 nm in diameter (Pierce and Adams, 2009).

However, there are locations where aerosols are in short supply and limit cloud formation. Small changes in the CCN abundance of such locations have been shown to alter clouds (e.g., Rosenfeld et al., 2006; Koren et al., 2012). The phenomenon of ship-tracks, which are frequently observed over aerosol-poor ocean environments, is a clear example of this: although we note that CCN changes required to generate ship tracks are much larger than 0.2%.

In addition to the 'clear air' mechanism, a completely separate theoretical pathway linking atmospheric ionization to clouds has also been suggested (Tinsley, 2000). It operates via the global atmospheric electric circuit (GEC) by changing the collision rate between electrically charged aerosol particles and cloud droplets. Charge is deposited on to aerosol particles, both in layer and convective cloud types, from current flow in the GEC which is modulated by the cosmic ray flux, as discussed in Chapter 4.8.

3 What do cloud observations show?

3.1 Decadal observations

Global cloud properties have been consistently measured by satellite-based instruments since the early 1980s. Long-term records of clouds have also been collected from the Earth's surface, although these data are less comprehensive. In the late 1990s, after accumulating around ten years of satellite data, a group of Danish researchers compared the cosmic ray flux and satellite cloud cover and found a positive correlation, suggesting cloud cover may change by ~2–4 % over the Schwabe cycle (Svensmark and Friis-Christensen, 1997). Later analyses suggested that this relationship was restricted to low-level clouds (<3.2 km) (Marsh and Svensmark, 2000). At that time, evidence suggested that there had been a long-term decline in the cosmic ray flux due to increasing solar activity over the 20th century. From this, the Danish researchers controversially concluded that the majority of global warming observed over the past century could be driven by solar-related decreases in cloud rather than by human emissions (Svensmark, 2007). This argument has become a key point for so-called 'climate skeptics' and is still widely used (e.g., Idso et al., 2013).

Further investigations into the reported cosmic ray–cloud correlations highlighted numerous issues, both with the analysis and the data. Regarding the analyses, the original correlation studies were criticized for poor-data handling and statistical errors, which resulted in an exaggerated correlation (e.g., Laut, 2003; Damon and Laut, 2004).

When additional solar parameters (e.g., total solar irradiance) were compared to cloud cover, similar (and even higher) correlations were found than were obtained with the cosmic ray flux (Kristjánsson et al., 2004). This is due to the co-temporal behavior of solar-related parameters over annual-to-decadal timescales, which makes it difficult to distinguish which solar properties (if any) may be important to clouds. Similarly, oscillations in the Earth's climate which operate over similar timescales to the Schwabe cycle (such as El Niño) compromise our ability to attribute cloud changes to a solar forcing (Farrar, 2000): they produce multi-year variations in climate, and make distinguishing the cause of observed cloud variations highly problematic.

Perhaps most importantly, there are significant issues with the cloud data itself. The most commonly used cloud data (from the International Satellite Cloud Climatology Project) is subject to large long-term errors and spurious trends. These errors come from numerous sources, such as changes in the number of contributing satellites over time, changes in the calibration satellites, and the degradation of instruments. Due to this, the creators of the data have explicitly stated that the records are not suitable for long-term analysis (Brest et al., 1997). To make matters worse, measurements of low-cloud cover – the variable of key importance in these results – are known to be unreliable, as the satellite instruments are non-cloud penetrating (hence high-clouds mask the view of low clouds).

Unsurprisingly, data errors played a strong role in affecting the conclusions of the original correlation studies. Newly available satellite data show little evidence of a cosmic ray–cloud link (Figure 2). In addition, the long-term trends which had previously been identified in solar activity were also found to be erroneous (Lockwood and Fröhlich, 2007): solar-activity has in fact shown no significant change over the second half of the 20th century. Consequently, even if a cosmic ray cloud relationship were true, it could not produce a trend in global temperatures during this time.

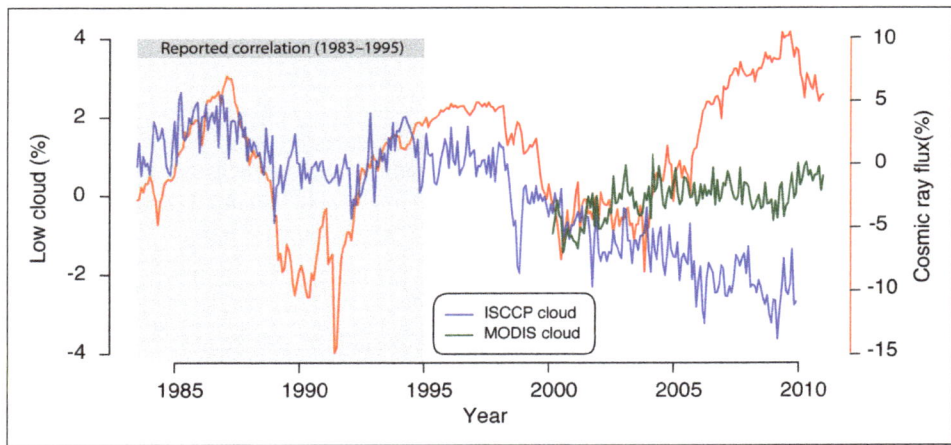

Fig. 2. The time-series of satellite cloud data (from two independent sources – ISCCP and MODIS), plotted with the cosmic ray flux. Image adapted from Laken *et al.* (2012).

3.2 Testing a link at daily-timescales

Another method of investigation has tried to overcome the weaknesses of long-term studies by exploiting short-term, sporadic, sudden reductions in the cosmic ray flux, so-called 'Forbush decreases'. These are connected with solar activity, and are mainly caused by coronal mass ejections from the Sun (Dumbović et al., 2011). In these tests, similar disturbances are identified and averaged together: this procedure is called a 'composite' or 'superposed epoch analysis' (further details in Laken and Čalogović, 2013a).

Far from clarifying questions of a cosmic ray–cloud link, composite studies have produced widely inconsistent results, with some finding statistical cosmic ray–cloud associations, and some not. Careful reinvestigation of studies that reported strong relationships have revealed that data treatment and statistical issues strongly affect the outcomes of composites, producing false positive results in many cases (e.g., Laken and Čalogović, 2013b).

Additionally, there are also limitations regarding the detection of solar-induced changes in clouds. These limits are mostly determined by the large meteorological variability of clouds and data quality. Consequently, it is possible that a small cosmic ray–cloud link may exist over limited locations and under certain conditions (being of the second order), but this link may be very difficult to detect in presently available cloud data (Laken et al., 2012).

4 Some final remarks

Although it is probable that the clear-air mechanism plays a role in enhancing aerosol nucleation, it seems unlikely that it is able to ultimately impact global cloud properties. However, it is possible that under aerosol-impoverished conditions this effect may become more influential, although the extent to which this occurs is currently unknown. Satellite observations of cloud over decadal-to-daily timescales have yielded a wide range of results, due to both data and analysis issues. Despite the disagreement however, it is clear that cosmic rays are unable to significantly alter global cloud cover, nor explain recent global warming.

Further reading

Carslaw, K.S., Harrison, R.G. & Kirkby, J.: 2002, Cosmic rays, clouds, and climate. Science 298(5599), 1732–1737

Laken B.A., Pallé E., Čalogović J., & Dunne, E.M.: 2012, A cosmic ray–climate link and cloud observations. Journal of Space Weather and Space Climate, 2(A18)

Pittock A.B.: 1978. A critical look at long-term Sun-weather reolationships. Reviews of Geophysics and Space Physics, 16(3), 400–420

Svensmark, H. & Calder, N.: 2007. The chilling stars: A new theory of climate change. Icon Books, London

CHAPTER 4.8

IMPACT OF SOLAR VARIABILITY ON THE GLOBAL ELECTRIC CIRCUIT

Colin Price[1], R. Giles Harrison[2] and Michael Rycroft[3]

1 Introduction to the global electric circuit (GEC)

Atmospheric electricity, and the global electric circuit (GEC) are amongst the oldest fields of research in the geophysical sciences. Soon after Benjamin Franklin's fundamental discovery of the equivalence between charge in the laboratory and the atmosphere in 1752, Lemonnier found that, even in fair weather regions of the planet, a persistent electric field existed at the Earth's surface of approximately 130 Vm^{-1} (Volt per meter) pointed downward to the Earth. While this vertical electric field E_0, or potential gradient (PG), remained fairly constant during short periods of measurements, it was shown in the 1920s that it had a systematic diurnal variation with Universal Time (UT) rather than with local time and location (Figure 1) (Harrison, 2013). From measurements taken aboard the Carnegie research vessel from global cruises made between about 1910 and 1930, the potential gradient (PG) data were shown to have the same Universal Time variations no matter where and when the observations were made. This universal diurnal variation of the surface PG is known as the "Carnegie Curve"; it shows a global minimum in PG around 03 UT and a global maximum around 19 UT.

This electric field pointed towards the Earth under fair weather conditions implies that the Earth has a permanent negative charge Q. This charge Q can be estimated as approximately half a million Coulombs, since $Q = 4\pi R_E^2 \epsilon_0 E_0$ where R_E is the radius of the Earth, ϵ_0 is the permittivity of free space (the electric constant), and E_0 is the magnitude of the PG field measured near the surface of the Earth (Figure 1)).

By 1887, Linss had discovered the existence of small charged particles – ions – in the atmosphere, implying that the atmosphere was not a perfect insulator,

---

[1] Department of Geosciences, Tel Aviv University, Tel Aviv, Israel
[2] Department of Meteorology, University of Reading, P.O. Box 243, Earley Gate, Reading, Berks, RG6 6BB in UK
[3] CAESAR Consultancy, 35 Millington Road, Cambridge CB3 9HW, U.K.

© EDP Sciences 2015
DOI: 10.1051/978-2-7598-1733-7.c134

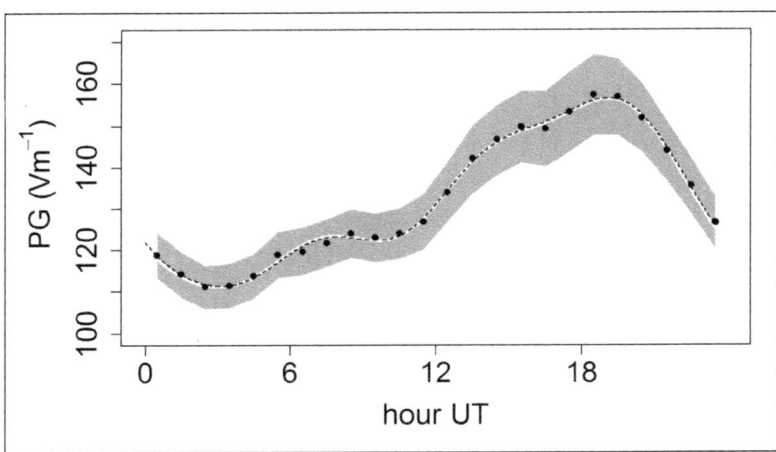

Fig. 1. Universal Time variations of the surface potential gradient (PG) as measured by the Carnegie research vessel in the 1920s. The grey envelope represents the variability of the data about the mean, from Harrison (2013).

rather that the atmosphere had a finite conductivity. Atmospheric ions are produced by galactic cosmic rays (GCRs) from above, which result in a maximum ionisation rate at around 15–20 km altitude, whilst from below the Earth's crust, there is ionising radiation from the radioactive decay of naturally occurring minerals, but only above land masses and not above the oceans. Another source is the Sun, which generates ionisation of the atmosphere above about 70 km, producing the highly conductive ionosphere.

Since the main sources of atmospheric ionisation occur from outside the Earth, atmospheric resistivity decreases with increasing altitude, with most of the atmospheric "resistance" being in the lowest 5 km of the atmosphere. In addition, the mobility of ions (i.e., their ability to move in an external electric field) is lowest near the surface where the atmospheric density is greatest. The larger the ions the slower they move. Near the surface, ions attach to larger molecules and particles reducing their mobility. Since the electric field is inversely proportional to the conductivity ($E \approx 1/\sigma$), the rapid increase of conductivity (σ) with altitude results in the PG showing a rapid decrease with altitude, with values less than 1 V m^{-1} when altitudes of 10 km are reached.

However, having electrically charged ions in the atmosphere with a finite electric field also implies there should be a "conduction current" flowing towards the Earth's surface at all times. The conduction current density is derived from Ohm's Law and is $J_z = \sigma E_z$ and, since σ increases with altitude, and E_z decreases with altitude, J_z is fairly constant with altitude z, with a value of a few pA m^{-2} (10^{-12} A m^{-2}). This was first measured directly by C.T.R. Wilson in the early 1900s, although it can be easily estimated using Ohm's Law from known values of σ (3.10^{-14} S m^{-1}) and E_0 (130 V m^{-1}). Integrated over the area of the

Earth ($4\pi R_E^2$), this amounts to a global current of 1200 A flowing to the Earth continuously.

Using instrumented balloons (radiosondes) to measure the vertical electric field as a function of height in the atmosphere, we see that the value of the vertically integrated field from the Earth's surface begins to change less and less until it reaches a steady value when the electric field no longer changes appreciably. This steady value is effectively the potential of the ionosphere with respect to the surface, known as the Ionospheric Potential (V_i); it has a value of about +250 kV (Markson, 2007). Given this value we can also calculate the fair weather atmospheric resistance for the whole planet, $R = V_i/I = 250$ kV$/1200$ A $\approx 200\ \Omega$.

Hence, we can describe the Earth-atmosphere system as an electric circuit with currents, potentials, and resistances. However, given the currents in the circuit, the charge on the Earth should disappear after some time. This timescale depends on the capacitance (C) of the electric circuit $C = Q/V_i \approx 2\ F$. Hence, the discharge time τ of the Earth's electric circuit is $\tau = RC \approx 400$ s or about 5 to 10 minutes. This means that all these currents and fields should disappear within a few minutes if the circuit generator was not maintained.

As these fields and currents are observed to be always present, something needs to be generating these currents continuously and, in 1920, Wilson suggested that the "batteries" in the electric circuit were thunderstorms and electrified shower clouds (i.e., extensive rain clouds) that are continuously active around the planet (Wilson, 1921). In 1929, Whipple showed that the diurnal variation of thunderstorm activity was very similar to the Carnegie Curve, providing evidence for the link between thunderstorm activity and the PG (Whipple, 1929). Today we know that, every second, there are about 50 lightning discharges somewhere on the planet, within 1000–2000 active thunderstorms. Hence, thunderstorms and other electrified clouds are continuously charging the electric circuit, which is continuously discharging in fair weather regions (Figure 2).

Near the geomagnetic poles there is an additional diurnal variation in the GEC parameters, caused by the interaction of the solar wind (which pulls the Sun's magnetic field away from the Sun) with the Earth's geomagnetic field. This generates an additional electric field which is superimposed on V_i. During geomagnetically quiet periods, the potentials on the dawn side of the polar cap are 20–30 kV greater than on the dusk side, which increases to a difference of 150–200 kV during geomagnetically disturbed periods. This effect generally only extends out to 30° from the North and South geomagnetic poles. Hence, the global electric circuit acts to couple the space environment to the lower atmosphere.

The description of the GEC given above describes what we call the direct current (DC) global circuit, which is constant on short time periods. However, there also exists an alternating current (AC) part to the global circuit, produced once again by the batteries in the circuit, and more specifically the lightning discharges.

Due to the high electrical conductivity of the Earth's surface and the ionosphere above 70 km altitude, the Earth-atmosphere-ionosphere can be thought of

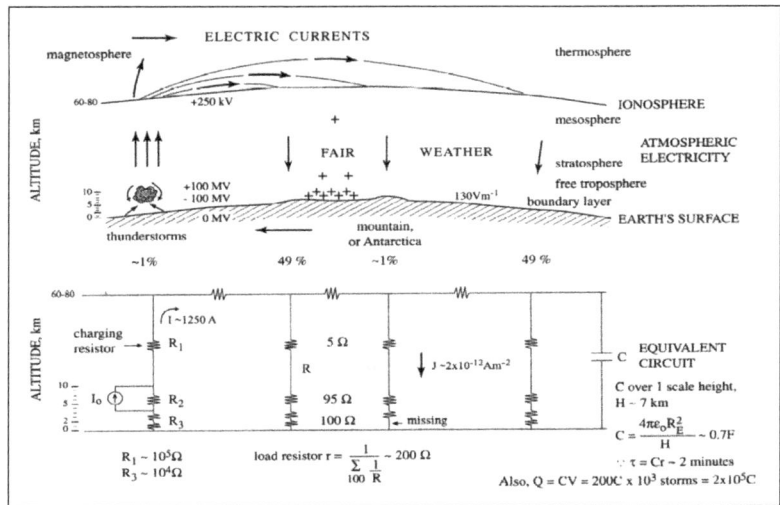

Fig. 2. A schematic representation of the global electric circuit for current flow from disturbed to fair weather regions, through the layers of the atmosphere, and the equivalent circuit diagram, after Rycroft et al. (2000).

as a waveguide for electromagnetic (EM) waves generated within the waveguide. Such EM waves are produced by lightning discharges, and the extremely low frequency (ELF) waves below 50 Hz frequency can actually propagate around the globe a number of times before dissipating into the background. Hence, these ELF waves result in constructive interference producing resonant waves at the Earth's eigen-frequencies (8Hz, 14Hz, 20 Hz, ...). These resonances are called the Schumann resonances after the theoretical prediction of these ELF resonant waves by W.O. Schumann in 1952 (Schumann, 1952). Since these Schumann resonance radio waves can be detected at any location on the globe, they also represent the GEC, and allow us to study changes in the GEC due to solar variability.

2 Impact of total solar irradiance on the global electric circuit

The total solar irradiance includes all electromagnetic radiation reaching the Earth from the Sun, principally the infrared, visible, ultraviolet and X-rays components. This EM radiation impacts two aspects of the GEC: the driving thunderstorms, and the ionosphere. Satellite measurements indicate that the total solar irradiance is about 1361 W m^{-2} (see Chapter 2.2).

The thunderstorms in the GEC are driven (like all weather phenomenon) by solar heating of the Earth's surface that results in convection, cloud formation, and cloud electrification. Thunderstorms occur primarily (70% of all thunderstorms) in the warm tropics, where thunderstorms form daily in the late afternoon hours, due to the direct heating of the surface by the Sun. In addition, 90% of global thunderstorms and lightning occur during the hot summer months when solar heating is at a maximum. Hence, on daily and seasonal timescales, the total solar irradiance is well correlated with thunderstorm activity. Consequently both the DC and AC parts of the global circuit will be impacted by any changes in the "batteries" of the circuit. There is some evidence that global thunderstorm activity also fluctuates on the ≈11-year solar cycle, but this topic needs a lot more research before we have conclusive results.

In the ionosphere, ultraviolet and X-rays ionise the so-called "neutral" atmosphere above 70 km, and determine the conductivity of the upper atmosphere, as well as the reflection height for the ELF (below 3 kHz) and VLF (3–30 kHz) radio waves producing the AC global circuit. While there are clear day/night changes in the ionospheric characteristics, and hence the AC global circuit, there is no evidence for a day/night impact on the DC global circuit. On other time scales (associated with solar flares or the ≈11-year sunspot cycle), changes in total solar irradiance ionisation in the lower ionosphere affects the Earth-ionosphere waveguide, resulting in changes in the Schumann resonance parameters, with less effects on the DC circuit. The impacts of total solar irradiance variations on the ionosphere are therefore likely to be of more importance in influencing the AC global circuit than the DC global circuit.

3   Impact of energetic charged particles

Energetic particles come from outside our solar system (GCRs, galactic cosmic rays), or from our own Sun (SEPs); they are scattered away from the Earth by interplanetary magnetic field irregularities, which are carried out from the Sun by the solar wind, the more so at solar maximum. Often associated with solar flares are CMEs, coronal mass ejections, enormous clouds of high speed plasma ejected by the Sun.

Energetic GCRs may directly trigger lightning through a mechanism known as "runaway breakdown", when electrons are accelerated to relativistic energies by the high electric fields in thunderstorms, and lose some of their energy by ionisation of the air (see Infobox 4.2). If the electrons generated are energetic enough, then an electron avalanche develops leading to electrical breakdown, or lightning. This mechanism needs a single electron of energy 1–10 keV to start the breakdown, which can be provided by GCR air showers and the cascade of particles. The electric fields needed to start runaway breakdown are an order of magnitude lower than the ≈ 1 MVm^{-1} needed to initiate conventional breakdown. This, and the paucity of observations demonstrating adequate electric fields in thunderclouds to cause conventional breakdown, has led to the suggestion that lightning could be triggered by electrons from GCRs initiating such a runaway breakdown.

If this mechanism is a significant source of the variation in lightning from suitably electrified clouds, there should be a positive relationship between GCRs (or other ionisation sources) and lightning. However, studies investigating this effect are difficult and often inconclusive due to the high levels of natural variability in

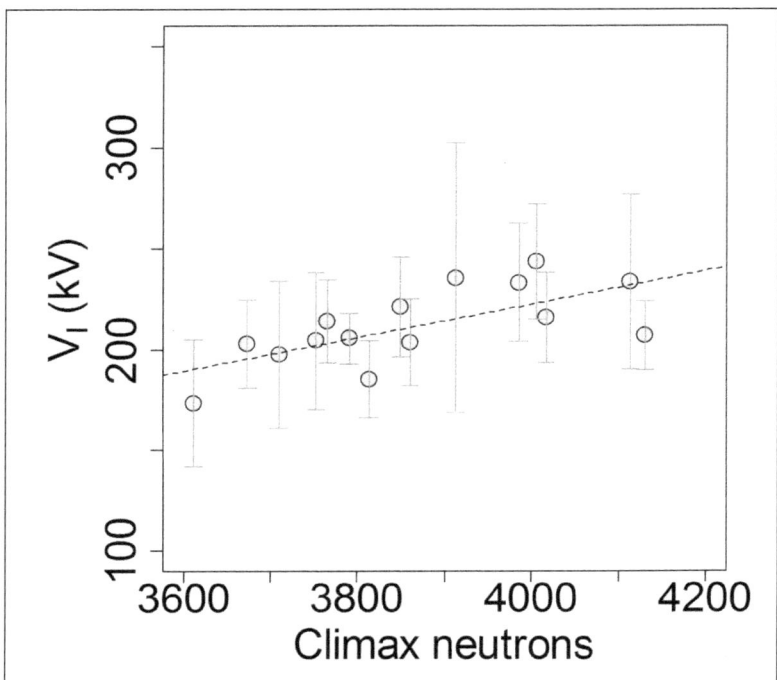

Fig. 3. Galactic Cosmic Ray (GCR, measured at the Climax neutron monitor) effects on the ionospheric potential (V_i) between 1967 and 1973, from Harrison and Usoskin (2010).

thunderstorm occurrence and lightning flash rates. There are also difficulties in obtaining reliable and consistently calibrated lightning data because of the continual improvement in the detection networks used for weather forecasting.

As GCRs entering the Earth's atmosphere are the main source of ionisation between the surface and the ionosphere, both periodic and transient GCR variations are expected to modulate the strength of the DC global circuit. As is apparent from Figure 3, an increase in the flux of GCRs at solar minimum increases the ionospheric potential. This may be due to changes in the columnar resistance which therefore affects the conduction current density and surface potential gradient (E_0) through the fair weather (non-thunderstorm) part of the global atmospheric electric circuit.

Solar energetic particles (SEPs), which often originate from explosive solar events, create an additional source of ionisation in the atmosphere, perturbing the vertical conductivity profile and the local electric field at high geomagnetic latitudes (see Chapter 2.3 and 4.5).

Coronal Mass Ejections (CMEs), which are often associated with solar flares, arrive at Earth a few days after the parent flare. If and when they hit the Earth's magnetosphere, CMEs create magnetic disturbances, such as geomagnetic storms;

the decrease in the incoming GCR flux is known as a Forbush decrease. Forbush decreases usually have an onset time of a day or so, with a slow recovery lasting up to about a week. They generally result in a decrease in the GCR ionisation rate measured by surface neutron monitors of 10% (although larger events, up to 30% have occurred). Their effects can extend from high latitudes to the middle latitudes.

The time difference between the arrival at Earth of SEPs from a solar flare and the CME shock impact might possibly allow the attribution of certain GEC changes to different particular processes. However, the additional influence of geomagnetic storms, which can occur throughout the duration of the event, and alter the geomagnetic cutoff energies for GCR penetration into the Earth's atmosphere complicate matters somewhat. Such magnetic variability has been shown to perturb the ionospheric potential V_i at high latitudes as well as the surface electric field, even down to mid latitude locations.

Although the effects of SEPs on atmospheric electrical parameters can be significant at high altitudes and at high latitudes, the magnitude of these effects on the global electric circuit is questionable. One coupling mechanism that could be influenced by SEP events is the response of the thunderstorm generator part of the GEC. The GEC current is dependent on stratospheric/mesospheric resistance above a thunderstorm, which varies with the ionisation rate. Increases in ionisation rate from SEPs (larger than 1 GeV) penetrating to low latitudes might decrease the electrical resistance above thunderstorms sufficiently to increase the upward charging current to the ionosphere.

The values of conductivity of the lower atmosphere do not impact the Schumann resonances, although changes in energetic charged particle fluxes may change the generator region by impacting the lightning activity itself. Hence, when talking about energetic particles, there is a strong link to the DC global circuit, but a weaker link to the AC global circuit.

4 Feedback on the climate system

External forcing of the global electric circuit may further feedback on the climate system, providing a solar-terrestrial-climate interaction linking solar variability to climate through the GEC. Again we can divide the impacts and feedbacks between the generator regions and the fair weather regions of the planet.

In the generator regions, thunderstorms act as huge ventilation shafts transporting boundary layer aerosols, water vapor and trace gases into the upper troposphere. Here, these aerosols and gases can alter the radiation balance of the planet, either through the production of cirrus clouds, or by an increase in greenhouse gas forcing. At the same time, lightning is a major source of nitrogen oxides (NO_x) in the atmosphere, a major precursor for the formation of tropospheric ozone (O_3). Hence, any changes in lightning or thunderstorm activity due to solar variability will directly impact the concentrations and distribution of tropospheric O_3, a strong greenhouse gas, but of much smaller influence than carbon dioxide (CO_2).

In the fair weather regions of the globe remote from thunderstorms, the fair weather currents (J_z) may have effects on clouds as they flow through and/or around them. Layer clouds, which are abundant globally, acquire positive ions at the upper portion of clouds and negative ions at the cloud base. How this attachment may impact cloud droplet nucleation, condensation, and freezing is not clear. However, recently some interesting results show that, in the dark polar night (no solar heating causing a diurnal variation), the cloud base height shows a very similar Universal Time diurnal variation to the Carnegie Curve. Could the fair weather currents be influencing cloud microphysics and dynamics, and hence possibly cloud coverage, cloud albedo or cloud lifetime, with implications for the Earth's radiation balance? More research is clearly required. The long-established geophysical topic of atmospheric electricity still retains many unanswered questions.

Further reading

Betz, H.D., U. Schumann, and P. Laroche (eds.): 2009, Lightning: Principles, Instruments and Applications, Springer.

Leblance, F., K.L. Aplin, Y. Yair, R.G. Harrison, J. P. Lebreton, and M. Leblanc (eds.): 2008, Planetary Atmospheric Electricity, Space Science Series of ISSI, Springer.

Rakov, V., and M. Uman: 2003, Lightning: Physics and Effects, Cambridge University Press.

Rust. D., and D. MacGorman: 1998, The Electrical Nature of Thunderstorms, Oxford University Press.

Williams, E., and E. Mareev: 2014, Atmospheric Research, 135-6, 208-227.

INFOBOX 4.1

MODELED IMPACT OF TOTAL SOLAR IRRADIANCE (TSI) FORCING

Kristoffer Rypdal[1] and Martin Rypdal[1]

Until recently, models used for projections of the future climate typically did not include modeling of processes above the troposphere. In such models, the radiation budget deals with the visible and infrared part of the spectrum, and the absorption, reflection, and re-emission of solar radiation from the surface and the lower atmosphere. For this purpose, the solar forcing can be characterised by one single quantity – the total solar irradiance (TSI). It is defined as the solar radiative energy flux per unit area at the distance from the Sun corresponding to the Earth's orbit. The TSI mechanism is the only component of the solar forcing that is usually included in simple energy-balance climate models (EBMs), but above-troposphere processes and spectral distribution of the solar radiation, the solar spectral irradiance (SSI), are included in many of the modern, complex general circulation models (GCMs). EBMs consider only the energy exchange between different parts of the climate system and the balance between radiative energy received from the Sun and radiation loss from the Earth to space. Moreover, GCMs model the dynamics of the fluids of the atmosphere and the oceans, and include sea ice and land surface processes. The more advanced Earth system models (ESMs) may include the effect of atmospheric aerosols, atmospheric and ocean chemistry, the carbon cycle, dynamic vegetation and ocean biology, and ice sheets. The most advanced models have been standardized and compared in the climate model intercomparison projects CMIP3 and CMIP5, which are presented in the later reports of the Intergovernment Panel of Climate Change (IPCC). Results for these model ensembles are shown in panels (a) and (c) of Figure 1. The TSI forcing could be considered a first-order approximation to the true solar forcing. It is reasonable to use it in conceptual and simplified climate models like the EBMs, but the most sophisticated ESMs should also include the kind of physics and chemistry described in this chapter.

Nevertheless, many features of global surface temperature, as observed in instrumental records, paleo-reconstructions, or ensembles of climate-model

---

[1] Department of Mathematics and Statistics, Faculty of Science and Technology, University of Tromsø, 9037 Tromsø, Norway

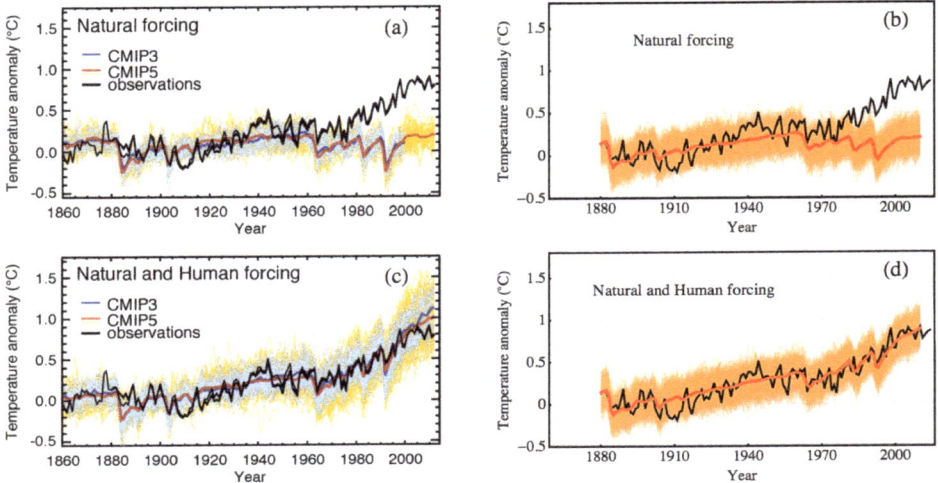

Fig. 1. Adapted from (IPCC, 2014) AR5, FAQ 10.1. Panel (a) shows individual runs and ensemble means of the CMIP3 and CMIP5 ensembles with natural forcing only. The black curves are global instrumental temperature records. Panel (c) is the same with natural + human forcing. Panel (b) and (d) show the same for ensembles of realizations of the simple response model. The black curves is the HadCrut4 instrumental global temperature record.

simulations, can be surprisingly accurately represented by simple linear response models subject to deterministic and stochastic forcing, where the solar forcing is represented by the TSI only. Examples of an ensemble of realizations of solutions of such a model is given in panel (b) and (d) in Figure 1. The results are strikingly similar to the CMIP3/5 ensemble, and suggests that the TSI may be a reasonably good representation of the solar forcing.

Such simple response models can be used to assess the impact of solar irradiance variability compared to other known drivers of the global temperature. The relative importance of the solar, volcanic, and anthropogenic components of the forcing is shown in Figure 2, as computed from such a model. We observe that the solar and anthropogenic component contribute approximately the same amount up to about year 1960, but after this the global warming is totally dominated by the anthropogenic forcing. In Figure 2, we also observe a weak wiggle of period about 11 years. This is the response to the Schwabe solar cycle in TSI. Observe that the response is quite weak, which is partly due the thermal inertia of the oceans and the long effective response time. The response to the solar cycle in this model is also discussed in Chapter 3.9.

These kinds of results have been tested in the more sophisticated climate models in a manner similar to Figure 1a and 1c, by separating the natural forcing into different components, but could also be tested further by separating the solar

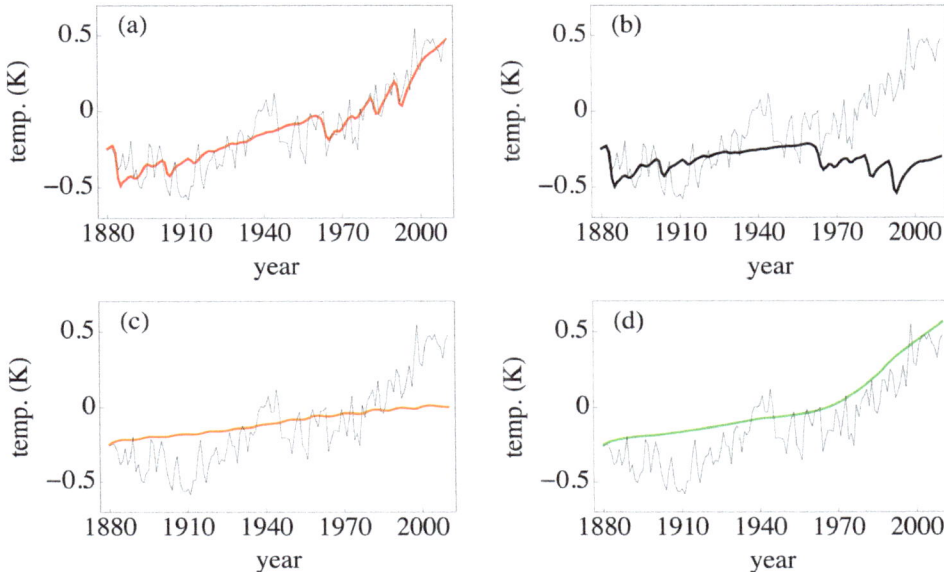

Fig. 2. Deterministic part of forced temperature change AD 1880–2010 according to the simple response model. (a) From total forcing. (b) From volcanic forcing. (c) From solar TSI forcing. (d) From anthropogenic forcing.

forcing into different spectral bands. The latter, however, requires reliable reconstructions of SSI back to AD 1860, and reliable modelling of the above-troposphere processes. Another example of application of the simple response model to reconstructed forcing for the last millennium is shown in Chapter 3.9.

INFOBOX 4.2

LIGHTNING, COSMIC RAYS AND ENERGETIC PARTICLES

Yoav Yair[1]

Thunderstorms and lightning are among the fiercest and most beautiful natural phenomena, feared, respected and worshiped by mankind since ancient times. As described in Chapter 4.8, at any given moment, there are a few thousand thunderstorms around the planet, with a global flash rate estimated at 50 strikes per second. Lightning activity depends on the season and geographical location: with the aid of lightning location networks and satellite data, we are now able to map the areas where lightning frequency is greatest. As Figure 1 shows, most lightning occur over the tropical continental regions in Africa, South-East Asia and South-America. Over the oceans, lightning is less frequent, and at high latitudes, it is almost entirely absent. The Inter-tropical Convergence Zone (ITCZ) migrates with the season from the Southern hemisphere summer to the Northern one, a fact that is clearly reflected in the location of lightning activity: almost 90% of global lightning occurs in the respective summer hemisphere.

The impressive lightning flash rate is a proof of the efficiency by which clouds are being electrically charged and discharged. But, even with our present-day capabilities of advanced computer simulations, sophisticated laboratory experiments and airborne measurements, we are still unsure about the nature of charging processes taking part within the tumultuous heart of a thundercloud. There, trillions of ions, water drops, ice crystals and hail particles collide with each other in sub-zero temperatures. On average, these collisions charge the particles such that smaller ones carry a net positive charge, and the larger ones negative charge. The difference in fall-speeds in air of particles of different phases, masses and shapes leads to the formation of charged regions with different polarities, making the thundercloud behave like a giant battery (with a "plus" and a "minus", separated by kilometers). When the resultant electric field that is present between the charge centers exceeds a certain threshold value, an electron avalanche starts at some point, and rapidly progresses to form a conductive channel that moves in the air by discrete steps (hence it is called "a stepped leader"). If the process occurs between two adjacent (but opposite) centers within the cloud, we would get

---

[1] School of Sustainability, Interdisciplinary Center Herzliya, Israel

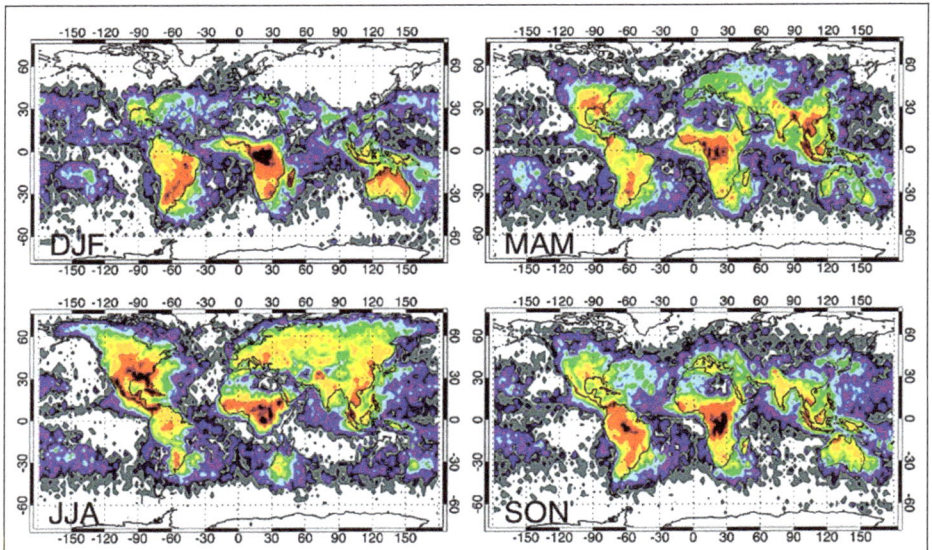

Fig. 1. The seasonal pattern of global lightning activity as observed by the Optical Transient Detector satellite. The different parts represent the three-monthly accumulated data, reflecting the seasonal pattern. Upper left, December-January-February (DJF); Upper right: March-April-May (MAM); Bottom left June-July-August (JJA); Bottom right, September-October-November (SON) ((Christian et al., 2003)). Copyright AGU, reprinted with permission).

an Intra-Cloud flash (IC). If the process moves downward, it forms the familiar chaotic zig-zag pattern until the leader tip attaches to an object on the ground, creating a cloud-to-ground flash (CG). The approaching leader short-circuits the charge center aloft to the surface, and huge amounts of electrons flow downwards, heating the surrounding air to over 30 000 °C, making it glow brightly. This is the lightning that we see, and it normally lasts less than half a second. The heated air around the lightning channel rapidly expands, a fact we all recognize as the sound of thunder. The contact point, where the leader attaches, can be a building, a tree or a person, a fact that makes lightning extremely dangerous and sometimes lethal.

There are still many mysteries surrounding the initial breakdown process that ignites a lightning flash. Balloon measurements performed inside developing thunderstorms show that the conventional breakdown field of about 1 MV m^{-1} is never reached, and that lightning occurs at fields weaker by an order of magnitude. This hints at another process taking place. One possible mechanism is the "Runaway Breakdown" process, in which acceleration of relativistic electrons originating from extensive air-showers (EAS) created by energetic cosmic rays trigger the avalanche, leading to a lightning discharge (Gurevich and Karashtin, 2013). Alternatively, the avalanche may be initiated by proton fluxes during solar proton events (SPE), especially those that are energetic enough to reach the lower troposphere (where clouds develop). These effects would likely show a latitudinal dependence, due to the shielding effect of the Earth's magnetic field that determines how many particles would actually penetrate down.

If fluxes of external energetic particles are indeed effective in the initiation of a leader that starts the discharge, then we might search for a signal of enhanced (or reduced) lightning activity in close conjunction with an increase (or decrease) in the flux of cosmic rays / solar energetic particles reaching the atmosphere. If, however, the influence of enhanced energetic particles (primary or secondary) is through slower changes in the ionisation and other charging processes affecting charge separation within thunderclouds, then the effect may be diluted by other microphysical processes or be delayed by hours to days.

Research conducted in the 1980s attempted to relate cosmic ray maxima (determined from minimum Kp index) with the number of lightning days, obtained from reports of meteorological stations (a "Lightning Day" is defined as a day when an observer sees a lightning flash or hears a thunder). Results suggest that there is a maximum in lightning activity 3 days after the maximum cosmic ray flux (Lethbridge, 1981). Although the exact mechanism is unclear, this correlation does not imply causality and cannot be explained by the direct seeding-ignition hypothesis. It may reflect changes in the ionisation rates, and by inference of the charging efficiency at cloud altitudes.

An opposite effect is expected following a Forbush Decrease (FD, and see Chapter 2.3), identified through neutron counts by the World-Wide Neutron Monitor Network. If there is a link, one can expect a decrease in the amount of lightning following such an event, because a large dip in the number of arriving cosmic ray particles would mean fewer seed electrons available for breakdown. Recent analysis of lightning data obtained from the World-Wide Lightning Location Network (wwlln.net) shows that in strong FD events, a minimum in lightning counts occurred a day after the onset of that event[1]. It should be noted that the WWLLN only detects the strongest cloud-to-ground flashes and has an overall global detection efficiency of about 25%, so the result may apply only to the strongest flashes. The delay is also hard to explain, because it means that the effect is not instantaneous as one would expect if the Runaway Breakdown hypothesis was true.

A study preformed for the continental US derived lightning from the National Lightning Detection Network (Chronis, 2009) and showed a minimum in lightning activity 4–5 days after a Forbush event. For longer time periods, e.g., monthly statistics of lightning occurrence, that study showed a positive trend between lightning activity and cosmic ray fluxes in Northern Hemisphere winter months (December-January-February), but a non-significant trend in summer. This may

---

[1] Okike, O and Collier AB, Testing the cosmic-ray lightning connection hypothesis. 30th URSI General Assembly and Scientific Symposium, Istanbul, Turkey, 13-20 August, 2011, 10.1109/URSIGAS.2011.6051175, IEEE, 2011.

be due to differences in the relative importance of various charging processes in winter and summer and to the nature of the clouds themselves. Another study, conducted in Europe (Schlegel et al., 2001) related lightning data from local networks in Germany and Austria to solar parameters (sunspot number R, geomagnetic activity Ap and radio flux index F10.7) and to cosmic ray fluxes. In general, they found only weak influence of solar activity on lightning frequency compared to the decisive meteorological factors, and found that the overall effect was opposite to the former study showing high frequency of lightning during times of low cosmic ray flux, i.e., when solar activity was high. The explanation for that result was the operation of planetary-wave dynamics. Still, it is unclear how exactly solar signals are transmitted to the lower atmosphere. In Brazil, a statistical analysis of thunder-day data from 1951 to 2009 in 7 major cities searched for a signal of the 11-year solar cycle in lightning activity (Pinto Neto et al., 2013). The results showed an anti-correlation between solar activity and lightning only in 3 cities (with a weaker one in the other 3, and none in the last), explained by a "shielding effect" of cosmic rays by the heliosphere in years with large sunspot numbers. A study conducted in south-east Asia failed to find any significant correlation between sunspot numbers, Ap index, the F10.7 index, and lightning (Siingh et al., 2013).

These different and sometimes contradictory results clearly show that there is a geographic variability in the observed correlations between solar variability (affecting cosmic ray fluxes) and lightning, and that it is different on different time scales. It also leaves open the questions on the exact nature of the coupling mechanism(s): is it via the ionization and charging process within the evolving cloud, or via the runaway-breakdown-leader process when the thunderstorms are mature. It may also be that both processes take place, or others still unknown. With the advance in global coverage of lightning flashes, better studies can be made, where new and improved results will help shed light on this intriguing topic.

In recent years, observations from satellites, ground-based sensors and laboratory experiments found intriguing new results showing that lightning discharges are related to energetic particles in a completely different manner. It appears that lightning emits and produces very energetic particles by themselves: X-rays, neutrons, positrons and gamma-rays are all generated by complex processes taking place in the tips of lightning leaders during various stages of the breakdown process. These new results imply that we are still far from a complete understanding of this wonderful natural phenomenon, and that there is a lot to be discovered.

INFOBOX 4.3

THE INFLUENCE OF SOLAR VARIABILITY ON EXTREME WEATHER

Maria Carmen Llasat[1]

The analysis of the influence of solar activity on climate extremes has the added difficulty of the definition of "extreme climate" and the limited number of data constituting the sample that should be analysed. Due to the lack of consensus to the definition of "extreme", it is usual to apply the ETCCDI (Expert Team on Climate Change Detection and Indices) indices defined by the World Meteorological Organisation (WMO, 2009[2]). The problem of working with the ETCCDI index or the extremes of the statistical distributions of climatic variables is that, usually, the required instrumental data are only available for a short period, so the analysis of the correlation with solar activity is unrepresentative. Then, taking into account that climate extremes are usually associated with climate risks, it is often preferable to work with hydrometeorological risks (UNISDR, 2009[3]), which besides providing a longer proxy data series, have also a major interest for the risk assessment and risk awareness. The main difference lies in the fact that the concept of "extreme" is based on the statistical distribution of the variables, while the "risk" is the combination of the probability of an event and its negative consequences (UNISDR, 2009), and can be considered the convolution of "hazard" (the probability of occurrence of an event of a certain intensity in a specific site over a specific time) and "vulnerability" (the characteristics and circumstances of a community, system or asset that make it susceptible to the damaging effects of a hazard). This would be the case of a flood series, where observed trends and variability could be due to factors related with the hazard, like climate (and, in its turn, to solar variability), but also to vulnerability changes. The major part of studies relating extremes with solar variability refer to floods, for which the reconstruction from historical data (documentary flood sources) and sediments

---

[1] Department of Astronomy and Meteorology, University of Barcelona, Spain
[2] Guidelines on analysis of extremes in a changing climate in support of informed decisions for adaptation, Tech. Rep. WCDMP No. 72 WMO/TDNo. 1500, WMO, 2009
[3] UNISDR Terminology on Disaster Risk Reduction, available at: http://www.unisdr.org/eng/terminology/terminology-2009-eng.html, 2009.

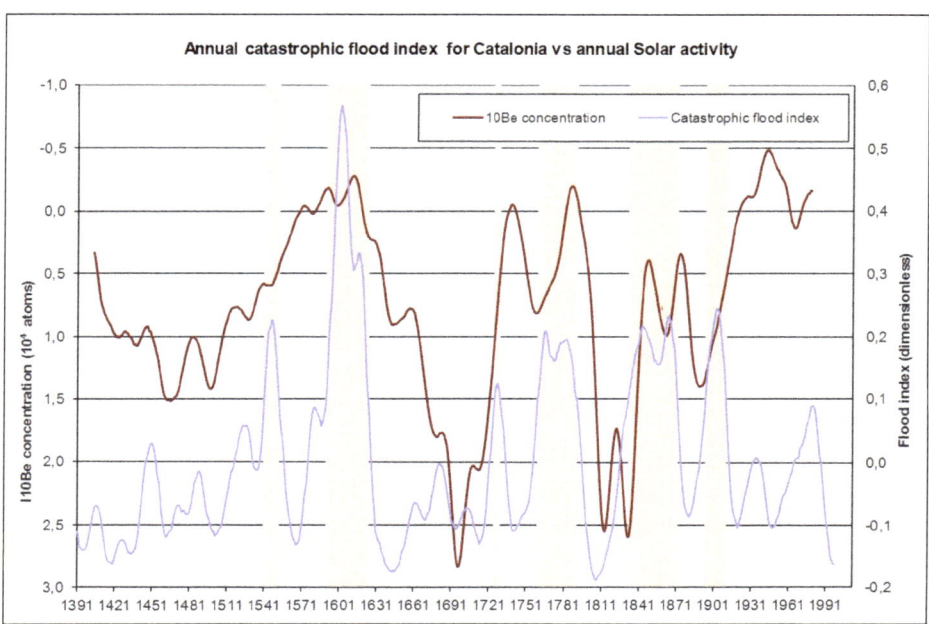

Fig. 1. Temporal evolution of solar activity taken from ^{10}Be versus annual catastrophic flood index series for Catalonia from 1389 to 2012. Data are smoothed by a 31-year low-pass Gaussian filter. The scale for ^{10}Be concentration is inverted. (from Barrera-Escoda and Llasat, 2015) .

(paleofloods) is possible (i.e., Benito et al. (2003); Vaquero (2004)). There is, however, a drawback that the detected variability may be due to non-climatic factors, and that if one wants to correlate it with solar activity, it is necessary to distinguish the climate signal from the other contributions. For this reason, it is necessary to separate catastrophic events (major floods and damages), mainly related with climate factors, from those considered ordinary or extraordinary (more frequent) that are usually more influenced by changes in vulnerability and hydraulic conditions (Llasat et al., 2005). Recent studies have shown that no significant common trend has been found for catastrophic flood events in Europe, in spite of the increase of the greenhouse effect, and that the influence of external factors, like solar variability could explain partially, the anomalous periods found over time.

Figure 1 shows the comparison between the annual evolution of solar activity taken from the ^{10}Be concentration and the catastrophic flood index series for NE of Spain for the period 1389–2012. Changes in production rate of ^{10}Be depend on solar activity and geomagnetic field strength, and although there are some dating uncertainty, the ^{10}Be concentration is usually higher for periods of minimum solar activity and lower for periods of maximum solar activity (Berggren et al., 2009). The long-term temporal correlation between ^{10}Be concentration and the catastrophic flood index series (data smoothed by a 31-year low-pass Gaussian filter) is -0.32, and -0.42, when only the autumn season (near 60% of the total catastrophic floods) is considered (Barrera-Escoda and Llasat, 2015). The oscillations of greatest magnitude on flood frequency have a common presence in different European basins which goes to reinforce their general character or origin and suggests that some outside factor is having a significant effect. The strongest links are recorded at the beginning of the Little Ice Age (approximately, 1560–1620), the last decades of the 18th century (1760–1800) and the central decades of the 19th century (1830–1870). Usually, anomalous periods of high catastrophic flood frequency occur in periods of increasingly greater solar activity or rapid change in it. The common flood-rich period 1560–1620 was recorded during a period of increasing solar activity, with a maximum near 1620. During the Maunder minimum, flooding activity decreased, and it increased again until the first part of the Dalton minimum that was a very unusual climatic period in southern Europe, characterised by high climatic irregularity accompanied by both types of hydrological extremes, catastrophic floods and drought periods. The last European common flood-rich period took place at the end of the Little Ice Age, near the maximum of solar activity recorded near 1880.

These marked clusters of historical floods would be associated with changes in the climatic pattern at both the regional and global scales, and, in their turn, they could be related to changes in solar activity. Some studies revealed a certain influence of solar activity on the winter-early spring North Atlantic circulation (both atmospheric and ocean circulations) and, consequently on the development and trajectory of Atlantic perturbations that could arrive to Europe (Moffa-Sánchez et al., 2014). In this case, low solar irradiance could promote positive sea level pressure changes over the northern part (i.e., Iceland) and negative in the subtropics blocking or diverting the usual flow of the westerly winds (negative NAO index). This winter configuration would shift the North Atlantic storm tracks further south carrying moist but colder air over some parts of central and southern Europe, but blocking the meridional transport of warm maritime winds. As heavy rainfall events and floods in Northeast of Spain are usually associated to strong warm and moist air advections from the south, this colder situation would be difficult for the floods production. Noting that floods produced by heavy rainfalls are associated with different circulation patterns, the effect of solar variability on them could change from one region and season to another. On the other hand, the influence in the differential warming of the stratosphere, as a consequence of changes in stratospheric ozone production due to the interaction with energetic particle precipitation and the variations in the ultraviolet spectrum related with solar variability, would influence the dynamics in the high troposphere which have an important role on circulation patterns, like North Atlantic Oscillation (Ermolli et al., 2013). This so-called "top-down" mechanism could explain some climate variability and associated conditions to produce heavy rainfall events, mainly at multidecadal scale. In conclusion, although some advances in the knowledge of the solar variability influence in climatic extremes have been made in the last decade, it remains as an important challenge to understand and simulate their past and future evolution that merits major research.

Further reading

Benito, G. and Thorndycraft, V.R. (ed.): 2004. Systematic Palaeoflood and Historical Data for the improvement of Flood Risk estimation. Methodological Guideliness. CSIC, Madrid, Spain.

Glade, T., Albini, P. and Francs, F.; 2001, The Use of Historical Data in Natural Hazard Assessments, Kluwer Academic Publishers, Dordrecht, The Netherlands

Hall, J., Arheimer, B., Borga, M., et al.: 2015, Understanding flood regime changes in Europe: a state-of-the-art assessment, Hydrology and Earth System Sciences 18, 2735.

References of Part IV

Almeida, J., S. Schobesberger, A. Kürten, I. K. Ortega, O. Kupiainen-Määttä, et al. Molecular understanding of sulphuric acid-amine particle nucleation in the atmosphere. *Nature*, **502**(7471), 359–363, 2013.

Baker, D. N., T. I. Pulkkinen, X. Li, S. G. Kanekal, J. B. Blake, R. S. Selesnick, M. G. Henderson, G. D. Reeves, H. E. Spence, and G. Rostoker. Coronal mass ejections, magnetic clouds, and relativistic magnetospheric electron events: ISTP. *Journal of Geophysical Research*, **103**, 17,279–17,292, 1998. 10.1029/97JA03329.

Baldwin, M. P., L. J. Gray, T. J. Dunkerton, K. Hamilton, P. H. Haynes, et al. The quasi-biennial oscillation. *Reviews of Geophysics*, **39**, 179–229, 2001. 10.1029/1999RG000073.

Barrera-Escoda, A., and M. C. Llasat. Evolving flood patterns in a Mediterranean region (1301-2012) and climatic factors - the case of Catalonia. *Hydrology and Earth System Sciences*, **19**, 465–483, 2015. 10.5194/hess-19-465-2015.

Beig, G., J. Scheer, M. G. Mlynczak, and P. Keckhut. Overview of the temperature response in the mesosphere and lower thermosphere to solar activity. *Reviews of Geophysics*, **46**, RG3002, 2008. 10.1029/2007RG000236.

Benito, G., A. Diez-Herrero, and M. Fernandez de Villalta. Magnitude and frequency of flooding in the Tagus basin (Central Spain) over the last millennium. *Climatic Change*, **58**, 171–192, 2003.

Berggren, A.-M., J. Beer, G. Possnert, A. Aldahan, P. Kubik, M. Christl, S. J. Johnsen, J. Abreu, and B. M. Vinther. A 600-year annual ^{10}Be record from the NGRIP ice core, Greenland. *Geophysical Research Letters*, **36**, L11801, 2009. 10.1029/2009GL038004.

Birkeland, K. The Norwegian aurora Polaris expedition 1902-1903. Aschhoug, 1908.

Brasseur, G. P., and S. Solomon. Aeronomy of the Middle Atmosphere: Chemistry and Physics of the Stratosphere and Mesosphere. Springer, 2005.

Brest, C. L., W. B. Rossow, and M. D. Roiter. Update of radiance calibrations for ISCCP. *Journal of Atmospheric and Oceanic Technology*, **14**(5), 1091–1109, 1997.

Butler, A. H., and L. M. Polvani. El Niño, La Niña, and stratospheric sudden warmings: A reevaluation in light of the observational record. *Geophysical Research Letters*, **38**, L13807, 2011. 10.1029/2011GL048084.

Calvo, N., M. A. Giorgetta, R. Garcia-Herrera, and E. Manzini. Correction to "Nonlinearity of the combined warm ENSO and QBO effects on the Northern Hemisphere polar vortex in MAECHAM5 simulations". *Journal of Geophysical Research (Atmospheres)*, **114**, D20117, 2009. 10.1029/2009JD013257.

Calvo, N., and D. R. Marsh. The combined effects of ENSO and the 11 year solar cycle on the Northern Hemisphere polar stratosphere. *Journal of Geophysical Research (Atmospheres)*, **116**, D23112, 2011. 10.1029/2010JD015226.

Camp, C. D., and K.-K. Tung. The Influence of the Solar Cycle and QBO on the Late-Winter Stratospheric Polar Vortex. *Journal of Atmospheric Sciences*, **64**, 1267, 2007. 10.1175/JAS3883.1.

CCMVAL, S. SPARC Report on the Evaluation of Chemistry-Climate Models. SPARC Report No. 5, WCRP-132, WMO/TD-No. 1526. Eyring, V., Shepherd, T. G., Waugh, D. W., 2010.

Chiodo, G., D. R. Marsh, R. Garcia-Herrera, N. Calvo, and J. A. García. On the detection of the solar signal in the tropical stratosphere. *Atmospheric Chemistry & Physics*, **14**, 5251–5269, 2014. 10.5194/acp-14-5251-2014.

Christian, H. J., R. J. Blakeslee, D. J. Boccippio, W. L. Boeck, D. E. Buechler, et al. Global frequency and distribution of lightning as observed from space by the Optical Transient Detector. *Journal of Geophysical Research (Atmospheres)*, **108**, 4005, 2003. 10.1029/2002JD002347.

Chronis, T. G. Investigating Possible Links between Incoming Cosmic Ray Fluxes and Lightning Activity over the United States. *Journal of Climate*, **22**, 5748, 2009. 10.1175/2009JCLI2912.1.

Clilverd, M. A., C. J. Rodger, T. Moffat-Griffin, E. Spanswick, P. Breen, F. W. Menk, R. S. Grew, K. Hayashi, and I. R. Mann. Energetic outer radiation belt electron precipitation during recurrent solar activity. *Journal of Geophysical Research (Space Physics)*, **115**, A08323, 2010. 10.1029/2009JA015204.

Crutzen, P. J., I. S. A. Isaksen, and G. C. Reid. Solar proton events - Stratospheric sources of nitric oxide. *Science*, **189**, 457–459, 1975. 10.1126/science.189.4201.457.

Damon, P. E., and P. Laut. Pattern of strange errors plagues solar activity and terrestrial climate data. *EOS, Transactions American Geophysical Union*, **85**(39), 370–374, 2004.

Dumbović, M., B. Vrsnak, J. Čalogović, and M. Karlica. Cosmic ray modulation by solar wind disturbances. *Astronomy and Astrophysics*, **531**, 91, 2011.

Dunne, E., L. Lee, C. Reddington, and K. Carslaw. No statistically significant effect of a short-term decrease in the nucleation rate on atmospheric aerosols. *Atmospheric Chemistry and Physics*, **12**(23), 11,573–11,587, 2012.

Ermolli, I., K. Matthes, T. Dudok de Wit, N. A. Krivova, K. Tourpali, et al. Recent variability of the solar spectral irradiance and its impact on climate modelling. *Atmospheric Chemistry and Physics*, **13**, 3945–3977, 2013. 10.5194/acp-13-3945-2013, 1303.5577.

Fairfield, D., and L. J. Cahill. Transition region magnetic field and polar magnetic disturbances. *Journal of Geophysical research*, **71**, 1966.

Farrar, P. D. Are cosmic rays influencing oceanic cloud coverage–or is it only El Nino? *Climatic Change*, **47**(1-2), 7–15, 2000.

Frame, T. H. A., and L. J. Gray. The 11-Yr Solar Cycle in ERA-40 Data: An Update to 2008. *Journal of Climate*, **23**, 2213–2222, 2010. 10.1175/2009JCLI3150.1.

Funke, B., A. Baumgaertner, M. Calisto, T. Egorova, C. H. Jackman, et al. Composition changes after the "Halloween" solar proton event: the High Energy Particle Precipitation in the Atmosphere (HEPPA) model versus MIPAS data intercomparison study. *Atmospheric Chemistry & Physics*, **11**, 9089–9139, 2011. 10.5194/acp-11-9089-2011.

Funke, B., M. López-Puertas, G. P. Stiller, and T. Clarmann. Mesospheric and stratospheric NO$_y$ produced by energetic particle precipitation during 2002-2012. *Journal of Geophysical Research (Atmospheres)*, **119**, 4429–4446, 2014. 10.1002/2013JD021404.

Gerber, E. P., A. Butler, N. Calvo, A. Charlton-Perez, M. Giorgetta, et al. Assessing and understanding the impact of stratospheric dynamics and variability on the earth system. *Bulletin of the American Meteorological Society*, **25**, 845–859, 2012. 10.1175/BAMS-D-11-00145.1.

Gray, L. J., J. Beer, M. Geller, J. D. Haigh, M. Lockwood, et al. Solar Influences on Climate. *Reviews of Geophysics*, **48**, RG4001, 2010. 10.1029/2009RG000282.

Gurevich, A. V., and A. N. Karashtin. Runaway Breakdown and Hydrometeors in Lightning Initiation. *Physical Review Letters*, **110**(18), 185005, 2013. 10.1103/PhysRevLett.110.185005.

Haigh, J. D. The Sun and the Earth's Climate. *Living Reviews in Solar Physics*, **4**, 2, 2007. 10.12942/lrsp-2007-2.

Hansen, K., K., C. Matthes, J. Petrick, and W. Wang. The influence of natural and anthropogenic factors on major stratospheric sudden warmings. *Journal of Geophysical Research*, **119**, 8117–8136, 2014.

Harrison, R., and I. Usoskin. Solar modulation in surface atmospheric electricity. *Journal of Atmospheric and Solar-Terrestrial Physics*, **72**, 176–182, 2010. 10.1016/j.jastp.2009.11.006.

Harrison, R. G. The Carnegie Curve. *Surveys in Geophysics*, **34**, 209–232, 2013. 10.1007/s10712-012-9210-2.

Holton, J. R. An Introduction to Dynamic Meteorology. Elsevier Academic Press, 4th edn., 2004. ISBN: 0-12-354016-X.

Holton, J. R., and H.-C. Tan. The Influence of the Equatorial Quasi-Biennial Oscillation on the Global Circulation at 50 mb. *Journal of Atmospheric Sciences*, **37**, 2200–2208, 1980. 10.1175/1520-0469(1980)037¡2200:TIOTEQ¿2.0.CO;2.

Idso, C. D., R. M. Carter, and S. F. Singer, eds. Climate Change Reconsidered II: Physical Science. Heartland Institute, 2013.

Ineson, S., and A. A. Scaife. The role of the stratosphere in the European climate response to El Niño. *Nature Geoscience*, **2**, 32–36, 2009. 10.1038/ngeo381.

Jackman, C. H., D. R. Marsh, F. M. Vitt, R. R. Garcia, C. E. Randall, E. L. Fleming, and S. M. Frith. Long-term middle atmospheric influence of very large solar proton events. *Journal of Geophysical Research (Atmospheres)*, **114**, D11304, 2009. 10.1029/2008JD011415.

Kirkby, J., J. Curtius, J. Almeida, E. Dunne, J. Duplissy, et al. Role of sulphuric acid, ammonia and galactic cosmic rays in atmospheric aerosol nucleation. *Nature*, **476**(7361), 429–433, 2011.

Kodera, K., and Y. Kuroda. Dynamical response to the solar cycle. *Journal of Geophysical Research (Atmospheres)*, **107**, 4749, 2002. 10.1029/2002JD002224.

Koren, I., O. Altaratz, L. A. Remer, G. Feingold, J. V. Martins, and R. H. Heiblum. Aerosol-induced intensification of rain from the tropics to the mid-latitudes. *Nature Geoscience*, **5**(2), 118–122, 2012.

Kristjánsson, J., J. Kristiansen, and E. Kaas. Solar activity, cosmic rays, clouds and climate–an update. *Advances in space research*, **34**(2), 407–415, 2004.

Labitzke, K., M. Kunze, and S. Brönnimann. Sunspots, the QBO and the stratosphere in the North Polar Region - 20 years later. *Meteorologische Zeitschrift*, **15**, 355–363, 2006. 10.1127/0941-2948/2006/0136.

Labitzke, K., and H. van Loon. Associations between the 11-Year Solar Cycle, the Quasi-Biennial Oscillation and the Atmosphere: A Summary of Recent Work. *Royal Society of London Philosophical Transactions Series A*, **330**, 577–589, 1990. 10.1098/rsta.1990.0039.

Laken, B. A., and J. Čalogović. Composite analysis with Monte Carlo methods: an example with cosmic rays and clouds. *Journal of Space Weather and Space Climate*, **3**, A29, 2013a.

Laken, B. A., and J. Čalogović. Does the diurnal temperature range respond to changes in the cosmic ray flux? *Environmental Research Letters*, **8**(4), 045,018, 2013b.

Laken, B. A., E. Pallé, J. Čalogović, and E. M. Dunne. A cosmic ray-climate link and cloud observations. *Journal of Space Weather and Space Climate*, **2**, A18, 2012.

Laut, P. Solar activity and terrestrial climate: an analysis of some purported correlations. *Journal of Atmospheric and Solar-Terrestrial Physics*, **65**(7), 801–812, 2003.

Lethbridge, M. D. Cosmic rays and thunderstorm frequency. *Geophysical Research Letters*, **8**, 521, 1981. 10.1029/GL008i005p00521.

Lilensten, J., T. Dudok de Wit, P.-O. Amblard, J. Aboudarham, F. Auchère, and M. Kretzschmar. Recommendation for a set of solar EUV lines to be monitored for aeronomy applications. *Annales Geophysicae*, **25**, 1299–1310, 2007. 10.5194/angeo-25-1299-2007.

Liu, W. The dynamic magnetospheres. IAGA Special Sopron Book Series 3, 2011.

Llasat, M.-C., M. Barriendos, A. Barrera, and T. Rigo. Floods in Catalonia (NE Spain) since the 14th century. Climatological and meteorological aspects from historical documentary sources and old instrumental records. *Journal of Hydrology*, **313**, 32–47, 2005. 10.1016/j.jhydrol.2005.02.004.

Lockwood, M., and C. Fröhlich. Recent oppositely directed trends in solar climate forcings and the global mean surface air temperature. *Proceedings of the Royal Society A: Mathematical, Physical and Engineering Science*, **463**(2086), 2447–2460, 2007.

Mann, I. R., T. P. O'Brien, and D. K. Milling. Correlations between ULF wave power, solar wind speed, and relativistic electron flux in the magnetosphere: solar cycle dependence. *Journal of Atmospheric and Solar-Terrestrial Physics*, **66**, 187–198, 2004. 10.1016/j.jastp.2003.10.002.

Markson, R. The Global Circuit Intensity: Its Measurement and Variation over the Last 50 Years. *Bulletin of the American Meteorological Society*, **88**, 223, 2007. 10.1175/BAMS-88-2-223.

Marsh, N. D., and H. Svensmark. Low cloud properties influenced by cosmic rays. *Physical Review Letters*, **85**(23), 5004, 2000.

Matthes, K., K. Kodera, R. R. Garcia, Y. Kuroda, D. R. Marsh, and K. Labitzke. The importance of time-varying forcing for QBO modulation of the atmospheric 11 year solar cycle signal. *Journal of Geophysical Research (Atmospheres)*, **118**, 4435–4447, 2013. 10.1002/jgrd.50424.

Meehl, G. A., and J. M. Arblaster. A Lagged Warm Event-Like Response to Peaks in Solar Forcing in the Pacific Region. *Journal of Climate*, **22**, 3647–3660, 2009. 10.1175/2009JCLI2619.1.

Meehl, G. A., J. M. Arblaster, K. Matthes, F. Sassi, and H. van Loon. Amplifying the Pacific Climate System Response to a Small 11-Year Solar Cycle Forcing. *Science*, **325**, 1114–, 2009. 10.1126/science.1172872.

Moffa-Sánchez, P., A. Born, I. R. Hall, D. J. R. Thornalley, and S. Barker. Solar forcing of North Atlantic surface temperature and salinity over the past millennium. *Nature Geoscience*, **7**, 275–278, 2014. 10.1038/ngeo2094.

Pierce, J., and P. Adams. Can cosmic rays affect cloud condensation nuclei by altering new particle formation rates? *Geophysical Research Letters*, **36**(9), 2009.

Pinto Neto, O., I. R. C. A. Pinto, and O. Pinto. The relationship between thunderstorm and solar activity for Brazil from 1951 to 2009. *Journal of Atmospheric and Solar-Terrestrial Physics*, **98**, 12–21, 2013. 10.1016/j.jastp.2013.03.010.

Ramanathan, V., R. Cess, E. Harrison, P. Minnis, B. Barkstrom, E. Ahmad, and D. Hartmann. Cloud-radiative forcing and climate: Results from the Earth Radiation Budget Experiment. *Science*, **243**(4887), 57–63, 1989.

Randall, C. E., V. L. Harvey, D. E. Siskind, J. France, P. F. Bernath, C. D. Boone, and K. A. Walker. NO_x descent in the Arctic middle atmosphere in early 2009. *Geophysical Research Letters*, **36**, L18811, 2009. 10.1029/2009GL039706.

Richter, J. H., K. Matthes, N. Calvo, and L. J. Gray. Influence of the quasi-biennial oscillation and El Niño-Southern Oscillation on the frequency of sudden stratospheric warmings. *Journal of Geophysical Research (Atmospheres)*, **116**, D20111, 2011. 10.1029/2011JD015757.

Robock, A. Volcanic eruptions and climate. *Reviews of Geophysics*, **38**, 191–219, 2000. 10.1029/1998RG000054.

Rodger, C. J., T. Raita, M. A. Clilverd, A. Seppälä, S. Dietrich, N. R. Thomson, and T. Ulich. Observations of relativistic electron precipitation from the radiation belts driven by EMIC waves. *Geophysical Research Letters*, **35**, L16106, 2008. 10.1029/2008GL034804.

Rosenfeld, D., Y. Kaufman, and I. Koren. Switching cloud cover and dynamical regimes from open to closed Benard cells in response to the suppression of precipitation by aerosols. *Atmospheric Chemistry and Physics*, **6**(9), 2503–2511, 2006.

Rostoker, G., S. Skone, and D. N. Baker. On the origin of relativistic electrons in the magnetosphere associated with some geomagnetic storms. *Geophysical Research Letters*, **25**, 3701–3704, 1998. 10.1029/98GL02801.

Rozanov, E., L. Callis, M. Schlesinger, F. Yang, N. Andronova, and V. Zubov. Atmospheric response to NO_y source due to energetic electron precipitation. *Geophysical Research Letters*, **32**, L14811, 2005. 10.1029/2005GL023041.

Rycroft, M. J., S. Israelsson, and C. Price. The global atmospheric electric circuit, solar activity and climate change. *Journal of Atmospheric and Solar-Terrestrial Physics*, **62**, 1563–1576, 2000. 10.1016/S1364-6826(00)00112-7.

Schlegel, K., G. Diendorfer, S. Thern, and M. Schmidt. Thunderstorms, lightning and solar activity-Middle Europe. *Journal of Atmospheric and Solar-Terrestrial Physics*, **63**, 1705–1713, 2001. 10.1016/S1364-6826(01)00053-0.

Schumann, W. O. Über die strahlungslosen Eigenschwingungen einer leitenden Kugel die von einer Luftschicht und einer Ionosphärenhülle umgeben ist. *Zeitschrift Naturforschung Teil A*, **7**, 149, 1952.

Seppälä, A., H. Lu, M. A. Clilverd, and C. J. Rodger. Geomagnetic activity signatures in wintertime stratosphere wind, temperature, and wave response. *Journal of Geophysical Research (Atmospheres)*, **118**, 2169–2183, 2013. 10.1002/jgrd.50236.

Siingh, D., P. R. Kumar, M. N. Kulkarni, R. P. Singh, and A. K. Singh. Lightning, convective rain and solar activity — Over the South/Southeast Asia. *Atmospheric Research*, **120**, 99–111, 2013. 10.1016/j.atmosres.2012.07.026.

Snow-Kropla, E., J. Pierce, D. Westervelt, and W. Trivitayanurak. Cosmic rays, aerosol formation and cloud-condensation nuclei: sensitivities to model uncertainties. *Atmospheric Chemistry and Physics*, **11**(8), 4001–4013, 2011.

Soukharev, B. E., and L. L. Hood. Solar cycle variation of stratospheric ozone: Multiple regression analysis of long-term satellite data sets and comparisons with models. *Journal of Geophysical Research (Atmospheres)*, **111**, D20314, 2006. 10.1029/2006JD007107.

Stenchikov, G., A. Robock, V. Ramaswamy, M. D. Schwarzkopf, K. Hamilton, and S. Ramachandran. Arctic Oscillation response to the 1991 Mount Pinatubo eruption: Effects of volcanic aerosols and ozone depletion. *Journal of Geophysical Research (Atmospheres)*, **107**, 4803, 2002. 10.1029/2002JD002090.

Stocker, T., and D. Qin, eds. Climate Change 2013 - The Physical Science Basis. Working Group I Contribution to the Fifth Assessment Report of the IPCC. Cambridge University Press, Cambridge, 2014.

Stubenrauch, C., W. Rossow, S. Kinne, S. Ackerman, G. Cesana, et al. Assessment of global cloud datsasets from satellites: Project and database initiated by the GEWEX Radiation Panel. *Bulletin of the American Meteorological Society*, **94**(7), 1031–1049, 2013.

Svensmark, H. Cosmoclimatology: a new theory emerges. *Astronomy & Geophysics*, **48**(1), 1–18, 2007.

Svensmark, H., and E. Friis-Christensen. Variation of cosmic ray flux and global cloud coverage—a missing link in solar-climate relationships. *Journal of Atmospheric and Solar-Terrestrial Physics*, **59**(11), 1225–1232, 1997.

Takahashi, K. Magnetospheric ULF: Synthesis and New Directions. AGU, 2006.

Tanskanen, E. I. A comprehensive high-throughput analysis of substorms observed by IMAGE magnetometer network: Years 1993-2003 examined. *Journal of Geophysical Research (Space Physics)*, **114**, A05204, 2009. 10.1029/2008JA013682.

Tanskanen, E. I., J. A. Slavin, A. J. Tanskanen, A. Viljanen, T. I. Pulkkinen, H. E. J. Koskinen, A. Pulkkinen, and J. Eastwood. Magnetospheric substorms are strongly modulated by interplanetary high-speed streams. *Geophysical Research Letters*, **32**, L16104, 2005. 10.1029/2005GL023318.

Tinsley, B. A. Influence of solar wind on the global electric circuit, and inferred effects on cloud microphysics, temperature, and dynamics in the troposphere. *Space Science Reviews*, **94**(1-2), 231–258, 2000.

Toohey, M., K. Krüger, M. Bittner, C. Timmreck, and H. Schmidt. The impact of volcanic aerosol on the Northern Hemisphere stratospheric polar vortex: mechanisms and sensitivity to forcing structure. *Atmospheric Chemistry & Physics*, **14**, 13,063–13,079, 2014. 10.5194/acp-14-13063-2014.

van Loon, H., and K. Labitzke. The Southern Oscillation. Part V: The Anomalies in the Lower Stratosphere of the Northern Hemisphere in Winter and a Comparison with the Quasi-Biennial Oscillation. *Monthly Weather Review*, **115**, 357, 1987. 10.1175/1520-0493(1987)115¡0357:TSOPVT¿2.0.CO;2.

van Loon, H., G. A. Meehl, and D. J. Shea. Coupled air-sea response to solar forcing in the Pacific region during northern winter. *Journal of Geophysical Research (Atmospheres)*, **112**, D02108, 2007. 10.1029/2006JD007378.

Vaquero, J. M. Solar signal in the number of floods recorded for the Tagus river basin over the last millennium. *Climatic Change*, **66**, 23–26, 2004.

Whipple, F. J. W. On the association of the diurnal variation of electric potential gradient in fine weather with the distribution of thunderstorms over the globe. *Quarterly Journal of the Royal Meteorological Society*, **55**, 1–18, 1929. 10.1002/qj.49705522902.

Wilson, C. T. R. Investigations on Lightning Discharges and on the Electric Field of Thunderstorms. *Royal Society of London Philosophical Transactions Series A*, **221**, 73–115, 1921. 10.1098/rsta.1921.0003.

Yu, F., and G. Luo. Effect of solar variations on particle formation and cloud condensation nuclei. *Environmental Research Letters*, **9**(4), 045,004, 2014.

Yu, F., and R. P. Turco. Ultrafine aerosol formation via ion-mediated nucleation. *Geophysical Research Letters*, **27**(6), 883–886, 2000.

Part V

CONCLUSION

© ISS, NASA

CHAPTER 5

CONCLUSIONS

Until the early 1980s, the only way the impact of solar variability on climate could be investigated was by performing statistical comparisons between climate records and proxies for solar activity, such as the sunspot number. A large number of correlations were found that way. However, many of them did not withstand closer scrutiny. Nor did such studies allow for a physical understanding of the mechanisms, which is the key to an assessment of the impact of solar variability on climate.

This situation has now progressed, as observations (of both solar variability, and the climate response), and also model simulations have gradually begun to unravel a highly complex picture, with multiple competing or synergistic mechanisms. This handbook has introduced several.

Understanding these mechanisms is also crucial for quantifying the contribution of solar variability to global warming observed since the 1950s. According to present knowledge, this solar contribution is most likely much smaller than that of man-made greenhouse gases (IPPC, 2014), although its actual extent remains an open and debated issue. The major socio-economic relevance of this question explains why so much attention has been given to the role of the Sun in climate change, as compared to the understanding of the influence of solar variability on climate. In this handbook, we have deliberately put the focus on the latter, investigating how the Sun may act on different climate and atmospheric parameters, on a variety of time scales.

To summarise the findings, we have split them into two parts. The first part deals with the scientific questions, and lists the key issues. For each of them, we start with solid facts, and end with open issues. The second part is more informal, and addresses some of the questions that came up during the 4-year project that led to this handbook.

1 Conclusions: a science perspective

Sun–climate connections: a multi-faceted system

The impact of solar output on the Earth's atmosphere is mediated by a large variety of coupled physical and chemical mechanisms. The signatures of the changing Sun are strongest in the upper atmosphere. They are considerably weaker in the lowermost layers of the atmosphere, which govern most of climate variability, and which are the main layers of interest for climate studies (see Chapters 1.1 and 1.2). In contrast to most climate studies, which focus on the conditions of the lower atmosphere only, a proper assessment of solar influences on climate requires the observation of the whole 3-dimensional region that extends from the lower atmosphere all the way to the solar surface.

One of the main issues in Sun–climate connections is the identification of physical mechanisms linking solar activity to the Earth's troposphere. The highly variable upper-atmosphere of the Earth is readily influenced by solar activity. However, the impact of this variability on the much denser lower-atmosphere is unclear: if these signals are translated to the troposphere they may be relatively small, and easily hidden in the natural variability of the troposphere. Since all solar effects are entrained by the same driver (i.e. the evolution of the solar magnetic field), it is often difficult to isolate the different mechanisms, even more so when they are coupled. Consequently, many claims for solar signatures in climate records are still based on correlation studies, and thus await a physical explanation. (see Chapter 1.3). Here, we stress again the importance of grounding these hypotheses in physical mechanisms that can be tested. In this context, models are precious allies, as they allow testing some hypotheses by investigating mechanisms individually.

Direct heating by solar radiation: an important effect, but not the only one

Solar electromagnetic radiation represents over 99.96% of the total energy that enters the Earth's atmosphere, and is frequently described as a single quantity, called TSI (Total Solar Irradiance). The variability of the TSI over a Schwabe cycle (with a period of typically 11 years) is of the order of 1 W m^{-2}, resulting in a top-of-the-atmosphere radiative forcing of about 0.2 W m^{-2} (see Chapter 4.1). This variation is causing a global mean surface temperature response of likely slightly less than 0.1 °C. This is of the same order of magnitude as the natural variability of the climate system on decadal time scales, thus making it very difficult to detect a solar signal in most climate observations.

Since the Maunder minimum that extended from approximately 1645 to 1715, solar activity (as expressed by the sunspot number) has progressively reached a peak of activity during the second half of the 20th century, called Grand maximum, followed since by much more moderate levels. This pattern is well accepted. However, how this translates into TSI changes remains a matter of intense debate, with differences ranging from 0.6–3 W m^{-2} since the Maunder minimum.

Conclusions

Narrowing this range depends on our ability to understand how the Sun radiates during periods of extreme quietness, such as the Maunder minimum. Comparisons with Sun-like stars provide further constraints. On yearly time scales and below, however, the agreement between the observed TSI and model reconstructions is within error bars (see Chapters 2.5 and 2.6).

The impact of ultraviolet radiation versus that of visible and infrared

Although the energy from UV wavelengths contributes a relatively small amount to the TSI (about 8%), it is much more variable than the TSI over the Schwabe cycle, and therefore cannot be neglected. In contrast to visible and infrared radiation, most of the UV radiation is absorbed in the middle and upper atmosphere. The main mechanism by which solar UV impacts climate is through ozone production and destruction, which alters the temperature of the middle atmosphere. Major progress has been achieved in the last decade, in incorporating these effects in climate models. While TSI variations impact global mean temperatures, UV variations are more likely to cause regional changes (see Chapters 4.1–4.3)

How changes in middle atmosphere can impact the much denser lower atmosphere is still a question of debate. Different mechanisms are likely to coexist, but the paucity of direct observations hinders further progress. A major challenge is the proper reconstruction of the solar spectral variability in the UV (see Chapter 4.2).

Energetic particles modulated by the solar wind: more important than expected

Variations in the solar wind permanently affect the state of the Earth's environment. One of their consequences is the intermittent precipitation of energetic particles into the atmosphere. There are various mechanisms by which these fluxes of particles can eventually affect the lower atmosphere, and climate, but their level of understanding is not as advanced as for radiative forcing (see Chapters 4.5 and 4.6). The primary mechanism involves the precipitation into polar regions of energetic particles (mainly protons and electrons) that produce chemically active species, such as HO_x and NO_x, which in turn affect the ozone balance in the middle atmosphere (at high latitudes, and in winter), similarly to what UV radiation does, but limited to high latitudes.

These mechanisms are still poorly documented due to the lack of observations, and also because the observed quantities are subject to strong internal variability that masks the solar signal. Recent results show that the effects of energetic particle impacts on middle atmosphere ozone are more frequent than previously assumed. In addition, their impact on regional climate may be comparable to that of the UV forcing. These particle precipitations are local and highly intermittent, and one of the challenges is to quantify their long-term climate impact.

Galactic cosmic rays: not a major player in climate change

Galactic cosmic rays (GCRs) refer to the flux of high-energy, extra-solar, particles which constantly impinge upon the Earth's atmosphere from all directions. By colliding with atoms in the atmosphere, they generate cascades of secondary particles that may extend to the Earth's surface, and consequently produce ionisation in the middle and lower atmosphere. GCRs are primarily modulated by the variable solar magnetic field and the geomagnetic field, and, as a result, vary inversely with solar activity (see Chapter 2.3). A long-debated hypothesis suggested that GCRs may provide a strong solar amplification mechanism, linking solar activity to the Earth's climate system. One suggested pathway by which this may occur, termed ion-mediated nucleation, operates via an ion-enhancement of aerosol formation and growth rates, leading to a modification of cloud properties. While laboratory-based studies have confirmed the existence of ion-mediated nucleation, it is now understood that conditions in the troposphere are usually unfavorable for this mechanism to exert a significant influence. Such conclusions are supported by laboratory experiments, climate model studies, and satellite observations. Although no evidence for a significant and widespread GCR–cloud link has been found, uncertainty still remains as to whether or not signals too small to be reliably detected remain buried within cloud observations. However, it is clear that the remaining uncertainty is not sufficient to account for 20th century climate change (see Chapter 4.7).

Increased certainty regarding a GCR–cloud link would be gained from improved aerosol and cloud datasets, in particular more reliable vertical profiles and cloud-type discrimination from satellite observations. Although there is no widespread globally significant GCR–cloud effect of importance to recent climate change, the possibility of locally significant, second-order, impacts on clouds has not yet been excluded: the uncertainty regarding such phenomena are high, to the extent that even the potential sign of cloud changes that may result from variations in the GCR flux is not reliably known. Various processes could account for local-scale relationships, the majority of which relate to the global electric circuit, which has recently received considerable interest (see Chapter 4.8). To reduce this uncertainty, a large coordinated effort of ground-based observations and experimentation will be required at multiple sites across the globe, with expertise from a range of disciplines including atmospheric electricity and cloud microphysics.

Internal climate variability blurs the whole picture

Random variability of the climate system is comparable in magnitude to that of the external drivers, which makes it very difficult to detect statistically meaningful correlations without having long-term climate records (see Chapter 3.9). In the stratosphere and above, the solar signature becomes the main cause of the observed variability, and so can often be more easily detected.

Long-term monitoring of the climate system is required to discriminate among concurrent mechanisms. Consequently, the key to a better understanding of the climate response resides in our ability to continuously monitor the state of the atmosphere, and the relevant solar forcings. These data are mainly obtained from spacecraft observations, and to a lesser extent from ground-based, or balloon-borne instruments. Unfortunately, the measurement of several key observables, such as the ozone content and the spectral irradiance, will be discontinued soon, thereby depriving us of the possibility to improve our understanding.

Global solar influence with significant regional manifestation

In recent years, considerable evidence has been accumulated in favour of regional effects of solar influences on climate. Recent observations suggest that regional temperature anomalies from solar associated changes may exceed global mean anomalies from global forcings, such as TSI changes. For example, changes in the UV flux, and particles precipitation both affect the ozone concentration in a way that eventually leads to an enhanced polar vortex, and a positive phase of the North Atlantic Oscillation, which in turn has a particular impact on European climate (see Chapter 4.3).

This role of regional effects is a topic of great interest, and again highlights the limitations of a reductionist position that only emphasises global averages. The possible coupling of natural oscillation modes with the solar cycle has recently called attention to the impact of regional effects. For example, the positive phase of the North Atlantic Oscillation seems to be preferred in the descending phase of the solar cycle.

Models are a precious ally to observations

Although observations are vital for understanding the climate impact of solar variability, models have become a key tool for testing hypotheses, and for investigating regions or regimes that are not properly covered by observations. Significant progress in understanding has been achieved using reanalysis datasets that enhance observations by blending them with model simulations and thereby obtaining realistic and complete high-resolution data (see Chapter 3.4). However, each model has its limitations and caveats, and understanding the uncertainties of models has become as important as determining the uncertainties associated with observations (see Chapter 3.5)

Future developments will be substantially influenced by the expected increase in computer power, which will eventually allow climate simulations with higher spatial resolution. This may allow to gradually overcome the necessity of parameterising small-scale dynamical disturbances, clouds, etc. Regarding the investigation of solar signatures, an important issue is the proper description of the impact of solar UV radiation and energetic particles on the middle and upper atmosphere. This is a challenging task, as it requires a description of physical and chemical processes involving many species. So far, much more attention has been given to the troposphere only, but climate models are gradually pushing their upper limit to higher altitudes, thereby enabling to study phenomena involving vertical coupling processes, such as solar signals (see Chapters 3.6 and 3.7).

2 Conclusions: a human perspective

This handbook was not only a scientific project, but also a human endeavour, involving about one hundred scientists with different scientific background. As such, it constitutes the sum of our present knowledge on multiple connections between solar variability and the Earth's climate. All the results shown are open to scrutiny; some may end up being incomplete, or may even be refuted. This is science at work.

These results, however, reveal a growing number of mechanisms by which solar variability can influence climate. So far we did not find any mechanisms, or sets of mechanisms that allow to attribute the global warming observed since the 1950s to solar activity. Any rebuttal, or claim that solar activity is the major cause of global warming, has to be accompanied by scientific evidence, and provide a true mechanism, physical or chemical, not just speculations that are left to scientists for later explanation.

Is the impact of solar activity on climate a mere hypothesis, or is it real?

This question has been an enduring source of confusion. What we are addressing here is the impact of solar variability, and not that of the Sun itself, whose influence on climate is of course overwhelming. Establishing the most likely solar cause for a change detected in climate is a completely different issue, which requires a careful investigation.

The borderline between a hypothesis and clear evidence gets crossed once we have full disclosure, i.e. once other scientists are able to reproduce our explanation and verify it by independent means, after putting it under careful scrutiny. From that point of view, there is today clear evidence for some aspects of solar variability to affect climate. One example is the modulation of the stratospheric ozone concentration by the UV flux, which then couples to the troposphere below by various mechanisms. Another example is the (sometimes considerable) regional impact of solar forcing, as opposed to a global one.

How is this solar impact perceived by the scientific community?

Most climate scientists today concentrate on the lowermost layer of the atmosphere only (i.e. the troposphere), as this is where most climate dynamics occur and most directly affects humanity. For many climate scientists, the contribution of solar variability boils down to one single quantity: the total solar irradiance (TSI, see Chapter 1.2). The prime effect of the TSI is a direct but weak heating of the troposphere, oceans and land. This picture is simplistic, but it still prevails in many textbooks.

There are reasons for the persistence of this simple TSI-based picture. Solar variability mainly affects the stratosphere, and layers above it. As it turns out, most interactions between atmospheric layers are one-directional, and perturbations tend to propagate upward. As a consequence, and quite understandably, most climate studies are inclined to ignore what is happening above the troposphere, as if it were deep space. This situation is evolving, and climate models are now increasingly often addressing the middle and upper atmosphere too (see Chapter 3.7).

Are there unexplored mechanisms?

Many questions remain unanswered, and there are still some mechanisms whose role is elusive. More detailed investigations are needed in order to understand whether, and how they may affect climate.

One illustration is given by the role of the interplanetary magnetic field, which is carried to the Earth by the solar wind, and primarily interacts with the ionised fraction of the Earth's upper atmosphere. These interactions can couple in several, and often subtle, ways to the lower atmospheric layers. For example, variations in the solar wind are known to affect the vertical electric field (see Chapter 4.8), which eventually influences chemical reactions, and thus, the lower-atmospheric composition.

Interestingly, atmospheric electricity has been studied for more than a century now, but the awareness for its possible connection with climate is very recent. Another largely unexplored aspect is the synchronisation of internal modes of climate variability with the solar cycle, see for example Chapter 4.3.

What obstacles do scientists working on these topics encounter?

There are several obstacles. The main one is the lack of interdisciplinary interactions, which are crucial for properly addressing this complex chain of mechanisms spanning from the Sun to the Earth. Incidentally, this is precisely what has motivated the TOSCA action that has led to this handbook.

In the aforementioned example of solar wind forcing, we typically need experts in the solar wind (to understand how the driver varies on the long term), of the terrestrial atmosphere (to understand the interaction of the solar wind with the Earth's environment), of the Earth's global electric circuit (i.e. one of the mechanisms by which solar variability may couple to the lower atmosphere), of the physics and chemistry of cloud nucleation processes (to understand how the microphysics of cloud formation is affected), not to forget cloud observation (i.e. the delicate interpretation of terrestrial and space-borne observations of cloud coverage). No single scientist can cover all these topics.

In contrast, the literature is replete with publications that address only one piece of the chain, and draw general conclusions. Today, anyone can download solar and climate data, correlate them, suggest an explanation, and draw far-reaching conclusions. Such correlations may raise public interest, but they rarely provide a deeper understanding of the underlying physics.

What we need are means for establishing causal connections between solar drivers and the climate response. Unfortunately, since all drivers are eventually modulated by solar magnetism, "smoking guns" are hard to find. In addition to this, most observations are sparse in time, and in space. Some regions, such as the mesosphere, are poorly known because they can be probed only by remote sensing. Most space-based data are available only since the 1980s, which results in a dramatic lack of hindsight for assessing long-term changes. This partly explains the important role of numerical models to help bridge these gaps of information, and the quest for proxies of past solar activity.

Is the controversy around the role of solar forcing also an obstacle?

The global warming issue has indeed sparked a passionate debate that has engulfed the field of Sun-climate connections. Until the 1980s, there was almost no interest in this field, which was considered as a non-rigorous and poorly documented field of research. Todays strong interest for it largely has to do with its societal consequences. This is the positive side of the debate.

There is also an adverse side to this debate, which has lost much of its scientific content as it became increasingly politicised. Today, this debate is intentionally being spread to the media chiefly by small numbers of politically motivated groups who give an outward appearance of scientific rigour, but in reality have no interest in science. This has short-circuited some of the key parapets of scientific research, which is independent peer review. Not surprisingly, it can be difficult to judge today which research claims should be taken seriously.

Interestingly, most major scientific controversies in the history of science (e.g. heliocentrism, relativity, and continental drift) took several decades before being finally accepted by the scientific community, and later on, by the broader public. In this sense, the climate debate is not an unusual one, except for its antiscience backslash, which highlights the importance of good communication.

What is needed to make progress on this issue?

An obvious answer is: more observations of the Earth's environment in order to better diagnose its development. There are also less obvious needs. One of them is traceability. Many pointless discussions could have been avoided by ensuring that the results are documented and verifiable. Today, a growing number of authors are making their results and computer codes publicly available when publishing their conclusions. This openness complies with one of the pillars of scientific research, which is the critical evaluation by peers. This kind of self-regulation should hopefully help maintain standards of high quality, and credibility.

Finally, there is a clear need for stronger interactions between different scientific communities. For historical reasons, they tend to concentrate on specific regions (e.g., the stratosphere) or processes (e.g., cloud formation), rather than on their broader interactions. We definitely need a shift towards the latter. This is not only a scientific challenge, but also a human endeavour. This takes time, but TOSCA is another step in that direction.

August 2015
Thierry Dudok de Wit, Jean Lilensten, Katja Matthes

Reference of Part V

Stocker, T., Qin, D., Plattner, G.-K., Tignor, M., Allen, S., Boschung, J., Nauels, A., Xia, Y., Bex, V., and Midgley, P.: 2013, *IPCC, 2013: Climate Change 2013: The physical science basis. Contribution of Working Group 1 to the Fifth Assessment Report of the Intergovernmental Panel on Climate Change*, Cambridge University Press

GLOSSARY

April 18, 2015

Source: some definitions come from the Glossary of Terms used in the IPCC Fourth Assessment Report (http://www.ipcc.ch)

Active region: a localised, transient volume of the solar atmosphere in which various solar features, such as sunspots and flares may be observed. Active regions are the result of enhanced magnetic fields; they are bipolar and may be complex if the region contains two or more bipolar groups. The solar radiative output in the UV increases with the number of active regions, which is modulated by the solar cycle. The sunspot number is often used as a proxy for the number of active regions.

Aeronomy: field of physics and chemistry dealing with the specific study of the Earth's (and also planetary) middle and upper atmospheres.

Aerosols: collection of airborne solid or liquid particles, with a typical size between 10^{-9} and 10^{-2} m that reside in the atmosphere for at least several hours. Aerosols may be of either natural or anthropogenic origin. They may influence climate in two ways: directly through scattering and absorbing radiation, and indirectly through acting as condensation nuclei for cloud formation or modifying the optical properties and lifetime of clouds.

Albedo: fraction of the light and energy received that is reflected or diffused by a non-luminous body. The albedo is always comprised between 0 and 1. It varies according to the wavelength. An albedo equal to zero at a given wavelength characterises a body that absorbs all this radiation perfectly. A value of 1 characterises a perfect mirror for that wavelength.

ap: 3-hourly index of geomagnetic activity. The ap index is measured in nano-Tesla, and is derived from the Kp index. It quantifies the variation of the geomagnetic field at ground level compared to quiet day conditions. Ap stands for a daily average of the ap index. The ap index is frequently used as a proxy for the flux of energetic particles precipitating into the Earth's atmosphere. See also Radiation belts.

Astronomical unit (AU): the mean Earth-Sun distance, equal to $1.496 \cdot 10^6$ km.

Atmosphere: gaseous envelope surrounding the Earth. The dry atmosphere consists almost entirely of nitrogen (78.1% volume mixing ratio) and oxygen (20.9% volume mixing ratio), together with a number of trace gases, such as argon (0.93% volume mixing ratio), helium and radiatively active greenhouse gases, such as carbon dioxide (0.035% volume mixing ratio) and ozone. In addition, the atmosphere contains the greenhouse gas, water vapour, whose amounts are highly variable but are typically around 1% volume mixing ratio. The atmosphere also contains clouds and aerosols.

Atmospheric tides: global-scale periodic oscillations of the atmosphere, by analogy with to ocean tides. They are typically excited by the alternation between day/night in the intense solar heating. Some are also driven by the pull of the gravitational field of the Moon.

Aurora: a visual phenomenon occurring during the night at high latitudes, called aurora borealis in the Northern hemisphere and aurora australis in the Southern hemisphere. Auroras are caused by the excitation of atmospheric species, followed by radiation of photons, due to collisions with ionised particles precipitating from the external magnetosphere along the geomagnetic field lines. Since the atmospheric composition changes with altitude, different colours originate from different heights. Yellow-green, for example, occurs from 90 to 250 km (atomic oxygen), while blue and red at lower heights (molecular nitrogen and molecular oxygen). Auroras are part of the complex phenomenon, known as magnetospheric substorm whose occurrence is generally related to solar activity. See Substorm.

Auroral oval: region where the auroras typically occur. The oval is an elliptical region around each geomagnetic pole, generally ranging in geomagnetic latitude from $\approx 65°$ at midnight to $\approx 75°$ at noon, but widens during magnetic storms and substorms. The oval is the region where reconnected magnetospheric field lines originate.

Chromosphere: region of the solar atmosphere, above the photosphere and below the corona, which is characterised by a sudden increase in temperature. During a total eclipse of the Sun, it shows up as a thin, bright pink layer, hence its name. The colour is due primarily to the emission of hydrogen at a wavelength of 656.3 nm. The chromospheres extends to about 10 000 km above the surface of the Sun.

Climate: in a narrow sense, it is usually defined as the average weather, or more rigorously, as the statistical description in terms of the mean and variability of relevant quantities over a period of time ranging from months to thousands or millions of years. The classical period for averaging these variables is 30 years, as defined by the World Meteorological Organisation. The relevant quantities are most often surface variables, such as temperature, precipitation, and wind. Climate in a wider sense is the state, including a statistical description, of the climate system.

Climate change: refers to a change in the state of the climate that can be identified (e.g., by using statistical tests) by changes in the mean and/or the variability of its properties, and which persists for an extended period, typically decades or longer. Climate change may be due to natural internal processes or external forcings, or to persistent anthropogenic changes in the composition of the atmosphere or in land use. Note that the Framework Convention on Climate Change (UNFCCC) defines climate change as: "a change of climate which is attributed directly or indirectly to human activity that alters the composition of the global atmosphere and which is in addition to natural climate variability observed over comparable time periods". The UNFCCC thus makes a distinction between climate change attributable to human activities altering the atmospheric composition, and climate variability attributable to natural causes.

Climate feedback: an interaction mechanism between processes in the climate system is called a climate feedback when the result of an initial process triggers changes in a second process that in turn influences the initial one. A positive feedback intensifies the original process, and a negative feedback reduces it.

Climate model: (spectrum or hierarchy) a numerical representation of the climate system based on the physical, chemical and biological properties of its components, their interactions and feedback processes, and accounting for all or some of its known properties. The climate system can be represented by models of varying complexity, that is, for any one component or combination of components a spectrum or hierarchy of models can be identified, differing in such aspects as the number of spatial dimensions, the extent to which physical, chemical or biological processes are explicitly represented, or the level at which empirical parameterisations are involved. Coupled Atmosphere-Ocean General Circulation Models (AOGCMs) provide a representation of the climate system that is near the most comprehensive end of the spectrum currently available. There is an evolution towards more complex models with interactive chemistry and biology. Climate models are applied as a research tool to study and simulate the climate, and for operational purposes, including monthly, seasonal and interannual climate predictions.

Climate sensitivity: in IPCC reports, equilibrium climate sensitivity refers to the equilibrium change in the annual mean global surface temperature following a doubling of the atmospheric equivalent carbon dioxide concentration. Due to computational constraints, the equilibrium climate sensitivity in a climate model is usually estimated by running an atmospheric general circulation model coupled to a mixed-layer ocean model, because equilibrium climate sensitivity is largely determined by atmospheric processes. Efficient models can be run to equilibrium with a dynamic ocean.

Climate system: the climate system is the highly complex system consisting of five major components: the atmosphere, the hydrosphere, the cryosphere, the land surface and the biosphere, and the interactions between them. The climate system evolves in time under the influence of its own internal dynamics and because of external forcings, such as volcanic eruptions, solar variations and anthropogenic forcings, such as the changing composition of the atmosphere and land use change.

Convection zone: external region of the inside of the Sun (representing the last 30% of the radius), where the energy produced by the nuclear core is transmitted by convection. This is the region where solar matter seethes. Active regions on the solar surface are generally considered to originate from a strong toroidal magnetic field generated at the base of the convection zone.

Corona: the outermost, most tenuous region of the solar atmosphere, characterised by very high temperatures ($> 10^6$ K) and very low density ($< 10^9$ cm^{-3}). The corona extends throughout the solar system, and transitions into the solar wind.

Coronal hole: extended region of the solar corona, which is characterised by low density and unipolar magnetic fields. In coronal holes, solar magnetic field lines open up into space, which allows a high speed solar wind to escape. These regions exhibit a deficit of EUV and XUV radiation.

Coronal mass ejection (CME): a sudden outflow of plasma from the solar corona, often associated with flares and eruptive prominences. CME that are directed earthwards usually reach the Earth in 1–2 days, and strongly perturb the geomagnetic field. This is known to impact the Global Electrical Circuit (GCE), and may be one of the mechanisms by which solar activity influences climate. See Global Electrical Circuit.

Corotating interaction regions: compression regions formed from the interaction of high and low speed solar wind streams. They are roughly aligned with the Archimedean spiral that characterises the large-scale structure of the solar wind, and appear to corotate with the Sun.

Corotation: joint rotation of a planet and its atmosphere.

Cosmic rays: see Galactic cosmic rays.

Cosmogenic isotope, or cosmogenic nuclide: isotopes created when a high-energy cosmic ray interacts with a nucleus, causing the break-up of the bombarded nucleus into several parts. Some of these isotopes, such as ^{10}Be and ^{14}C, are radioactive and are produced in the Earth's atmosphere. Their production rate is weakly modulated by the level of solar magnetic activity. Because of this, their concentration in natural archives, such as ice cores can be used as a tracer for past solar activity. Cosmogenic isotopes today provide access to solar activity up to several 10 000 years back.

Dalton minimum: a period of low solar activity, lasting from about 1790 to 1830, and named after the meteorologist John Dalton. As for the Maunder and Spörer minima, the Dalton minimum coincided with a period of relatively lower temperatures, especially in the Northern hemisphere.

Detection and attribution: climate varies continually on all time scales. Detection of climate change is the process of demonstrating that climate has changed in some defined statistical sense, without providing a reason for that change. Attribution of causes of climate change is the process of establishing the most likely causes for the detected change with some defined level of confidence.

Dst index: or disturbance storm time index, is the weighted average of the north-south component of the geomagnetic field, measured by four stations that are located near the geomagnetic equator. The Dst index gives information about the strength of the ring current around Earth, and is used in space weather for detecting geomagnetic storms. See Radiation belts.

Dynamical system: a process or set of processes whose evolution in time is governed by a set of deterministic physical laws. The climate system is a dynamical system. See Abrupt climate change; Chaos; Nonlinearity; Predictability.

El Niño-Southern Oscillation (ENSO): the term El Niño was initially used to describe a warm-water current that periodically flows along the coast of Ecuador and Perú, disrupting the local fishery. It has since become identified with a basin-wide warming of the tropical Pacific Ocean east of the dateline. This oceanic event is associated with a fluctuation of a global-scale tropical and subtropical surface pressure pattern called the Southern Oscillation. This coupled atmosphere-ocean phenomenon, with preferred time scales of two to about seven years, is collectively known as the El Niño-Southern Oscillation (ENSO). It is often measured by the surface pressure anomaly difference between Darwin and Tahiti and the sea surface temperatures in the central and eastern equatorial Pacific. During an ENSO event, the prevailing trade winds weaken, reducing upwelling and altering ocean currents such that the sea surface temperatures warm, further weakening the trade winds. This event has a great impact on the wind, sea surface temperature and precipitation patterns in the tropical Pacific. It has climatic effects throughout the Pacific region and in many other parts of the world, through global teleconnections. The cold phase of ENSO is called La Niña. Together they constitute one of the major modes of internal climate variability on decadal time scales.

Ensemble: a group of parallel model simulations used for climate projections. Variation of the results across the ensemble members gives an estimate of uncertainty. Ensembles made with the same model but different initial conditions only characterise the uncertainty associated with internal climate variability, whereas multimodel ensembles including simulations by several models also include the impact of model differences. Perturbed-parameter ensembles, in which model parameters are varied in a systematic manner, aim to produce a more objective estimate of modelling uncertainty than is possible with traditional multi-model ensembles.

EUV or Extreme Ultraviolet: spectral band, which is part of the UV, with wavelengths ranging from 10 to 121 nm. Solar spectral irradiance emitted in the EUV is highly variable, and is the main source of ionisation in the Earth's upper atmosphere.

Exosphere: outermost layer of the Earth's atmosphere, defined as the region where collisions between particles are rare enough to be considered as negligible.

Atoms thus move freely, and some may even escape into space without undergoing collisions. The bottom of the exosphere (called exobase) is located between 350 and 800 km altitude, depending on its temperature, which is strongly modulated by the solar UV flux.

External forcing: refers to a forcing agent outside the climate system causing a change in the climate system. Volcanic eruptions, solar variations and anthropogenic changes in the composition of the atmosphere and land use change are external forcings.

F10.7 index: also called decimetric index, is the solar radio flux measured at a wavelength of 10.7 cm (expressed in 10^{22} Wm^{-2}Hz^{-1}). This solar proxy closely matches the variability of the solar UV flux. Unlike the latter, however, it can be conveniently measured from the ground, hence its frequent use as a gauge of UV flux. The quiet Sun has an F10.7 index of approximately 70, whereas levels in excess of 300 are observed at solar maximum. This index does not include enhancements that are due to solar flares.

Facula: a bright region on the solar photosphere, which is best seen in white light. Faculae correspond to regions with enhanced magnetic field, and can be a precursor of a sunspot. Their counterpart in the chromosphere is called facula. The modulation of plages and faculae by the solar cycle is one of the primary causes of the variations observed in the solar spectral irradiance.

Fingerprint: the climate response pattern in space and/or time to a specific forcing is commonly referred to as a fingerprint. Fingerprints are used to detect the presence of this response in observations and are typically estimated using forced climate model simulations.

Flare: an explosion on the Sun usually releasing large amounts of energy and particles, and generally occurring within an active region. Flares are more likely to occur at solar maximum. Flares last from minutes to hours.

FUV or Far Ultraviolet: spectral band, which is part of the UV, with wavelengths ranging from 121 to 200 nm. Solar spectral irradiance emitted in the FUV is mostly absorbed in the mesosphere and above.

Galactic Cosmic Rays (GCR) or Cosmic rays: high-energy particles, originating outside the Solar system. GCR are composed of protons ($\approx$ 90%) and atomic nuclei, with energies that can be of orders of magnitude larger than that of radiation belt particles, i.e. in the MeV–GeV range, and beyond. When penetrating the Earth's atmosphere, they generally produce showers of secondary particles, some of which reach the ground surface. High-energy particles produced by solar flares and/or by coronal mass ejections are called solar cosmic rays or Solar Energetic Particles. See also Solar Energetic Particle (SEP) event.

Geomagnetic field: the magnetic field observed in and around the Earth. The geomagnetic field can be approximated by a centered dipole field, with the axis of the dipole inclined to the Earth's rotational axis by about 11.5°. At the Earth's surface, its intensity is approximately 32 000 nT at the equator, and 62 000 nT at the north pole.

Geomagnetic storm: a temporary disturbance of the Earth's magnetosphere, which occurs when a solar wind perturbation interacts with the geomagnetic field. Geomagnetic storms cause an increase in movement of plasma through the magnetosphere (driven by increased electric fields inside the magnetosphere) and an increase in electric current in the magnetosphere and ionosphere. They last for several days and their main consequences are the injection of large quantities of ions in the radiation belts, which then partly precipitate into the upper atmosphere, where they release their energy and create NO_x. Storms are distinct from substorms in that the latter are much less energetic, and last for a few hours only.

Geostationary (orbit): satellite orbit that always flies over the same point of the terrestrial equator. Geostationary satellites are all located on the same circular orbit, which is located in the equatorial plane, at an altitude of 35 784 km.

GeV: unit of energy, which is frequently used for ionised particles: 1 GeV = 10^9 eV (electron-volts), which is equivalent to $1.6 \cdot 10^{-10}$ J (Joule). Particles with energies in the GeV range are relativistic because their speed is only a fraction of the speed of light. See also MeV.

Global Electrical Circuit (GEC): refers to the continuous electrical current that flows vertically between the ionosphere and the Earth's surface. This flow is powered by thunderstorms, which cause a build-up of positive charge in the ionosphere. During fair weather conditions, this positive charge slowly flows back to the Earth's surface. Cosmic rays are the principal source of atmospheric ions in the lower atmosphere; their intensity is modulated by solar activity, which may thus influence the GEC.

Gravity waves: generally occur in the atmosphere at interfaces between layers, and are primarily governed by the Earth's gravity. Gravity waves are well known to occur near the tropopause. They are generated from below by airflow over mountains. As they move upward and reach more rarefied air, their amplitude increases, until non-linear effects causes them to break, transferring their energy to the mean flow. See planetary waves.

Greenhouse effect: greenhouse gases effectively absorb thermal infrared radiation, emitted by the Earth's surface, by the atmosphere itself due to the same gases, and by clouds. Atmospheric radiation is emitted to all sides, including downward to the Earth's surface. Thus, greenhouse gases trap heat within the surface–troposphere system. This is called the greenhouse effect. Thermal infrared radiation in the troposphere is strongly coupled with the temperature of the atmosphere at the altitude at which it is emitted. In the troposphere, the temperature generally decreases with height. Effectively, infrared radiation emitted to space originates from an altitude with a temperature of, on average, $-19\,°C$, in balance with the net incoming solar radiation, whereas the Earth's surface is kept at a much higher temperature of, on average, $14\,°C$. An increase in the concentration of greenhouse gases leads to an increased infrared opacity of the atmosphere, and therefore to an effective radiation into space from a higher altitude at a lower temperature. This causes a radiative forcing that leads to an enhancement of the greenhouse effect, the so-called enhanced greenhouse effect.

Group sunspot number: index of solar activity that is similar to the sunspot number. The group sunspot number, however, is defined from the number of sunspot groups only, which makes it easier to infer from historic drawings.

Hadley Circulation: a direct, thermally driven overturning cell in the atmosphere consisting of poleward flow in the upper troposphere, subsiding air into the subtropical anticyclones, return flow as part of the trade winds near the surface, and with rising air near the equator in the so-called Inter-Tropical Convergence Zone.

Heliosphere: region in space that undergoes the influence of the solar wind. It spreads from approximately 50 to 100 Astronomical Units (AU) from the Sun.

Heterosphere: layer of the atmosphere that is typically located above 80 km. In the heterosphere, each molecular constituent has its own scale height because for each of them, the pressure decreases differently with altitude. The layer below is called homosphere.

Holocene: the Holocene geological epoch is the latter of two Quaternary epochs, extending from about 11 600 years to and including the present.

Homosphere: designates the lower layers of the terrestrial atmosphere in which the scale height is the same for all constituents: for all, the pressure and the concentration decrease in the same way with altitude. The homosphere ends at an altitude of about 80 km, beyond which starts the heterosphere.

HO_x: a generic term used for hydrogen oxide radicals. These highly reactive, and consequently short-lived radicals are produced during UV-light dissociation of H_2O_2, typically above 50 km altitude. These radicals play a role in the middle stratosphere by depleting it from its ozone. See Radical and NO_x.

Hydrosphere: the component of the climate system composed of liquid surface and subterranean water, such as oceans, seas, rivers, freshwater lakes, underground water, etc.

Insolation: the amount of solar radiation reaching the Earth by latitude and by season. Usually insolation refers to the radiation arriving at the top of the atmosphere. Sometimes it is specified as referring to the radiation arriving at the Earth's surface. See also: Total Solar Irradiance.

Interplanetary magnetic field (IMF): the solar magnetic field, carried by the solar wind into interplanetary space and twisted into an Archimedean spiral by the Sun's rotation.

Inter-Tropical Convergence Zone: area encircling the Earth near the equator where the northeast and southeast trade winds come together. Variation in its location zone drastically affects rainfall in many equatorial nations, accentuating the risk of having droughts and floods.

Ion: a charged molecule or an atom.

Ionisation: is a process by which an atom or a molecule acquires a negative or positive charge by gaining or losing electrons.

Ionosphere: region of the Earth's upper atmosphere consisting of charged particles (ions and electrons) that mingle with the thermosphere to form the upper atmosphere. The ionosphere is the ionised counterpart of the thermosphere and typically extends from 85 km to 500–1000 km. The ionosphere is often subdivided into 4 regions, called D-region (between 60 and 90 km), E-region (between 90 and 140 km), and F1 and F2-region (between 140 and 200 km).

Jet streams: meandering-shaped and fast flowing air currents that are located near the altitude of the tropopause. Each hemisphere has both a polar jet and a subtropical jet. The Northern hemisphere polar jet flows over the middle to northern latitudes of North America, Europe, and Asia. Jet streams are caused by a combination of the Earth's rotation, and atmospheric heating by solar radiation. There is growing evidence for their gradual weakening and equatorward drift with global warming.

keV: unit of energy, which is frequently used for ionised particles : $1 \text{ keV} = 10^3 \text{ eV}$ (electron-volts), which is equivalent to $1.6 \cdot 10^{-16}$ J (Joule). Most solar wind protons and electrons have energies between 0.1 and 10 eV. See also MeV and GeV.

Kp: index of geomagnetic activity, which measures disturbances of the horizontal component of the Earth's magnetic field at ground level. The Kp index is a weighted average of 13 stations that are located between 46 and 60° of southern and northern geomagnetic latitude. Its values are expressed on a semi-logarithmic scale from 0 to 9+, with 1 being calm and > 5 indicating a geomagnetic storm. The Kp index is delivered on a 3-hourly basis. See also ap and Dst indices.

Lagrange point: zone in space where the gravity and centrifugal force of two bodies balances out. The L1 point is on the Sun-Earth line (at about 1% of that distance from the Earth). Several Sun-observing spacecraft are located there because they rotate around the Sun in exactly one year.

Lithosphere: the upper layer of the solid Earth, both continental and oceanic, which is composed of all crustal rocks and the cold, mainly elastic, part of the uppermost mantle. Volcanic activity, although part of the lithosphere, is not considered part of the climate system, but acts as an external forcing factor.

Limb (solar limb): luminous edge of the solar disk, or of a heavenly body.

Longwave radiation: radiation emitted by the Earth's surface, the atmosphere and the clouds. See thermal infrared radiation.

Magnetic cloud: a region of the solar wind extending over about 0.15 AU, characterised by a strong magnetic field, a large and smooth rotation of the magnetic field direction and a low proton temperature. Magnetic clouds represent a subset of interplanetary coronal mass ejections.

Magnetic storm: a worldwide perturbation of the geomagnetic field due to an enhanced ring current. The intensity of the storm can be characterised by the minimum of the Dst (disturbance storm time) index, such that during intense storms the global field decreases at least by a hundred nanoTesla. A storm occurs when the interplanetary magnetic field turns southward for a prolonged period of time producing reconnection at the magnetopause. During the storm's main phase, which can last as long as two to two and a half days in the case of a severe storm, charged particles in the near-Earth plasma sheet are energised and injected deeper into the inner magnetosphere, producing the storm-time ring current

Magnetopause: boundary between the magnetosphere and the solar wind determined by the pressure balance between the solar wind on one side, and the magnetic pressure of the planetary field on the other. It is typically located at a geocentric distance of about 10 Earth radii on the upstream side.

Magnetosphere: is the volume of space around an astronomical object that is controlled by that object's magnetic field. The Earth's magnetosphere is the cavity formed by the Earth's magnetic field in the flow of plasma from the Sun known as the solar wind.

Magnetospheric substorm: see Substorm.

Magnetosphere: is the volume of space around an astronomical object that is controlled by that object's magnetic field. The Earth's magnetosphere is the cavity formed by the Earth's magnetic field in the flow of plasma from the Sun, known as the solar wind. The terrestrial magnetosphere is highly dynamic, because of the continuously changing solar wind.

Maunder Minimum: a period of exceptionally low solar activity, named after the astronomer Edward Maunder, and which lasted from approximately 1645 to 1715. This period of low solar activity also coincides with one of the cold and wet climatic periods in the Northern Hemisphere, also called "Little Ice Age".

Mesopause: boundary between the mesosphere and the thermosphere. The mesopause is one of the coldest layers of the atmosphere, with temperatures as low as 130 K.

Mesosphere: region of the atmosphere situated above the stratosphere, at an altitude of approximately 50 to 90 km, where the temperature decreases with height. The upper boundary of the mesosphere is the mesopause, which can be the coldest naturally occurring place on Earth with temperatures below 130 K.

MeV: unit of energy, which is frequently used for ionised particles: 1 MeV $= 10^6$ eV (electron-volts), which is equivalent to $1.6 \cdot 10^{-13}$ J (Joule). Particles with energies in the MeV range are called high-energy particles. Most solar wind protons and electrons in contrast have energies between 0.1 and 10 eV. See also keV and GeV.

Middle atmosphere: name given to the stratosphere, mesosphere and lowest part of the thermosphere, which span heights from approximately 10 km to 100 km.

Modes of climate variability: natural variability of the climate system, in particular on seasonal and longer time scales, predominantly occurs with preferred spatial patterns and time scales, through the dynamical characteristics of the atmospheric circulation and through interactions with the land and ocean surfaces. Such patterns are often called regimes, modes or teleconnections. Examples are the North Atlantic Oscillation (NAO), the Pacific-North American pattern (PNA), the El Niño-Southern Oscillation (ENSO), the Northern Annular Mode (NAM; previously called Arctic Oscillation, AO) and the Southern Annular Mode (SAM; previously called the Antarctic Oscillation, AAO).

MUV or Medium Ultraviolet: spectral band, which is part of the UV, with wavelengths ranging from 200 to 300 nm. Solar spectral irradiance emitted in the FUV is predominantly absorbed in the stratosphere and in the mesosphere. In the stratosphere, MUV radiation is responsible for the generation of ozone, and as such represents an important connection between solar variability and climate.

Nonlinearity: a process is called nonlinear when there is no simple proportional relation between cause and effect. The climate system contains many such nonlinear processes, resulting in a system with a potentially very complex behaviour. Such complexity may lead to abrupt climate change. See also Chaos; Predictability.

North Atlantic Oscillation (NAO): the North Atlantic Oscillation consists of opposing variations of barometric pressure near Iceland and near the Azores. It therefore corresponds to fluctuations in the strength of the main westerly winds across the Atlantic into Europe, and thus to fluctuations in the embedded cyclones with their associated frontal systems. The NAO is one of the main contributors to natural climate variability on the decadal time scale.

NO_x: a generic term used for mono-nitrogen oxides NO (nitric oxide) and NO_2 (nitrogen dioxide). Man-made NO_x is mainly produced by combustion processes. Natural sources of NO_x are lightning, but also energetic particles precipitating into the atmosphere. In atmospheric chemistry, the term means the total concentration of NO and NO_2. The term NO_y (reactive, odd nitrogen) refers to both NO_x and compounds that are produced from the oxidation of NO_x which include nitric acid.

Ozone: the triatomic form of oxygen (O_3) is a gaseous atmospheric constituent. In the troposphere, it is created both naturally and by photochemical reactions involving gases resulting from human activities (smog). Tropospheric ozone acts as a greenhouse gas. In the stratosphere, ozone is created by the interaction between solar ultraviolet radiation and molecular oxygen (O_2). Stratospheric ozone, whose main source is solar UV radiation, plays a dominant role in the stratospheric radiative balance.

Ozone layer: the stratosphere contains a layer in which the concentration of ozone is greatest, the so-called ozone layer. The layer extends from about 12 to 40 km above the Earth's surface. The ozone concentration reaches a maximum between about 20 and 25 km. This layer is being depleted by human emissions of chlorine and bromine compounds. Every year, during the Southern Hemisphere spring, a very strong depletion of the ozone layer takes place over the Antarctic region, caused by anthropogenic chlorine and bromine compounds in combination with the specific meteorological conditions of that region. This phenomenon is called the ozone hole.

Pacific decadal oscillation: coupled decadal-to-inter-decadal variability of the atmospheric circulation and underlying ocean in the Pacific Basin. It is most prominent in the North Pacific, where fluctuations in the strength of the winter Aleutian Low pressure system co-vary with North Pacific sea surface temperatures, and are linked to decadal variations in atmospheric circulation, sea surface temperatures and ocean circulation throughout the whole Pacific Basin. Such fluctuations have the effect of modulating the El Niño-Southern Oscillation cycle. Key measures of Pacific decadal variability are the North Pacific Index (NPI), the Pacific Decadal Oscillation (PDO) index and the Inter-decadal Pacific Oscillation (IPO) index.

Paleoclimate: climate during periods prior to the development of measuring instruments, including historic and geologic time, for which only indirect (i.e. proxy) climate records are available. See Proxy.

Photolysis: chemical process by which molecules are broken down into smaller units through the absorption of light. Atmospheric chemistry is heavily driven by photolysis.

Photosphere: the lowest visible layer of the solar atmosphere that corresponds to the solar surface viewed in white light. The major part of solar radiation comes from the photosphere, and a very small part from the corona.

Plage: bright region of the solar chromosphere that exhibits enhanced emissions in the UV and in the visible bands of the spectrum. Plages are typically found near sunspots, and map closely to the faculae in the photosphere below. Their occurrence is modulated by the solar magnetic cycle, and is one of the main causes of the changing solar spectral irradiance.

Planetary waves (or Rossby waves): large-scale meanders in high-altitude winds that have a major influence on wind speeds, temperature, distribution of ozone, and other characteristics of the middle atmosphere structure. Planetary waves, together with gravity waves, are responsible for quasi-periodic oscillations in the stratosphere, such as the the Quasi-Biennial Oscillation (QBO). They are typically generated by orographic and adiabatic heating in the troposphere. See Gravity waves.

Plasma: gas in which charged particles (ions, electrons) are free. In geophysical plasmas, the charges are neutral, i.e. there are just about as many electrons as there are monovalent ions. A plasma is a good electrical conductor and is strongly affected by magnetic fields.

Plasmapause: external boundary of the plasmasphere.

Plasmasphere: a magnetospheric region of cool and dense plasma that may be considered an outer extension of the ionosphere with which it is coupled. Like the ionosphere, the plasmasphere tends to corotate with the Earth.

Polar vortex: a persistent, large-scale cyclone that circles the geographical north and south poles. Their bases are located in the middle and upper troposphere, and extend into the stratosphere. The rotation of these vortices is caused by the Coriolis effect. Polar vortices are weaker during summer and strongest during winter; when they are strong, the westerlies also increase in strength. Sudden stratospheric warming events, when temperatures within the stratosphere warm dramatically over a short time, are associated with weaker polar vortices. See also Sudden stratospheric warming, and Westerlies.

Predictability: the extent to which future states of a system may be predicted based on knowledge of current and past states of the system. Since knowledge of the climate system's past and current states is generally imperfect, as are the models that utilise this knowledge to produce a climate prediction, and since the climate system is inherently nonlinear and chaotic, predictability of the climate system is inherently limited. Even with arbitrarily accurate models and observations, there may still be limits to the predictability of such a nonlinear system

Proxy: a proxy climate indicator is a local record that is interpreted, using physical and biophysical principles, to represent some combination of climate-related variations back in time. Climate-related data derived in this way are referred to as proxy data. The F10.7 index is a proxy for the solar UV flux.

Quasi-biennial oscillation (QBO): one of the most important modes of variability of the tropical stratosphere. This quasi-periodic oscillation of the equatorial zonal wind between easterlies and westerlies has a period of typically 28 to 29 months. The phase of the QBO is correlated with the solar cycle, which impacts the variability of the arctic stratosphere.

Radiation belts (van Allen radiation belts): annular shell in the magnetosphere surrounding the Earth at 1.2 to 6 Earth radii, in which highly energised particles are trapped by closed geomagnetic field lines. There are two belts. The inner one is part of the plasmasphere and corotates with the Earth; its maximum proton density lies near 5000 km. Inner belt protons have high energies (10–50 MeV range) and originate from the decay of secondary neutrons created during collisions between cosmic rays and upper atmospheric particles. The outer belt extends on to the magnetopause on the sunward side and to about 6 Earth radii on the nightside; it mostly consist of electrons. The altitude of maximum proton density is near 16 000–20 000 km. Outer belt protons have lower energies (about 200 eV to 1 MeV) and come from the solar wind and ionosphere. The population of radiation belt part particles gets enhanced by geomagnetic storms, and their precipitation into the upper atmosphere (at high latitudes) affects its chemistry.

Radiative forcing: the change in the net, downward minus upward, irradiance (expressed in Wm^{-2}) at the tropopause due to a change in an external driver of climate change, such as, for example, a change in the concentration of carbon dioxide or the output of the Sun. Radiative forcing is computed with all tropospheric properties held fixed at their unperturbed values, and after allowing for stratospheric temperatures, if perturbed, to readjust to radiative-dynamical equilibrium. Radiative forcing is called instantaneous if no change in stratospheric temperature is accounted for. For the purposes of this report, radiative forcing is further defined as the change relative to the year 1750 and, unless otherwise noted, refers to a global and annual average value. Radiative forcing is not to be confused with cloud radiative forcing, a similar terminology for describing an unrelated measure of the impact of clouds on the irradiance at the top of the atmosphere.

Radiative transfer: in the context of atmospheric physics, refers to the transfer of electromagnetic radiation through the atmosphere. Numerical models are needed to simulate the complex interaction of radiation with the atmosphere as it propagates.

Radiative zone: internal region of the Sun, between the nuclear oven and the convection zone, from 0.2 to 0.7 solar radii. Here, the energy produced by the nuclear core is transmitted by radiation.

Radical: in chemistry, radicals denote groups of atoms behaving as a single unit in a number of compounds. Hydroxyl (HO*) is an important radical for atmospheric chemistry.

Reanalysis: reanalyses are atmospheric and oceanic analyses of temperature, wind, current, and other meteorological and oceanographic quantities, created by processing past meteorological and oceanographic data using fixed state-of-the-art weather forecasting models and data assimilation techniques. Using fixed data assimilation avoids effects from the changing analysis system that occurs in operational analyses. Although continuity is improved, global reanalyses still suffer from changing coverage and biases in the observing systems.

Reconnection: magnetic reconnection refers to the breaking and reconnecting of oppositely directed magnetic field lines in plasma regions, which come in contact with each other. In the process, magnetic field energy is converted to plasma kinetic and thermal energy. It is the key mechanism in the solar wind-magnetosphere coupling and in the solar flare process.

Ring Current: in the magnetosphere, a toroidal electric current flowing westward around the Earth in a region near the geomagnetic equator, at geocentric distances between 2 and 9 Earth radii. It is generated by the drift of trapped charged particles with energies of 10–300 keV. This current is responsible of global decreases in the Earth's surface magnetic field during geomagnetic storms, when it is strongly enhanced by the injection of ions originated in the solar wind and the terrestrial ionosphere.

Schwabe cycle: see Solar cycle.

Shortwave radiation: radiation emitted by the Sun, in contrast to longwave radiation. See also Solar radiation, and Thermal infrared radiation.

Solar activity: the Sun exhibits periods of high activity observed in numbers of sunspots, as well as radiative output, magnetic activity and emission of high-energy particles. These variations take place on a range of time scales from millions of years to minutes. See Solar cycle.

Solar cycle (Schwabe cycle): a quasi-regular modulation of solar activity with varying amplitude, and a period of typically 10–11 years. One of the main manifestations of the solar cycle is the variation in the sunspot number. The true solar cycle (or Hale cycle) is actually a 20–22-year one, with two alternate periods of opposite magnetic polarity.

Solar corona: see Corona

Solar maximum: the time at which the Sun reaches its highest level of activity, as defined by the 12-month averaged sunspot number.

Solar minimum: the time at which the Sun reaches its lowest level of activity, as defined by the 12-month averaged sunspot number.

Solar energetic particle (SEP) event: is a burst of high-energy charged particles (dominated by protons) that are emitted by solar flares or accelerated in interplanetary space by a shock wave associated with a coronal mass ejection. These particles have energies in the MeV/nucleon and sometimes up to GeV/nucleon range. SEP events may occur anytime during a solar cycle, although avoiding minima, and typically last for hours to days. These particles penetrate deep into the atmosphere at high latitudes. They are often referred to as solar cosmic rays. See also Galactic cosmic rays.

Solar proton event (SPE): see Solar energetic particle (SEP) event.

Solar radiation: electromagnetic radiation emitted by the Sun. It is also referred to as shortwave radiation. Solar radiation has a distinctive range of wavelengths (spectrum) determined by the temperature of the Sun, peaking in visible wavelengths.

Solar spectral irradiance (SSI): the spectrally-resolved amount of solar radiation (or irradiance per unit wavelength) received outside the Earth's atmosphere on a surface normal to the incident radiation, and at the Earth's mean distance from the Sun. Reliable measurements of the SSI can only be made from space, and the first full coverage of the solar spectrum (UV to infrared) extends back only to 2003 only. The variability of the SSI is highly-wavelength-dependent, and ranges from 0.1% in the visible to over 100% in the extreme ultraviolet. See also Total solar irradiance.

Solar storm: refers to various types of explosive events that occur at the surface of the Sun. Best known are solar flares, which are sudden flashes of brightness that can release up to 10^{26} J in just a few minutes. Flares also lead to the acceleration of mostly protons and electrons, which propagate away from the Sun at speeds close to that of light. Flares occur intermittently, but are more frequent when the Sun is active and has large, complex sunspots. They are frequently accompanied by the release of large amounts of plasma, called Coronal Mass Ejections (CME).

Solar wind: stream of charged particles (i.e., a plasma), which are ejected from the upper solar atmosphere. The solar wind consists mostly of high-energy electrons and protons (about 1 keV) that are able to escape the Sun's gravity in part because of the high temperature of the corona and the high kinetic energy particles gain through a process that is not well understood at this time. Several perturbations of the Earth's environment, such as geomagnetic storms, are driven by the solar wind. Fast solar winds mostly originate from solar coronal holes or from high latitude regions, with speeds in excess of about 700 km s^{-1}. Slow winds mostly originate from the quiet Sun, with speeds that are typically below 450 km s^{-1}. Alternating fast and slow solar winds disturb the Earth's magnetosphere, generate geomagnetic storms, and may lead, for example, to the precipitation of energetic particles from the radiation belts into the atmosphere. See Radiation belts.

Space climate: denotes the long-term variations in solar activity, as well as the related long-term changes in the heliosphere, the solar wind and the heliospheric magnetic field, and their effects in the near-Earth environment, including the magnetosphere and ionosphere, the upper and lower atmosphere, climate and other related systems.

Space weather: the physical and phenomenological state of natural space environments. The associated discipline aims, through observation, monitoring, analysis and modelling, at understanding and predicting the state of the Sun, the interplanetary and planetary environments, and the solar and non-solar driven perturbations that affect them; and also at forecasting and nowcasting the possible impacts on biological and technological systems.

Spörer minimum: a period of low solar activity, from approximately 1420 to 1570, named after the astronomer Friedrich Spörer. Like the Maunder and Dalton minima, this episode coincided with a period of lower-than-average global temperatures.

Sprite: brief optical flash of light that emanates from the upper atmosphere (mesosphere) of the Earth occurring above thunderstorm clouds.

Stratopause: atmospheric level, which separates the stratosphere and the mesosphere, and is located around 50–55 km altitude. The temperature, which gradually increases with height in the stratosphere, starts decreasing again above the stratopause.

Stratosphere: highly stratified region of the atmosphere above the troposphere, extending from about 10 km (ranging from 9 km at high latitudes to 16 km in the tropics on average) to about 50 km altitude. The ozone layer is located in the upper stratosphere.

Substorm: a localised and brief disturbance in the Earth's magnetosphere by which energy is released from the tail of the magnetosphere, and injected into the high latitude ionosphere. Its main manifestation is the brightening of auroras in polar regions. Substorms are distinct from geomagnetic storms in that the latter take place over a period of several days, are observable from anywhere on Earth, inject a large number of ions into the outer radiation belt. See also Magnetic storm.

Sudden stratospheric warming (SSW): event where the polar vortex of westerly winds in the winter hemisphere slows down or even reverses direction over the course of a few days. The change is accompanied by a rise of stratospheric temperature by several tens of kelvins. SSWs are mainly forced by planetary scale waves that propagate up from the lower atmosphere. The resulting reverse of high altitude winds weakens the jet stream, eventually resulting in dramatic reductions in temperature in the Eastern U.S. and in Europe.

Sunspot: dark zone of the solar photosphere, with a mean diameter approaching a few thousand kilometres. Sunspots are regions whose enhanced magnetic field block upwelling plasma motion, and thus causes their temperature to be somewhat lower than their surroundings. Their lifetime ranges from hours and months.

Sunspot Number: index of solar activity defined from the number of sunspots, and sunspot groups present on the Sun. The sunspot number is the most widely used gauge of solar activity; its recordings go back to 1610 but homogeneous data are available since the late 19th century only.

Synodic rotation: rotation of the Sun observed from the Earth, i.e. taking into account the Earth's own rotation (on its axis, and around the Sun).

Teleconnection: a connection between climate variations over widely separated parts of the world. In physical terms, teleconnections are often a consequence of large-scale wave motions, whereby energy is transferred from source regions along preferred paths in the atmosphere.

Thermal infrared radiation: radiation emitted by the Earth's surface, the atmosphere and the clouds. It is also known as terrestrial or longwave radiation, and is to be distinguished from the near-infrared radiation that is part of the solar spectrum. Infrared radiation, in general, has a distinctive range of wavelengths (spectrum) longer than the wavelength of the red colour in the visible part of the spectrum. The spectrum of thermal infrared radiation is practically distinct from that of shortwave or solar radiation because of the difference in temperature between the Sun and the Earth-atmosphere system.

Thermosphere: neutral component of the upper atmosphere. This layer is located between the mesosphere and the exosphere, and extends from about 85 km to 500–1000 km. Solar UV radiation causes the thermosphere to become electrically charged, thus producing its ionised counterpart, called ionosphere.

The dynamics of the thermosphere is dominated by atmospheric tides, which are driven by the intense diurnal heating.

Total solar irradiance (TSI): the amount of solar radiation received outside the Earth's atmosphere on a surface normal to the incident radiation, and at the Earth's mean distance from the Sun. Reliable measurements of solar radiation can only be made from space and the precise record extends back only to 1978. The generally accepted value is 1361 W m^{-2}, with an accuracy of about 0.2%. Variations of a few tenths of a percent are common, usually associated with the passage of sunspots across the solar disk. The solar cycle variation of TSI is of the order of 0.1%. The TSI is still occasionally (and incorrectly) called "solar constant".

Transition region: a very thin region that separates the solar chromosphere from the solar corona. The transition region is characterised by a sudden increase in temperature (from a few thousand to a few million degrees).

Tropopause: the interface between the troposphere and the stratosphere. The tropopause is located around 9 km altitude (at high latitude), and up to to 16 km (in tropical regions).

Troposphere: the lowermost part of the atmosphere, and also the most dynamic one, where clouds and weather phenomena occur. The troposphere is bounded by the tropopause. In the troposphere, temperatures generally decrease with height.

Upper atmosphere: often refers to that part of the atmosphere, which is above the troposphere. More generally, the upper atmosphere refers to the mesosphere, and above, and thus to altitudes beyond 50–55 km. See also middle and lower atmosphere.

Westerlies: prevailing winds from the west toward the east in the middle latitudes between 30 and 60° latitude.

XUV or Soft X-ray: spectral band with wavelengths ranging from 0.1 to 10 nm, i.e. with wavelengths smaller than in the UV band. Solar spectral irradiance emitted in the XUV is highly variable, and this energetic radiation is predominantly absorbed in the ionosphere, where it dissociates atoms and molecules.

Zonal: latitudinal, i.e. easterly or westerly, as opposed to meridional.

THE AUTHORS

- Galina A. Bazilevskaya (LPI, Nuclear Physics and Astrophysics Division, Russia), specialist in solar-terrestrial physics with a keen interest in cosmic rays generation, propagation and modulation in the heliosphere and atmosphere.

- Rasmus Benestad (MET Norway, Oslo, Norway), physicist and specialist in climate analysis, statistics and sun-climate connections.

- Roxana Bojariu (National Meteorological Administration, Bucharest, Romania), specialist in climate variability and predictability, with a focus on North Atlantic/Arctic Oscillation.

- Jaša Čalogović (Hvar Observatory, University of Zagreb, Croatia), specialist in solar-terrestrial connections (e.g. cosmic ray-cloud link) and space weather.

- Thomas von Clarmann (Karlsruhe Institute of Technology, Karlsruhe, Germany), specialist in remote sensing and stratospheric chemistry and physics.

- Thierry Dudok de Wit (University of Orléans, France), solar-terrestrial physicist with a keen interest in the statistical analysis of solar spectral variability

- Ilaria Ermolli (INAF, Osservatorio Astronomico di Roma, Italy), specialist in solar magnetism, with a focus on its effects on irradiance variations.

- Patrizia Francia (University L'Aquila, Italy), magnetospheric physics, ULF waves, polar cap activity.

- Bernd Funke (IAA-CSIC, Granada, Spain), specialist in middle atmosphere science and satellite data analysis, with a particular interest in particle-induced solar variability.

- Hans Gleisner (Danish Meteorological Institute), specialist on retrieving climate data from GPS radio occultation measurements.

- Margit Haberreiter (PMOD/WRC, Switzerland), solar physicist with strong interest in the radiative transfer in the solar atmosphere, and the causes and consequences of solar variability.

- Joanna Haigh (Imperial College London, UK), specialist in atmospheric radiative transfer and radiative–dynamical coupling.

- Arnold Hanslmeier (University of Graz, Austria), specialist in solar cycle variation studies and space weather; has published several books for specialists and non-specialists in the fields of astrophysics.

- Giles Harrison (University of Reading, UK), specialist in atmospheric electricity, and novel meteorological instrumentation.

- Sverre Holm (Dept. of informatics, University of Oslo, Norway), specialist on spectral estimation.

- Maarit J. Käpylä (Aalto University, Finland), solar dynamo theory and modelling of turbulent magnetoconvection.

- Ingo Kirchner (Freie Universität Berlin, Institute for Meteorology, Germany), climate modeller and meteorologist, with a special interest in climate dynamics and it's driving processes.

- Natalie Krivova (MPS – Göttingen, Germany), specialist in solar variability, solar irradiance and solar influence on Earth's climate

- Markus Kunze (Freie Universität Berlin, Germany), specialist in middle atmosphere data analysis and modeling.

- Erkki Kyrölä (Finnish Meteorological Institute, Helsinki, Finland), specialist in radiative transfer, remote sensing and Earth observation, with a focus on the middle atmosphere.

- Benjamin A. Laken (University of Oslo, Norway), a Physical Geographer, with a focus on applying modern computer analysis methods to investigate climate processes and impacts.

- Mai Mai Lam (Department of Meteorology, University of Reading, UK), solar–terrestrial physicist with expertise in radiation belts and solar wind–troposphere coupling.

- Ulrike Langematz (Freie Universität Berlin, Berlin, Germany), specialist in modeling chemistry–climate interactions in the lower and middle atmosphere and solar influence on climate.

- Maria Carmen Llasat (University of Barcelona, Barcelona, Spain), specialist on meteorology and climatology, and natural hazards.

- Jean Lilensten (IPAG – Grenoble, France), specialist of the planetary upper atmospheres (thermospheres and ionospheres) and space weather.

- Franz-Josef Lübken (Director of IAP, Kühlungsborn, Germany), specialist in ground-based and in-situ measurements in the polar mesosphere/lower thermosphere.

- Katja Matthes (GEOMAR & CAU Kiel, Germany), specialist in modeling solar influence on climate, with a focus on stratosphere–troposphere–ocean coupling.

The authors

- Amanda Maycock (University of Cambridge, UK), specialist in composition climate interactions, with a focus on stratosphere–troposphere interaction.
- Irina A. Mironova (St. Petersburg State University, Department of Earth Physics, Russia), specialist in solar–terrestrial physics with a keen interest in solar cosmic ray propagation and their effects into polar atmosphere.
- Stergios Misios (Aristotle University of Thessaloniki, Greece), specialist in modelling solar–climate linkages.
- Dann Mitchell (Atmospheric physics, Oxford University, UK), specialist on the stratospheric polar vortex.
- Christian Muller (B.USOC- Brussels, Belgium), data manager of the SOLAR payload on the International Space Station and ICTP expert on natural forcings on climate.
- Kalevi Mursula (Oulu University, Finland), space climate, solar wind and heliospheric magnetic field, long-term solar effects to near Earth space and climate.
- Donal Murtagh (Chalmers University of Technology, Göteborg, Sweden), specialist in remote sensing and Earth observation techniques.
- Keri Nicoll (University of Reading, UK), specialist in atmospheric measurements and instrumentation with a focus on solar–terrestrial effects on clouds and atmospheric electricity.
- Tine Nilsen (Applied mathematics, The Arctic University of Norway), specialist on statistical analysis of paleoclimatic reconstruction and climate model data.
- Colin Price (Tel Aviv University, Israel), specialist in atmospheric electricity, thunderstorms, lightning and climate change.
- Kerstin Prömmel (Freie Universität Berlin, Institute for Meteorology, Germany), climate modeller and meteorologist, with a special interest in palaeoclimate and decadal climate variability.
- Craig J. Rodger (Department of Physics, University of Otago, New Zealand), specialist on the detection, analysis and impact of energetic electron precipitation into the atmosphere.
- Eugene Rozanov (Institute of Atmospheric and Climate Science, ETHZ and PMOD/WRC, Switzerland), specialist in the climate and atmospheric chemistry modelling.
- Michael Rycroft (University of Bath and CAESAR Consultancy, UK), specialist in the global electric circuit, with a broad range of interests in atmospheric and space science.

- Kristoffer Rypdal (Applied mathematics, The Arctic University of Norway), specialist on the mathematics and statistics of climate variability.

- Martin Rypdal (Applied mathematics, The Arctic University of Norway), specialist on the mathematics of climate and finance.

- Hauke Schmidt (Max Planck Institute for Meteorology, Germany), climate modeller specialised in effects of natural and anthropogenic forcings, and atmospheric vertical coupling.

- Werner Schmutz (PMOD/WRC, Switzerland), an astrophysicist who is investigating the solar irradiance and its impact on the terrestrial climate.

- Annika Seppälä (Finnish Meteorological Institute, Finland & British Antarctic Survey, United Kingdom), specialist in solar and magnetospheric particle precipitation impacts on the middle atmosphere chemistry and dynamics, satellite observation expert.

- Sami K. Solanki (Max Planck Institute for Solar System Research, Germany), solar and heliospheric physics, solar magnetism and Sun–Earth relations

- Timofei Sukhodolov, (Institute of Atmospheric and Climate Science, ETHZ and PMOD/WRC, Switzerland), specialist in the modelling of atmospheric chemistry.

- Eija I. Tanskanen (Finnish Meteorological Institute, Finland), geomagnetic activity and its solar and solar wind sources such as high-speed streams and Alfvén waves.

- Peter Thejll (Climate and Arctic Research, Danish Meteorological Institute). Investigated Sun–climate link, now specialising in statistics of extreme weather and climate events, and earth observations.

- Rémi Thiéblemont (GEOMAR Helmholtz Centre for Ocean Research Kiel, Germany), specialist in middle atmosphere dynamics and its influence on climate.

- Matthew Toohey (GEOMAR Helmholtz Centre for Ocean Research Kiel, Germany), specialist in the climate impacts of major volcanic eruptions.

- Kleareti Tourpali (Aristotle University of Thessaloniki, Greece), atmospheric physicist with expertise in ozone–climate interactions and solar activity effects, specialist in the atmospheric environment and climate.

- Ricardo M. Trigo (Universidade de Lisboa, Portugal), specialist in climate variability, climate extremes and climate change mechanisms.

- Ilya Usoskin (Sodankylä Geophysical Observatory, Finland), cosmic rays, ground level enhancements, grand solar maxima and minima.

- José M. Vaquero (Universidad de Extremadura, Spain), physicist with a keen interest in the reconstruction of solar activity from historical documents.

- Astrid Veronig (University of Graz, Austria), solar physicist with research focus on flares, coronal mass ejections and their impact on the Earth's space weather.

- Pekka T. Verronen (Finnish Meteorological Institute, Finland), specialist on middle atmospheric physics and chemistry, and effects of solar energetic particle precipitation.

- Yoav Yair (Interdisciplinary Center, Herzliya, Israel), An atmospheric physicist specialising in thunderstorms and lightning on Earth and other planets.

www.ingramcontent.com/pod-product-compliance
Ingram Content Group UK Ltd.
Pitfield, Milton Keynes, MK11 3LW, UK
UKHW061222180426